Nanoelectronics for Next-Generation Integrated Circuits

The incessant scaling of complementary metal-oxide semiconductor (CMOS) technology has resulted in significant performance improvements in very-large-scale integration (VLSI) design techniques and system architectures. This trend is expected to continue in the future, but this requires breakthroughs in the design of nano-CMOS and post-CMOS technologies. Nanoelectronics refers to the possible future technologies beyond conventional CMOS scaling limits. This volume addresses the current state-of-the-art nanoelectronic technologies and presents potential options for next-generation integrated circuits.

Nanoelectronics for Next-Generation Integrated Circuits is a useful reference guide for researchers, engineers, and advanced students working on the frontier of the design and modeling of nanoelectronic devices and their integration aspects with future CMOS circuits. This comprehensive volume eloquently presents the design methodologies for spintronics memories, quantum-dot cellular automata, and post-CMOS FETs, including applications in emerging integrated circuit technologies.

Nanoelectronics for Next-Generation Integrated Circuits

Edited by
Rohit Dhiman

CRC Press
Taylor & Francis Group
Boca Raton London New York

CRC Press is an imprint of the
Taylor & Francis Group, an informa business

Front cover image: sdecoret/Shutterstock

First edition published 2023
by CRC Press
6000 Broken Sound Parkway NW, Suite 300, Boca Raton, FL 33487-2742

and by CRC Press
4 Park Square, Milton Park, Abingdon, Oxon, OX14 4RN

CRC Press is an imprint of Taylor & Francis Group, LLC

ISBN: 978-0-367-72652-2 (HBK)
ISBN: 978-0-367-72656-0 (PBK)
ISBN: 978-1-003-15575-1 (EBK)

DOI: 10.1201/9781003155751

Typeset in Times
by Deanta Global Publishing Services, Chennai, India

To my respected Parents for their affection and untiring efforts in my upbringing; my loving family, Anjali, and our loving daughter, Shipra Dhiman, for their precious time and patience.

–Rohit Dhiman

Contents

Preface

The continuous scaling of complementary metal-oxide semiconductor (CMOS) technology has resulted in significant performance improvements in very-large-scale integration (VLSI) circuit design techniques and system architectures. In 1965, Gordon Moore, one of the eminent scientists in Silicon Valley and co-founder of the Intel Corporation, anticipated the growing trend of transistors in integrated circuits (ICs). He speculated that the active transistor count in IC would be magnified twice every 18–24 months. Since then, the number of transistors has risen dramatically in an IC according to Moore's trend, and tremendous growth has been seen wherein the technology has escalated from small scale to large, ultra-large, and eventually to the giga-scale integration. In order to develop more compressed, budget-friendly, and sophisticated ICs, *'More Moore'* (M_m) and *'More than Moore'* (M_{tm}) concepts have evolved in the past few years. To construct advanced CMOS technologies and scale numerous digital functions, the M_m concept is utilized. M_{tm} largely emphasizes non-digital functions via heterogeneous system integration. According to the *International Technology Roadmap for Semiconductors*, Intel's next-generation billion transistor processors have set out to achieve industry leading performance in the order of GHz. This trend is expected to continue in the future, but this will require breakthroughs in the design of VLSI technologies, generally known as nanoelectronics. With the development of novel materials and nano-scale devices, research is being directed to gain better physical insights into the parameters that influence device, circuit, and system characteristics. This book, *Nanoelectronics for Next-Generation Integrated Circuits*, is written by researchers in the respective areas of nanoelectronics, such as graphene-based electronics, spintronics, quantum-dot cellular automata (QCA), NWFETs, and other relevant areas. The 14 chapters of the book are classified under four sections that cover modeling, simulation, and applications of electronic, magnetic, and compound semiconductors in nanoelectronic devices, circuits, and systems. This comprehensive volume eloquently presents the design methodologies for spintronics memories, quantum-dot cellular automata, potential NWFETs, TFETs, and their applications from the viewpoint of nano-scale ICs and their application. The book shall serve as an invaluable reference for graduate students, Ph.D./M.S./M.Tech. scholars, researchers, and practicing engineers working on the frontier of nano-scale VLSI designs, circuits, systems, and their applications.

The first section of the book provides an overview and emphasizes the trends of nanoelectronics. Chapter 1: 'Emerging Graphene-Based Electronics: Properties to Potentials' deals with the self-healing phenomenon in graphene and its corresponding applications in the phenomenon of sub-nano sensor design. The chapter gives an overview of the interesting phenomenon of magnetic phase transitions (MPTs) detected under the collective impact of an electric field and temperature leading to the thermo-electromagnetic effect and provides insights into the intrinsic MPTs in graphene. The chapter also gives an overview of such emerging applications of graphene, i.e., sub-nano sensors, graphene-based spin-transfer torque magneto-resistive random-access memory, quantum computing devices, magnonics, and spintronic memory applications. Chapter 2: 'Models for Modern Spintronics Memories with Layered Magnetic Interfaces' explores spintronic technology-based non-volatile magneto-resistive random-access memory (MRAM). The spin interface and performance enhancement of MRAM cells due to the voltage-controlled magnetic anisotropy (VCMA) and Rashba spin–orbit interaction (SOI) effect are also discussed. Chapter 3: 'Evaluation of Magnetic Anisotropy via Intrinsic Spin Infusion' introduces a compact physical model to examine spin infusion and spin-transfer torque in magnetic multilayers, which leads to the magneto-resistance effect and tunnel magneto-resistance (TMR) effect in magnetic tunnel junctions. The applications and performance enhancement of MRAM cells due to the spin infusion mechanism are also discussed briefly. QCA can replace the CMOS technology in an ultra-nano-scale regime. Chapter 4: 'Quantum-dot Cellular Automata (QCA) Nanotechnology for the Next-Generation Systems' describes the implementation of QCA-based full adders with dots of 20 nm diameter. The results demonstrate that the proposed design provides reduced latency and design complexity.

With the advancements in CMOS scaling, the power density constraint puts a limit on the number of transistors that can be simultaneously switched on. Therefore, to exploit the full benefits of scaling, this book extensively investigates novel post-CMOS devices in the second section, Modeling and Design of Post-CMOS Field-effect Transistors (FETs). GAA nanowire (NW) architecture is considered as the most promising solution for the aggressive scaling of CMOS transistors. The electrical characteristics of different NW FETs such as ZnO/CuO/PbS NW FETs, III-V NW FETs, III-V NW TFETs, JL NW FETs, NW NCFETs, reconfigurable NW FETs, and Si NW FETs for their potential applications in post-CMOS electronics are provided in Chapter 5: 'An Overview of Nanowire Field-Effect Transistors for Future Nano-scale Integrated Circuits.' The reliability issues of NW FETs are also discussed in this chapter. Extreme thermal management crises caused by a surge in power dissipation are limited in the next-generation digital systems, since supply voltage is reduced. For such low-power systems, transistors with smaller sub-threshold slopes are desired. Tunnel FETs (TFETs) are appropriate alternatives to cope with the continuous scaling down of device dimensions. The incorporation of dual gate-source electrode dielectrically modulated (DG-SE-DM) and hetero-material source electrode dielectrically modulated (HM-SE-DM) into TFET-based biosensors that have outstanding carrier transport properties have opened up new vistas to the device designers with faster and better device performance. Thus, in Chapter 6: 'Investigation of Tunnel Field-Effect Transistors (TFETs) for Label-Free Biosensing,' the electrical performance of TFET-based biosensors for high performance and low-power IC technology is presented. The impressive potential of cylindrical NW GAATFETs for reduced short-channel effects at a 20 nm gate length is explored in Chapter 7: 'Analog and Linearity Analysis of Vertical Nanowire TFET.' This chapter also covers the comprehensive description of linearity and analog/RF parameters of the NW GAATFET and signifies the impetus behind the use of this device for low-power circuit applications. GaN-based HEMTs have proven to be excellent devices for applications requiring high frequency, high strength, and high temperature. Chapter 8: 'Effect of Variation in Gate Material on Enhancement mode P-GaN AlGaN/GaN HEMT' elucidates the variation in gate material on the performance of the enhancement mode in P-GaN-based AlGaN/GaN HEMT.

The third section of the book explores the possibilities of IC design with high performance memory architectures, and addresses the challenges that still need to be met. Recently, resistive random-access memory (RRAM) has gained intense attention in terms of its reliability, nano-scale dimensions, energy per bit, and fast read/write times. Also, it is promising for use in neuromorphic computing and digital applications. This part of the book, in Chapter 9: 'Electrical Modeling of One Selector-One resistor (1S-1R) for Mitigating the Sneak Path Current in a Nano-Crossbar Array,' provides insights into the sneak current problem and techniques to mitigate it in a nano-crossbar array. This chapter describes the electrical modeling of selector device (1S), RRAM device (1R), integration of selector model with RRAM (1S-1R), and formation of 1S-1R crossbar arrays. The memory design solutions for modern IoT edge devices and highly accurate artificial neural network architectures are described in Chapter 10: 'SRAM: An Essential Part of Integrated Circuits.' A low-voltage highly stable 12T SRAM cell was also designed for the battery-operated IoT-enabled wireless sensor network nodes. In today's sophisticated nano-era and densely packed IC designs, leakage reduction plays a crucial role in VLSI circuit design. The lector technique is found to be effective for leakage current reduction, and readers can explore some of its research aspects utilized in SRAM arrays in Chapter 11: 'Implementation of 512bit SRAM Tile using Lector Technique for Leakage Power Reduction.'

The last section of the book explores the possibilities for IC design with some emerging technologies and addresses the challenges that still need to be met. The advancements in the technology sector, witnessed nowadays, have mostly been driven by state-of-the-art VLSI devices and circuits. This all has been possible due to the miniaturization of devices and components up to the nano-scale and beyond. So, any new configuration of a signal processing module, such as an analog filter that employs nano-scale transistors, will always be welcomed. Chapter 12: 'Characterization of Stochastic Process Variability Effects on Nano-scale Analog Circuits' examines, in detail, the statistical design characterization for nano-scale analog ICs. Chapter 13: 'Versatile Single Input Single Output Filter Topology Suitable for Integrated Circuits' addresses a novel voltage-mode filter topology. The topology has a single input and a single output, and its versatility lies in the realization of different filtering functionalities on the appropriate selection of the generic impedances. Digital integrated circuits or intellectual property (IP) cores are widely used as computation engines or processors (application-specific and/or hardware accelerators) in sophisticated electronics products such as smartphones, cameras, camcorders, set-top boxes, smartwatches, etc. Recently, there has been a paradigm shift from the conventional IC/IP design flow to a secured IC/IP design flow that considers security awareness during the design process. The section concludes with

Chapter 14: 'Secured Integrated Circuit (IC/IP) Design Flow,' which throws light on some of the latest hardware security and IC/IP core protection methodologies integrated within high-level synthesis design flow, such as structural obfuscation, functional obfuscation, and hardware watermarking, including case study analysis of DSP applications. This book is a unique agglomeration of topics covering the recent advances in the field of nanoelectronics and other potential research areas for the efficient design exploration of post-CMOS based devices, circuits, and systems.

Acknowledgments

I would like to express my sincere gratitude to the authors of the individual chapters who have devoted their significant time and contributed their expertise to shaping this book. I express my earnest appreciation to all the authors for their excellent insights, as I am sure that this edited volume will be a useful text to many readers interested in nanoelectronics and its applications. I am grateful to the editorial team of the CRC Press, Taylor & Francis, for its tremendous support through the stages of preparation and for finally bringing out this book as an excellent academic treasure for its readers. I also express with sincere gratitude the technical and financial support received from the Science and Engineering Research Board through Core Research Grant (File No.: SERB/CRG/2021/000780).

Rohit Dhiman

List of Contributors

Amit Acharyya
Department of Electrical Engineering
Indian Institute of Technology, Hyderabad
Hyderabad, India

Swati Ghosh Acharyya
School of Engineering Sciences and Technology
University of Hyderabad
Hyderabad, India

Suneet Kumar Agnihotri
Department of Electronics and Communication Engineering
PDPM Indian Institute of Information Technology, Design and Manufacturing
Jabalpur, India

J. Ajayan
Department of Electronics and Communication Engineering
SR University
Warangal, India

Shamshad Alam
Department of Electronics and Electrical Engineering
Indian Institute of Technology
Guwahati, India

Rajeevan Chandel
Department of Electronics and Communication Engineering
National Institute of Technology, Hamirpur
Hamirpur, India

Bhartendu Chaturvedi
Department of Electronics and Communication Engineering
Jaypee Institute of Information Technology
Noida, India

Tarun Chaudhary
Electronics and Communication Engineering Department
Dr. B.R. Ambedkar National Institute of Technology, Jalandhar
Jalandhar, India

Rahul Chaurasia
Department of Computer Science and Engineering
Indian Institute of Technology, Indore
Indore, India

Sanghamitra Debroy
Department of Electrical Engineering
Indian Institute of Technology, Hyderabad
Hyderabad, India

Rohit Dhiman
Department of Electronics and Communication Engineering
National Institute of Technology, Hamirpur
Hamirpur, India

Shivani Godha
Electronic Professional
Delhi, India

A.F. Haider
Department of Electrical and Electronic Engineering
University of Nottingham Malaysia
Selangor, Malaysia

Divyansh Jain
Department of Electronics and Electrical Communication Engineering
Indian Institute of Technology, Kharagpur
Kharagpur, India

K.B. Jinesh
Department of Physics
Indian Institute of Space Science and Technology
Thiruvananthapuram, India

Jitender
Department of Electronics and Communication Engineering
Ajay Kumar Garg Engineering College
Ghaziabad, India

Prasad Pradeeprao Kanhegaonkar
Department of Computer Science and Engineering
Indian Institute of Technology, Indore
Indore, India

Mamta Khosla
Electronics and Communication Engineering Department
Dr. B.R. Ambedkar National Institute of Technology, Jalandhar
Jalandhar, India

K. Madhu Kiran
Department of Electronics and Communication Engineering
National Institute of Technology, Hamirpur
Hamirpur, India

T. Nandha Kumar
Department of Electrical and Electronic Engineering
University of Nottingham Malaysia
Selangor, Malaysia

J. Arya Lekshmi
Department of Electrical and Electronic Engineering
University of Nottingham Malaysia
Selangor, Malaysia

Anil Lodhi
Department of Electronics and Communication Engineering
PDPM Indian Institute of Information Technology, Design and Manufacturing
Jabalpur, India

Jitendra Mohan
Department of Electronics and Communication Engineering
Jaypee Institute of Information Technology
Noida, India

P. Mohankumar
Department of Mechatronics Engineering
Sona College of Technology
Salem, India

D. Nirmal
Department of Electronics and Communication Engineering
Karunya Institute of Technology and Science
Coimbatore, India

Soumya Pandit
Institute of Radio Physics and Electronics
University of Calcutta
Kolkata, India

Chithraja Rajan
Department of Electronics and Communication Engineering
Shri Ramdeobaba College of Engineering and Management (RCOEM)
Nagpur, India

Ashish Raman
Electronics and Communication Engineering Department
Dr. B.R. Ambedkar National Institute of Technology, Jalandhar
Jalandhar, India

Ravi Ranjan
Electronics and Communication Engineering Department
Dr. B.R. Ambedkar National Institute of Technology, Jalandhar
Jalandhar, India

Dip Prakash Samajdar
Department of Electronics and Communication Engineering
PDPM Indian Institute of Information Technology, Design and Manufacturing
Jabalpur, India

Sahil Sankhyan
Electronics and Communication Engineering Department
Dr. B.R. Ambedkar National Institute of Technology, Jalandhar
Jalandhar, India

Anirban Sengupta
Department of Computer Science and Engineering
Indian Institute of Technology, Indore
Indore, India

Vijay Kumar Sharma
School of Electronics & Communication Engineering
Shri Mata Vaishno Devi University
Katra, India

Vishal Sharma
Centre for Integrated Circuits and Systems
VIRTUS
School of EEE
Nanyang Technological University (NTU)
Singapore

Aniket Singha
Department of Electronics and Electrical Communication Engineering
Indian Institute of Technology, Kharagpur
Kharagpur, India

Santhosh Sivasubramani
Department of Electrical Engineering
Indian Institute of Technology, Hyderabad
Hyderabad, India

Shubham Tayal
Department of Electronics and Communication Engineering
SR University
Warangal, India

About the Editor

Rohit Dhiman received his Bachelor of Technology degree in Electronics and Communication Engineering from HP University Shimla, India, in 2007. He did his Master of Technology (M.Tech.) degree in VLSI Design at the National Institute of Technology (NIT) Hamirpur in 2009. He was awarded a Doctor of Philosophy (Ph.D.) degree from NIT Hamirpur in 2014. Presently, Dr. Rohit Dhiman is working as an Assistant Professor in the Department of Electronics and Communication Engineering at NIT Hamirpur and is the author/co-author of reputed publications in journals and conference proceedings of repute. He has been awarded the Young Scientist Award from the Department of Science and Technology, Science and Engineering Research Board, GoI, New Delhi. He has also been bestowed with the prestigious Young Faculty Research Fellowship from the Ministry of Electronics and Information Technology (MeitY), Government of India, and has three sponsored research projects to his credit. His major research interest is in device and circuit modeling for low-power VLSI design.

1 Emerging Graphene-Based Electronics
Properties to Potentials

Sanghamitra Debroy, Santhosh Sivasubramani, Swati Ghosh Acharyya, and Amit Acharyya

CONTENTS

1.1 INTRODUCTION

Since the discovery by Andre Geim et al. in 2004 of a novel method to isolate single atomic layers of graphene from graphite [1], graphene has gained significant attention and become the focus of widespread research owing to its mechanical, electronic, thermal, and optical properties [1–9]. The extraordinary band structure of graphene provides it with exceptional electrical transport properties, giving it huge potential as a candidate for next-generation electronics [10–17].

Defects in graphene, particularly cracks, substantially alter its mechanical and electrical properties [18], as shown in several studies on the failure behavior of graphene [19–24]. Thus studying the mechanical behavior of graphene in the presence of defects is vital for the design of nano-graphene-based structural (nano-composites) and nano-electromechanical systems (NEMS).

It has been reported, using experimental and molecular dynamics simulations, that graphene is able to heal its vacancies and other topological defects with the help of metal doping [25–32]. It has also been shown by researchers that this self-healing phenomenon can take place in single-layer graphene without an external aid [33]; it was demonstrated that the cracks nucleated 'in-situ,' owing to the application of tensile load, and self-healed within pico-seconds (ps) of load relaxation. This

DOI: 10.1201/9781003155751-1

self-healing was found to be independent of crack length, though a critical crack opening displacement range of 0.3–0.5 nm was required for self-healing to occur. The concept of self-healing was further used for the design of a sub-nano sensor using graphene that was able to sense a crack as soon as it started nucleation or even if it had propagated over a certain distance, thus making it a potential candidate for next-generation electronics applications.

We report here on our study of the intrinsic magnetic phase transitions (MPTs) in a pristine single-layer zigzag graphene nanoribbon (szGNR). Interestingly, by tuning the electric field (E) and temperature (T), three distinct magnetic phase behaviors (para, ferro, and antiferromagnetic) were exhibited in a pristine szGNR. The combined influence of the external electric field (E) and temperature (T) paved the way for the thermo-electromagnetic effect in pristine graphene, and also revealed the unrivaled positional parameters of these intrinsic magnetic phase transitions in szGNRs. The MPTs occurring in the system also followed a positional trend, and a change in these positional parameters, concerning the size of the szGNR, the quantitative application of external electric field (E), and temperature (T), were also reported in this study. These fundamental insights into intrinsic magnetic phase transitions in graphene provide an important step forward in developing graphene-based spin-transfer torque magneto-resistive random-access memory (STT-MRAM), quantum computing devices, magnonics, and spintronic memory applications.

1.2 SIMULATION METHODOLOGY FOR SELF-HEALING STUDIES

MD simulations were performed using LAMMPS (large-scale atomic/molecular massively parallel simulator) [34] and the AIREBO potential (adaptive inter-molecular reactive bond order potential) was used to define the force field [35, 36].

A quantized fracture mechanics (QFM) modified Griffith's model was implemented, and based on the QFM modeling fracture strength was largely related to crack tip radius and crack length. This method was accurate for studying fracture behavior, previously proved by [37, 38]. Quantized fracture strength is given by the following equation:

$$\sigma_f(l,\rho) = \sigma_c\sqrt{\frac{1+\rho/2L_0}{1+2l/L_0}} \tag{1.1}$$

Where σ_c is the fracture strength of a pristine sheet, ρ is the crack tip radius, l is the crack length, and L_0 is the minimum propagation distance of the crack.

The simulations were presented on a (5×5 nm) graphene sheet containing 1008 carbon atoms with PBC boundary conditions in two in-plane directions, and a vacuum space of 100 nm was considered along the z-direction to avoid edge effects. All simulations were presented at 300 K room temperature by a Nosé–Hoover thermostat. A velocity-Verlet time integrating scheme was applied with a time step of 0.0005 ps and held at equilibrium for 30 ps. Once equilibrium was attained, the tensile test was conducted using the deformation control method by applying strain loading with an increment of 0.001 ps until the ultimate tensile strength of the material was achieved. To study the self-healing phenomenon, all the forces acting on the deformed graphene sheet were eliminated, and the sheet was allowed to relax for a period of 150 ps.

1.3 SELF-HEALING OF PRISTINE AND PRE-EXISTING DEFECTIVE GRAPHENE SHEETS

Graphene sheets were tensile tested for pristine conditions in a longitudinal loading direction. Cracks formed and propagated in the pristine sheets after the load exceeded the ultimate tensile strength during uniaxial loading. At this point, all the forces acting on the deformed graphene sheets were removed, and the sheets were allowed to relax. The cracks were found to heal when a critical crack opening displacement of 0.3–0.5 nm was met. Figure 1.1 shows the different stages of self-healing in pristine sheets loaded in a longitudinal direction.

In the case of a pristine graphene sheet loaded in a longitudinal direction at a relaxation time of $t = 0$ ps, when the crack opening length was less than or equal to 0.5 nm, the carbon atoms displayed dangling bonds in the cracked region that were dynamically unstable [39]. At point $t=0$ ps (the point of relaxation of the sheet), 91% of the carbon bond lengths were found to be elongated in the range of 1.43–1.78 Å, as compared to a bond length of 1.42 Å; at $t=0.1$ ps the carbon atoms started reconstructing through dangling bond saturation, forming non-hexagonal rings, i.e., a pentagon–heptagon (5–7–7–5) structure, on the graphene sheet, known as Stone–Wales construction [40]. The Stone–Wales construction acts as a spontaneous topological change for the repair of the defective sheet, which is actively preferred in self-healing processes [41]. At $t = 0.2$ ps, the carbon atom coalesced due to atomic rearrangements, leading to the healing of the broken bonds, although 33% of the bonds still remained elongated in the range of 1.43–1.52 Å; 67% of the bonds were in the equilibrium bond length of 1.42 Å.

The strength of the healed graphene sheet was investigated via tensile testing, and the fracture strength observed was 16% less when compared to the pristine graphene sheet, owing to the formation of the Stone–Wales defect in the healed sheet; this demonstrated that the sheet could be reused repetitively but with a certain degradation in the strength of the sheet each time.

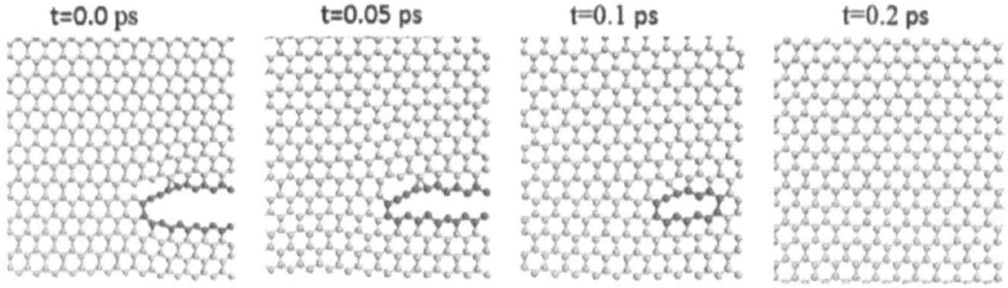

FIGURE 1.1 The different stages of self-healing in a pristine sheet after relaxation in the case of loading in a longitudinal mode.

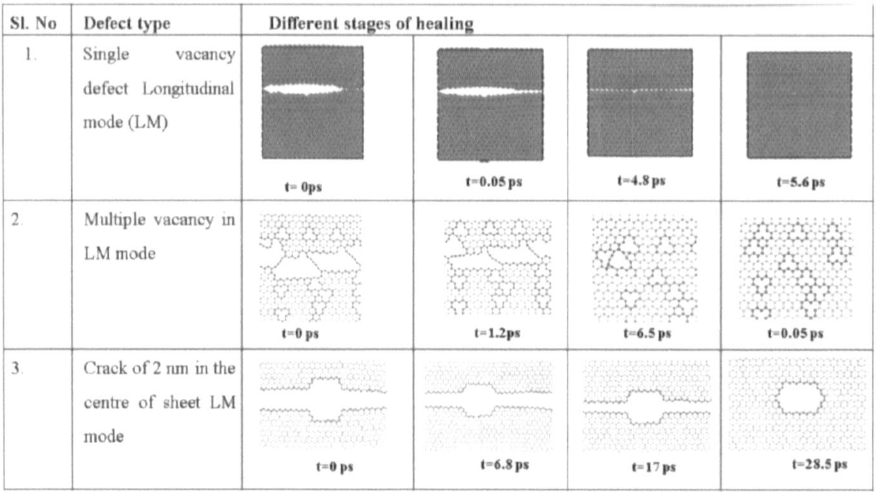

FIGURE 1.2 Effect of defects on the self-healing phenomenon.

Different types of imperfections and defects in the self-healing of cracks were reported here using MD simulations. The defect densities in the graphene sheet were in varying concentrations of missing atoms: 0.09, 0.2, 0.4, 0.6, 2, and 5% (single vacancy, multiple vacancies, cracks, etc.); this occurred in both the longitudinal and transverse modes assessed with tensile loading. Self-healing was present in all cases, though the defects created initially existed as shown in Figure 1.2, as initially deleted atoms cannot heal because there is no extra atoms to fill the defect in the simulation box. Crack healing took place over a span of pico-seconds whenever the critical gap is less than or equal to 0.5 nm. However, the healing time of single to multiple vacancies also increased.

We studied crack length and crack opening displacement to arrive at the critical crack length and crack opening displacement required for self-healing to occur. We found that self-healing was independent of the length of the crack. However, we observed that critical crack opening displacement must be within 0.3 to 0.5 nm for the crack to heal.

1.4 POSSIBLE APPLICATION OF THE SELF-HEALING PHENOMENON OF GRAPHENE

We envisage a sub-nano sensor design using the self-healing phenomenon of graphene that will sense a crack as crack nucleation starts or has even propagated a certain distance, making it a potential candidate for next-generation electronics. The sensor design is applicable only for graphene sheets loaded in a longitudinal direction, as the sensor requires that tears in the graphene sheet have a totally open circuit. But in Figure 1.2 we can observe that for pristine and defective sheets in a transverse mode, the sheets don't totally tear apart upon uniaxial loading, though self-healing is observed.

Figure 1.3 demonstrates the workings of a sub-nano sensor. A voltage source was attached to the graphene sheet to generate a current that passed through the graphene sheet, which was loaded in a horizontal direction, as shown in Figure 1.3(a). Cracks were generated in the graphene sheet due to tensile loading of the ultimate tensile strength at room temperature. When the voltage output was equal to the voltage input it acted as a closed-loop circuit. As the strain increased, and the resistivity of the graphene increased [42, 43], some output voltage was still obtained. However, as crack propagation led to the total tear of the sheet, the voltage output was equal to 0, as an open circuit was obtained (Figure 1.3b). A microcontroller can be used in conjunction with a circuit designed to receive a zero voltage signal and act instantaneously, sending a signal to release the load or stop further mechanical stretching or bending activity. The graphene sheet will immediately heal again (if the crack opening displacement is less than 0.5 nm) and the sensor becomes ready to measure any further

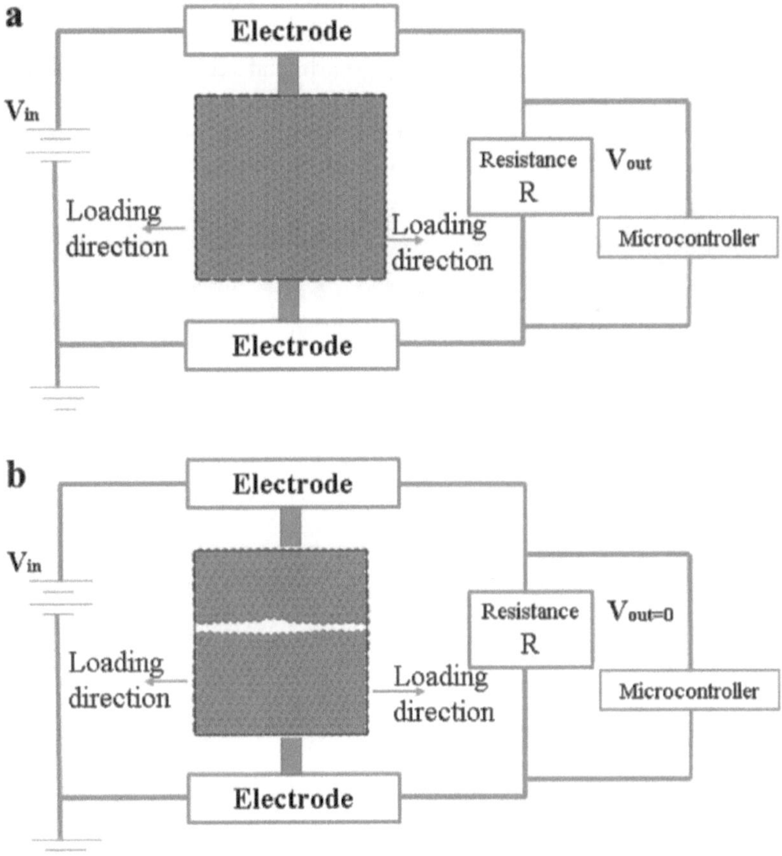

FIGURE 1.3 Schematic view of a nano-graphene sensor: (A) without crack; (B) with crack.

surges of the deforming load. Thus when the critical crack opening displacement is less than 0.3–0.5 nm, the graphene tends to heal the crack by itself. Thus even if a crack nucleates, it can be tracked as well as healed. Healing happens at a very fast rate of nearly 0.2 ps, as shown in Figure 1.4. A crack opening displacement above 0.5 nm can be detected, but healing cannot be achieved due to the open circuit, as shown in Figure 1.5. Thus if the microcontroller sends a signal to release the deforming load or pressure, the crack opening displacement can be controlled and thus the graphene sheet can be healed; this confirms perennial use of the sensor and thus that this technology can be extended to numerous other applications, such as for relief valves in pressure vessels, boilers, and cranes, etc. Crack healing was found to take place regardless of the nature of the original defect in the sheet. Thus, the envisaged sub-nano sensor can be efficiently used for pristine or defective parent graphene sheets loaded in a longitudinal direction.

These findings are believed to be significant for the development of artificial skin. Skin is the largest organ in the human body and is mechanically self-healing. If such behavior can be replicated in artificial skin, then it will signal a huge leap forward for robot technology [44–46]. Due to unprecedented stretching or bending, rupture of this electronic artificial skin may occur in robots, and thus by using the envisaged sub-nano crack sensing methodology the artificial skin could be healed, which would provide a significant application for the near future.

In the next section, we give an overview of the phenomenon of magnetic phase transitions detected under the collective impact of electric field (E) and temperature (T),

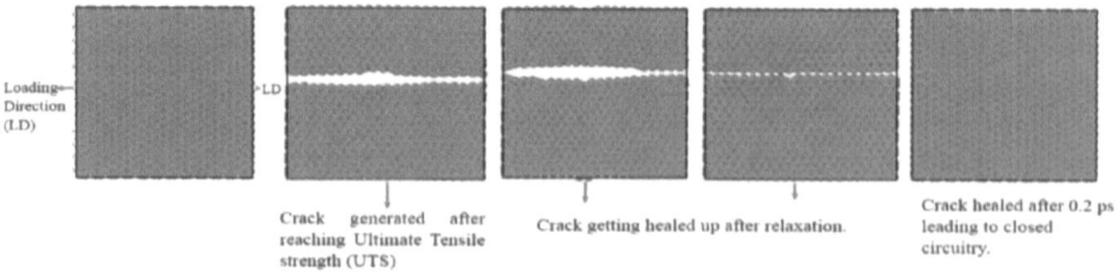

FIGURE 1.4 A loaded graphene sheet leading to detection of a crack as well as healing of the crack. Sub-nano sensing of the crack by applying voltage and measuring the output current.

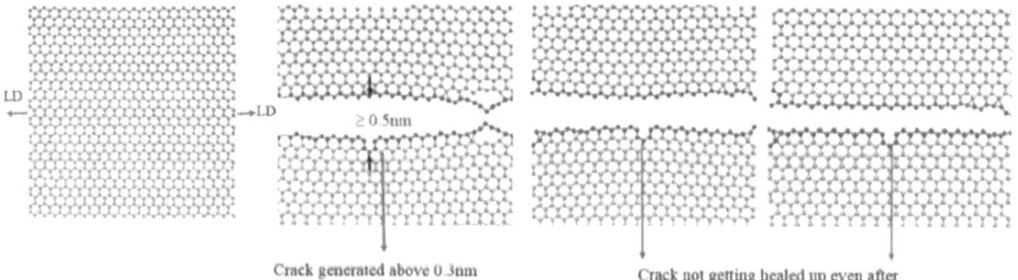

FIGURE 1.5 A loaded graphene sheet leading to detection of a crack with a voltage output of 0 V.

leading to a thermo-electromagnetic effect on a pristine single-layer zigzag graphene nanoribbon, which provides an insight into the intrinsic MPTs in graphene.

1.5 MAGNETIC PHASE TRANSITIONS

The magnetic properties of low dimensional materials have gained importance by virtue of the exceptional electron mobility, increased coulomb attraction, and their lightweight nature. Magnetism in graphene [47] also gained attention after the revelation of the nanoscale edge effect, the *p* electron energy spectrum, and self-healing phenomenon, which are capable of driving self-sustained, energy-efficient, ultra-low power devices in spintronic and magnonic applications [48–51]. Adding to the low dimensionality, if the intrinsic magnetic nature of pristine graphene is harnessed, this could bridge the gap between computing data signals using its spin properties and remote transmission of the processed big data in Internet of Things (IoT) sensor devices and nano-processors [52].

The combination of electric field and temperature was applied transversely to szGNRs (Figure 1.6) of 1, 2.5, 5, 7.5, and 10 nm, respectively. It was found that a similar property of MPTs (Figure 1.7) was exhibited irrespective of the size of the graphene nanoribbon, which in turn played a role in the position of the emerging magnetic phase. The PM, FM, and AFM phases appear at particular E and T values. With this control of the tunability of the electric field and the temperature, in real-time applications the value of other variables can be calculated, if either E or T is given. Reducing or increasing E and T will enable the re-configurability of pristine szGNR-based devices and make them ultra-low-power [53].

A magnetic phase diagram capturing phase transitions is detailed in Figure 1.7. The magnetization values were used to present the contour plot against the magnetic intensity and the product of the electric field and temperature (represented as ET). Numbers 1, 2, 3, and 4 were superimposed on the contour plot, distinguishing the

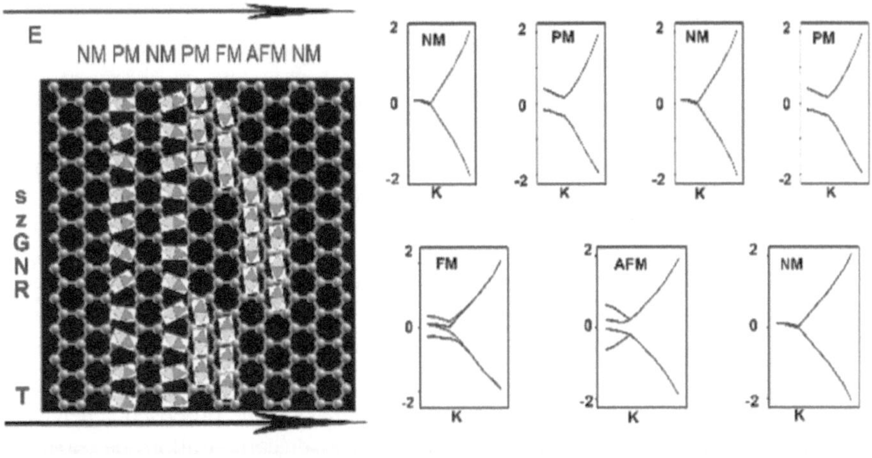

FIGURE 1.6 Pictorial representation of the concept [53, 54].

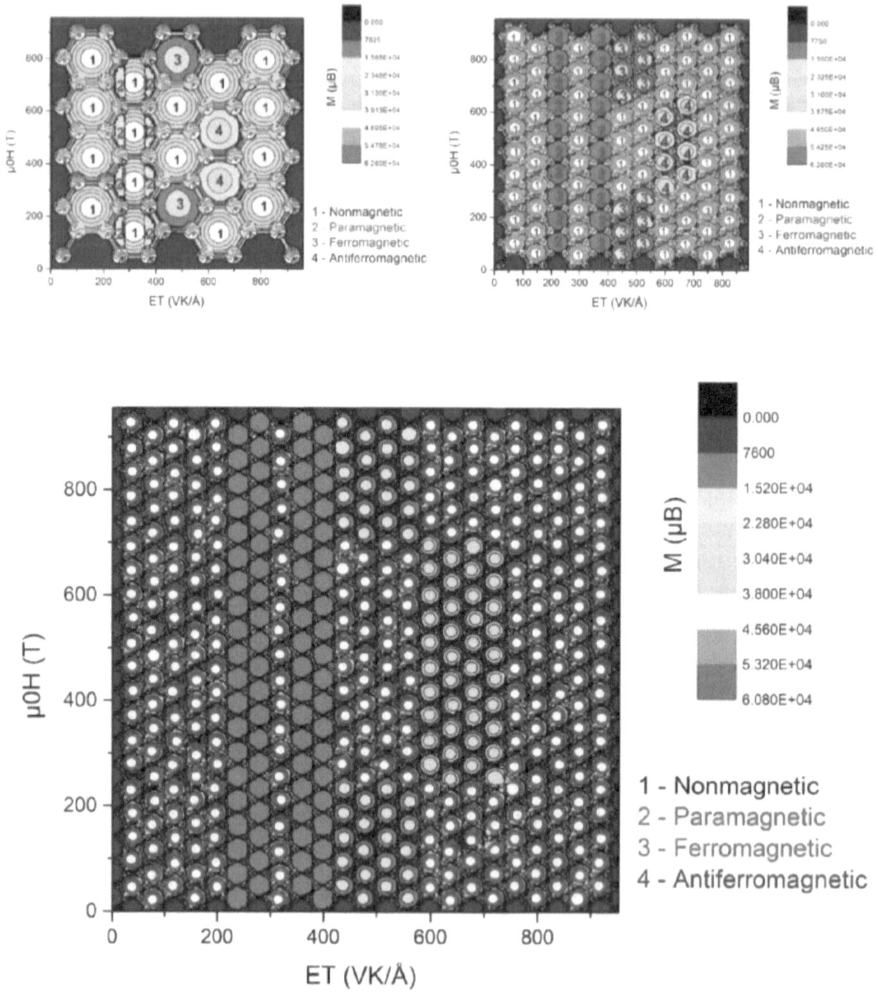

FIGURE 1.7 Magnetic phase transitions shown in various dimensions [54].

phases NM, PM, FM, and AFM, respectively. Pristine single-layer zigzag graphene nanoribbons of 1×1, 2.5×2.5 and 5.5×5.5 nm were studied [53, 54].

The proposed Bow-Tie scheme shown in Figure 1.8 is expressed by

$$M = f(X)f(Z) + a \tag{1.2}$$

where $X = \{E, P, O\}$ and $Z = \{T, S, G\}$ (M denotes magnetism, X and Z are set variables, a represents spin current, E is the electric field, P is pressure, O is optical force, T is temperature, S is strain, and G is geometry). The proposed scheme for the thermo-electromagnetic effect, its general form for carbon allotropes, and the envisaged device application in principle with the findings reported in this chapter

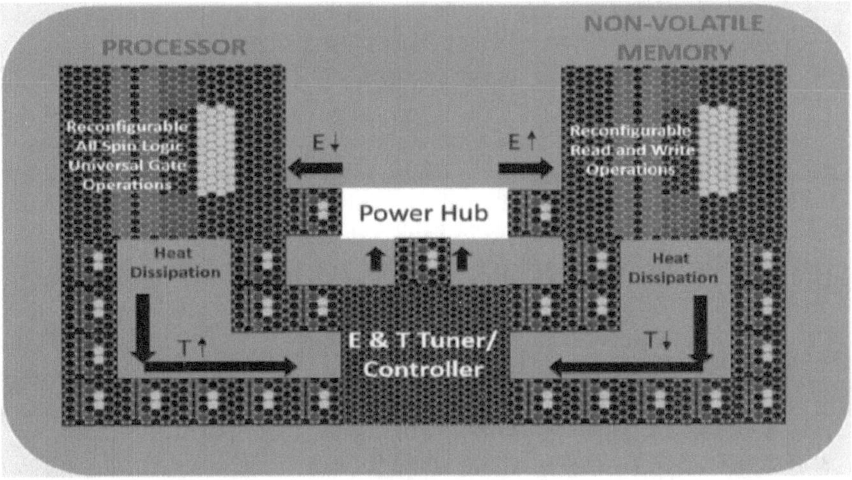

FIGURE 1.8 A Bow-Tie schematic to achieve magnetic phase transitions and the envisaged processor application using the proposed scheme [54].

are discussed in Figure 1.8. A Bow-Tie schematic is proposed to induce intrinsic magnetism in defect-free carbon allotropes, where the two tail ends count for the set variables. Magnetism is commonly induced using dopants, vacancies, and defects. The aim of this study was to induce it by applying energy to a pristine szGNR as defined by the set variables X and Z [53, 54]. Thus, the intrinsic magnetic phase transitions in szGNRs with their corresponding positional parameters were studied using first-principle calculations.

1.6 CONCLUSIONS

In this chapter, we discussed:

i. The self-healing nature of single-layer graphene for pristine and pre-existing defective graphene sheets (single and multiple vacancies and cracks of different orientations).
ii. Self-healing within pico-seconds of load relaxation was found for defective and pristine graphene sheets when the distance between crack surfaces was ≤ 0.5nm both in the presence and in the absence of a pre-existing defect.
iii. We envisage the self-healing phenomena of graphene for artificial skin in the future designs of robots, making graphene a next-generation electronic sensor for sub-nano sensing applications.
iv. Intrinsic magnetic phase transitions in szGNRs, under the combined influence of E and T (the thermo-electromagnetic effect), was reported.
v. We showed that the novel scheme of applying E and T in combination drastically tuned the magnetic properties of pristine szGNRs. The system

changes from the NM phase to the PM phase to the FM phase to the AFM phase, and then finally to the NM phase with changes in E and T.

vi. The obtained results also concluded that E and T are inversely proportional to each other for achieving the intrinsic tuneable magnetism reported here, which can lead to the novel electronic devices as envisaged in this chapter.

vii. These fundamental insights to intrinsic magnetic phase transitions in graphene provide an important step forward in developing graphene-based spin-transfer torque magneto-resistive random-access memory, quantum computing devices, magnonics, and spintronic memory applications.

REFERENCES

1. A. K. Geim, Graphene: status and prospects. *Science* 324, 1530–1534 (2009).
2. A. K. Geim, & K. S. Novoselov, The rise of graphene. *Nat. Mater.* 6, 183–191 (2007).
3. K. S. Novoselov, A. K. Geim, S. V. Morozov, D. Jiang, M. I. Katsnelson, I. V. Grigorieva, S. V. Dubonos, & A. A. Firsov, Two-dimensional gas of massless Dirac fermions in graphene. *Nature* 438, 197–200 (2005).
4. S. V. Morozov, K. S. Novoselov, M. I. Katsnelso, F. Schedin, D. C. Elias, J. A. Jaszczak, & A. K. Geim, Giant intrinsic carrier mobilities in graphene and its bilayer. *Phys. Rev. Lett.* 100, 016602 (2008).
5. U. Kürüm, O. Ö. Ekiz, H. G. Yaglioglu, A. Elmali, M. Ürel, H. Güner, A. K. Mızrak, B. Ortac, & A. Dâna, Electrochemically tunable ultrafast optical response of graphene oxide. *Appl. Phys. Lett.* 98, 141103 (2011).
6. J. Liu, A. R. Wright, C. Zhang, & Z. Ma, Strong terahertz conductance of graphene nanoribbons under a magnetic field. *Appl. Phys. Lett.* 93, 041106 (2008).
7. K. Saito, J. Nakamura, & A. Natori, Ballistic thermal conductance of a graphene sheet. *Phys. Rev. B* 76, 115409 (2007).
8. A. A. Balandin, S. Ghosh, W. Bao, I. Calizo, D. Teweldebrhan, F. Miao, & C. N. Lau, Superior thermal conductor graphene. *Nano Lett.* 8, 902–907 (2008).
9. I. W. Frank, D. M. Tanenbaum, A. M. van der Zande, & P. L. McEuen, Mechanical property of suspended graphene sheets. *J. Vac. Sci. Technol. B* 25, 2558 (2007).
10. K. I. Bolotin, K. J. Sikes, J. Hone, H. L. Stormer, & P. Kim, Temperature-dependent transport in suspended graphene. *Phys. Rev. Lett.* 101, 096802 (2008).
11. L. Chenguang, Y. Zhenning, N. David, Z. Aruna, & J. Bor, Graphene-based supercapacitor with an ultrahigh energy density. *Nano Lett.* 10, 4863–4868 (2010).
12. A. H. Castro Neto, F. Guinea, N. M. R. Peres, K. S. Novoselov, & A. K. Geim, The electronic properties of graphene. *Rev. Mod. Phys.* 81, 109 (2009).
13. W. Qiong, X. Yuxi, Y. Zhiyi, L. Anran, & S. Gaoquan, Super capacitors based on flexible graphene/ polyaniline nanofiber composite films. *Nano Lett.* 4, 1963–1970 (2010).
14. Y. M. Lin, et al., Operation of graphene transistors at gigahertz frequencies. *Nano Lett.* **9**, 422–426 (2009).
15. L. Lei, et al., High-speed graphene transistors with a self-aligned nanowire gate. *Nature* 467, 305–308 (2010).
16. S. Roman, T. Floriano, & R. L. Valeria, Logic gates with a single graphene transistor. *Appl. Phys. Lett.* 94, 073305 (2009).
17. J. W. Kang, & K. W. Lee, Molecular dynamics study of carbon-nanotube shuttle-memory on graphene nanoribbon array. *Comput. Mater. Sci.* 93, 164–168 (October 2014).
18. N. Gorjizadeh, A. Farajian, & Y. Kawazoe, The effects of defects on the conductance of graphene nanoribbons. *Nanotechnology*. 20, 015201 (2009).

19. H. Zhao, & N. R. Aluru, Temperature and strain-rate dependent fracture strength of graphene. *J. Appl. Phys.* 108, 064321 (2010).
20. M. Q. Le, & R. C. Batra, Crack propagation in pre-strained single-layer graphene sheets. *Comput. Mater. Sci.* 84, 238–243 (March 2014).
21. M. Q. Le, & R. C. Batra, Single-edge crack growth in graphene sheets under tension. *Comput. Mater. Sci.* 69, 381–388 (March 2013).
22. T. C. Theodosiou, & D. A. Saravanos, Numerical simulation of graphene fracture using molecular mechanics based nonlinear finite elements. *Comput. Mater. Sci.* 82, 56–65 (1 February 2014).
23. Y. Y. Zhang, & Y. T. Gu, Mechanical properties of graphene: Effects of layer number, temperature and isotope. *Comput. Mater. Sci.* 71, 197–200 (April 2013).
24. Z. Peng, et al., Fracture toughness of graphene. *Nat. Commun.* 5, 3782 (2014).
25. R. Zan, Q. M. Ramasse, U. Bangert, & K. S. Novoselov, Graphene reknits its holes. *Nano Lett.* 12, 3936–3940 (2012).
26. L. Tsetseris, & S. T. Pantelides, Adatom complexes and self-healing mechanisms on graphene and single-wall carbon nanotubes. *Carbon.* 47, 901–908 (2008).
27. V. O. Özçelik, H. H. Gurel, & S. Ciraci, Self-healing of vacancy defects in single-layer graphene and silicene, *Phys. Rev. B.* 88, 045440 (2013).
28. L. Yung- Chang, Y. Chao- Hui, H. Ju-Chun, & C. Po- Wen, High mobility flexible graphene field-effect transistors with self-healing gate dielectrics, *Nano Lett.* 6, 4469–4474 (2012).
29. J. Chen, et al., Self healing of defected graphene. *Appl. Phys. Lett.* 102, 103107 (2013).
30. D. Y. Kim, et al., Self-healing reduced graphene oxide films by supersonic kinetic spraying. *Adv. Funct. Mater.* 24, 4986–4995 (2014).
31. X. Xingcheng, X. Tao, & C. Yang-Tse, Self-healable graphenepolymer composites. *J. Mater. Chem.* 20, 3508–3514 (2010).
32. C. Hou, et al., A strong and stretchable self-healing film with self-activated pressure sensitivity for potential artificial skin applications, *Sci. Rep.* 3, 3138 (2013).
33. V. Sanghamitra Debroy P. K. Miriyala, K. Vijaya Sekhar, S. G. Acharyya, & A. Acharyya, Graphene heals thy cracks. Ms. Ref. No.: COMMAT-D-15-00423R1. (In press).
34. S. Plimpton, Fast parallel algorithms for short-range molecular dynamics. *J. Comput. Phys.* 117, 1–19 (1995).
35. S. J. Stuart, A. B. Tutein, & J. A. Harrison, A reactive potential for hydrocarbons with inter molecular interactions. *J. Chem. Phys.* 112, 6472–6486 (2000).
36. J. Tersoff, New empirical approach for the structure and energy of covalent systems. *Phys. Rev. B.* 37, 6991 (1988).
37. D. W. Brenner, Empirical potential for hydrocarbons for use in simulating the chemical vapor deposition of diamond films. *Phys. Rev. B.* 46, 1992 (1948).
38. S. A. Nose, A unified formulation of the constant temperature molecular dynamics methods, *J. Chem. Phys.* 81, 511–519 (1984).
39. C. L. Zhang, & H. S. Shen, Self-healing in defective carbon nanotubes: a molecular dynamics study. *J. Phys. Condens. Matter.* 19, 386212 (2007). doi: 10.1088/0953-8984/19/38/386212
40. M. B. Nardelli, B. I. Yakobson, & J. Bernholc, Brittle and ductile behavior in carbon nanotubes. *Phys. Rev. Lett.* 81 4656 (1998). doi: http://dx.doi.org/10.1103/PhysRevLett.81.4656
41. C.L.Zhang,&H.S.Shen,Self-healingindefectivecarbonnanotubes:amoleculardynamics study. *J. Phys.: Condens. Matter.* 19, 386212, 2007. doi:10.1088/0953-8984/19/38/386212
42. X. W. Fu, Z. M. Liao, J. X. Zhou, Y. B. Zhou, H. C. Wu, R. Zhang, G. Y. Jing, J. Xu, X. S. Wu, W. L. Guo, & D. P. Yu, Strain dependent resistance in chemical vapor deposition grown graphene. *Appl. Phys. Lett.* 99, 213107 (2011).

43. Z. Jing, Z. Guang Yu, & S. Dong-Xia, Review of graphene-based strain sensors. *Chin. Phys. B.* 22 (5), 057701 (2013).
44. T. Someya, et al., A large-area, flexible pressure sensor matrix with organic field effect transistors for artificial skin applications. *Proc. Natl. Acad. Sci.* 101, 9966–9970 (2004).
45. I. V. Yannas, & J. F. Burke, Design of and artificial skin. I. Basic design principles. *J. Biomed. Mater. Res.* 14, 65–81 (1980).
46. V. Maheshwari, & R. F. Saraf, Tactile devices to sense touch on a par with a human finger. *Angew. Chem. Int. Ed.* 47, 7808–7826 (2008).
47. H. Lee, H.-J. Lee, & S. Y. Kim, Room-temperature ferromagnetism from an array of asymmetric zigzag-edge nanoribbons in a graphene junction. *Carbon.* 127, 57–63 (2018).
48. L. A. Gonzalez-Arraga, J. L. Lado, F. Guinea, & P. San-Jose, Electrically controllable magnetism in twisted bilayer graphene. *Phys. Rev. Lett.* 119 (10), 107201 (2017).
49. S. Nigar, Z. F. Zhou, H. Wang, & M. Imtiaz, Modulating the electronic and magnetic properties of graphene. *RSC Adv.* 7 (81), 51546–51580 (2017).
50. G. Z. Magda, X. Jin, I. Hagymasi, P. Vancso, Z. Osvath, P. Nemes-Incze, C. Hwang, L. P. Biro, & L. Tapaszto, Room-temperature magnetic order on zigzag edges of narrow graphene nanoribbons. *Nature.* 514 (7524), 60811 (2014).
51. C. Hyun et al., Graphene edges and beyond: Temperature-driven structures and electromagnetic properties. *ACS Nano.* 9 (5), 4669–4674 (2015).
52. M. P. López-Sancho, & L. Brey, Charged topological solitons in zigzag graphene nanoribbons. *2D Mater.* 5 (1), 015026 (2017).
53. S. Sivasubramani, & A. Acharyya, *Investigation on electronic transport and magnetic properties of graphene for its applications in nanomagnetic computing*, Master's Thesis Indian Institute of Technology, Hyderabad, India (2018).
54. S. Sivasubramani, S. Debroy, S. G. Acharyya, & A. Acharyya, Tunable intrinsic magnetic phase transition in pristine single-layer graphene nanoribbons. *Nanotechnology.* 29 (45), 455701 (2018).

2 Models for Modern Spintronics Memories with Layered Magnetic Interfaces

Divyansh Jain and Aniket Singha

CONTENTS

2.1 INTRODUCTION

Magneto-resistive random-access memory (MRAM) is a non-volatile memory where data is stored in a magnetic domain orientation [1]. It was developed in the mid-1980s using spin electron perspectives. Advocates of this technology claim that magneto-resistive RAM will eventually overtake competitor technologies to become a universal memory technology [2]. Currently, MRAM uses traditional memory technologies, i.e., SRAM, DRAM, and flash memory, and plays a crucial role in the semiconductor market. MRAM stores information in magnetic moments but does not supply an electric charge or current flow to the junction. A lot of effort is currently geared toward the commercialization of spin-transfer torque magnetic random-access memory (STT-MRAM) in standard consumer-level systems due to their nonvolatility, unlimited endurance, low-power consumption, better CMOS adaptability, lower read–write access time, and high-speed operation. The main building block of an STT-MRAM cell is a magnetic tunnel junction (MTJ), which comprises two magnetic layers separated by a thin insulating layer. One of the two

DOI: 10.1201/9781003155751-2

ferromagnetic plates is a permanent magnet set to a specific polarity. The other plate's magnetization is controlled by an external field for storing memory bits. This configuration is called a magnetic tunnel junction and is the primary structure of an MRAM cell. A storage device is made from a grid of memory cells. Conventional MTJs contain in-plane magnetic anisotropy (IMA) for the magnetic layers. Different phenomena, like the spin-Hall-effect (SHE) and magnetic anisotropy effect, etc., are proposed for enhancing the performance of STT-MRAM [3]. Due to the finite TMR ratio, the high write current (~ 100 μA) and low sense edge [4, 5] are the two most significant concerns for the commercialization of STT-MRAM cells. The drawback of high write current lies in the increased energy demand for data manipulation [6], thereby constraining the transistor's scalability [7]. Due to the poor spin-infusion efficiency, and the least-resistance write method, SHE model-based memory bit-cells are used. We end this chapter with a brief note on the spin interface and performance enhancement of MRAM cells due to voltage-controlled magnetic anisotropy (VCMA) and the Rashba spin-orbit interaction (SOI) effect.

2.2 ADVANCED SPINTRONICS MEMORIES

2.2.1 Modern Spintronics Memories

Next-generation MRAM cells are modeled to generate toggle spin torque in the spin channel using the VCMA and Rashba SOI effect. The VCMA effect characterizes the connection between the electric field and interfacial anisotropy. It increases the precessional switching probability at large voltages [8, 9]. Anti-ferromagnetic (AFM) materials are immune to magnetic field fluctuations [10, 11]. The magnetic properties of magnetic layers in a MRAM cell are governed by the Stoner–Wohlfarth principle. The switching properties are dependent on the external field pulse strength and direction, pulse length, and pulse shape. It determines the types of torque, Spin-transfer torque (STT) operation, and free layer optimization in Stoner physics particles. STT is an effect in which the orientation of a magnetic layer in MTJ can be modified by means of the spin-polarized current. Spin-orbit torque is induced through the spin current and mediates the transfer of angular momentum from the lattice to the magnetic layer. It leads to sustained magnetic oscillations or switching of the magnetic structures [10]. The toggle spin torque (TST) follows the interplay effect between the STT and spin-orbit-torque (SOT) mechanism.

In Figure 2.1, the MRAM cell uses toggle-like operations, which rely on the STT and SOT mechanism [12]. The current pulses (I_{STT} and I_{SOT}) and magnetic field H are marked for the write operation. The free layer, pinned layer, and tunnel barrier of MTJ are denoted by *FL*, *PL*, and *TB*, respectively. The heavy metal, source line, bit line, and write/read word line are indicated by *HM*, *SL*, *BL*, and *WWL*/*RWL*, respectively.

2.2.2 VCMA Model

Magnetic anisotropy describes the directional dependence of the magnetic properties of a material. The easy axis is in the crystal direction, along which a small magnetic

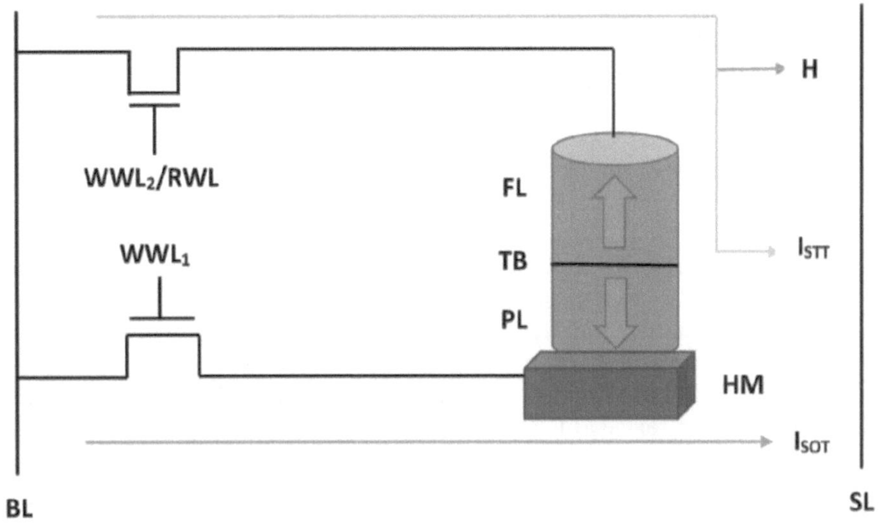

FIGURE 2.1 Typical bit-cells of the TST-MRAM memory unit.

field is enough to achieve the saturated magnetization state. The hard axis is in the crystal direction, along which a high field is required to orient magnetic polarization. A small magnetic fluctuation can reorient the magnetic polarization vector from the hard toward the easy axis. Magnetic anisotropy is divided into PMA and IMA. In PMA, the easy axis is perpendicular to the thin-film surface, and the direction of the hard axis is in-plane to the thin film; this is a crucial parameter for enhancing a nanomagnet's thermal cohesion. Generally, effective PMA occurs at the interface [13] between a ferromagnetic material (Co, Al) and non-magnetic material (Pt, Pd, W, and Au). The Co/Pt, Co/Pd, and Co/Au heterostructures demonstrate PMA and have a high Gilbert damping ratio [14]. Hence, due to high switching time, they are not suitable for current-induced magnetization switches in geophysical applications. The thin films lie on the hard axis due to the internal lines of magnetic force. Due to the identical thermal stability factor (TSF), the switching current of the PMA-based device is lower than the IMA-based device. PMA is classified into an interfacial PMA (IPMA) and a crystalline PMA (CPMA). IPMA exists in the CoFeB composite and generates a perpendicular anisotropic field along the planes perpendicular to the thin films. The thickness of this composite is lower than the critical thickness for IPMA. However, CPMA exists in high crystalline anisotropy materials, i.e., Co/Pt and Fe/Pd layers, to develop the largest magnetization field. The TST-MRAM cell's primary advantage is its higher signal stability and robustness against magnetic fluctuations than conventional MRAM cells. An anti-ferromagnetic thin film is often used as a replacement for a non-magnetic thin film to enhance MRAM performance. The anti-ferromagnetic film causes a shift in the soft magnetization vector of a ferromagnetic layer. The operation of the TST-MRAM cell is based on the VCMA effect, described in the following paragraph.

VCMA employs magnetization states in nanomagnets and is used for low-power spintronic applications. The physical phenomenon behind the VCMA effect is that it involves an electric-field-induced change in the d-orbital occupation at the interface between the transition metal and non-magnetic material (such as CoFeB/MgO, Al_2O_3/MgO, etc.). The VCMA effect is actuated under negative bias conditions. The anisotropy coefficient in the VCMA effect, known as the VCMA coefficient, depends on magnetic layer thickness. Reduced PMA causes the transfer of voltage-induced oxygen ions into a thin ferromagnetic metal layer [15] for non-linear operations. This process is controlled by the oxidation procedure [16] of electron ions. It induces the electromigration/charge-trapping effect in MgO barriers [17, 18]. This electromigration induces a disturbance in MgO film formation [19] and is transported by a spin electron's critical location.

The thermal field fluctuations, given by the zero-mean Gaussian density function (GDF), can be represented as follows:

$$\sigma_{h_{th}} = \frac{2K_b \alpha T}{\mu \gamma V_f M_s \Delta t}, \tag{2.1}$$

where α is the Gilbert damping constant, μ *is* the permeability in a vacuum, γ is the gyromagnetic ratio, V_f is the free layer enclosed volume, and $\Delta\ t$ is the step time. The injected electrons are polarized along the magnetization vector, thereby activating either groping [20] or delays switching for the MRAM cell.

VCMA-MTJ polarization [21] describes the charge miniaturization in the magnetic layer along the x-axis. The energy state is altered by an induced voltage (V_E) and external magnetic field simultaneously. These polarization dynamics are governed by the Landau–Lifshitz–Gilbert (LLG) equation. The LLG equation [22] for magnetic torque is represented as:

$$\frac{\partial \overrightarrow{M}_0}{\partial t} = -\gamma\left(\overrightarrow{M}_0 \times \overrightarrow{H}_{eff}\right) - \frac{\gamma\lambda}{M_s}\left\{\overrightarrow{M}_0 \times \left(\overrightarrow{M}_0 \times \overrightarrow{H}_{eff}\right)\right\}, \tag{2.2}$$

where $\overrightarrow{M}_0$ is the magnetization, λ is the phenomenological damping boundary, γ is the gyromagnetic ratio, and g is the Lande g-factor.

According to Equation (2.1), a small damping ratio is suitable for MRAM operations. In 1955, Gilbert expressed the large damping [23] equation in thin films as:

$$\frac{\partial \overrightarrow{M}}{\partial t} = \gamma\left\{\overrightarrow{M} \times \left(\overrightarrow{H}_{eff} - \frac{\alpha}{\gamma M_s} \times \frac{\mathrm{d}\overrightarrow{M}}{\mathrm{d}t}\right)\right\}, \tag{2.2A}$$

where α and M_s represent the Gilbert damping parameter and the saturation magnetization [24], respectively.

The VCMA model is derived from the LLG equation under the assumptions of the critical thickness of free and fixed magnetic layers and takes account of the precession, damping, and STT mechanism. The VCMA model equation is represented as follows:

$$\frac{1+\alpha^2}{\gamma}\cdot\frac{\partial\vec{M}}{\partial t}=-\vec{M}\times\vec{H}_{eff}(V)-\alpha\cdot\vec{M}\times\left(\vec{M}\times\vec{H}_{eff}(V)\right)+\frac{\hbar PJ}{2et_FM_s}\vec{M}\times\left(\vec{M}\times\vec{M}_P\right), \quad (2.3)$$

where $\vec{M}$ is the magnetization vector of the free layer, $\vec{H}_{eff}(V)$ is the voltage-dependent effective magnetic field, $\hbar$ is the reduced Plank's constant, P is the spin polarization, J is the switching current density, e is the electron charge, and $\vec{M}_P$ is the magnetization vector of the pinned layer.

Generally, $\vec{H}_{eff}(V)$ takes into account the different internal and external magnetic field components affecting the free layer [25] and is expressed as:

$$\vec{H}_{eff}(V)=\vec{H}_{ext}+\vec{H}_d+\vec{H}_{th}+\vec{H}_{K\perp eff}(V),$$
$$\vec{H}_{K\perp eff}(V)=\left(\frac{2K_i(V)}{\mu_0M_st_F}\right)\vec{m}_z\hat{z}, \quad (2.4)$$

where $\vec{H}_{ext}$ is the external magnetic field, $\vec{H}_d$ is the demagnetization field, $\vec{H}_{th}$ is the thermal field, $\vec{H}_{K\perp eff}(V)$ is the voltage-dependent effective magnetic anisotropy field, μ_0 is the permeability, and $\vec{m}_z$ is the magnetization vector.

The TSF for MTJ is an essential parameter determining the FM layers' data retention capability [9]. This model depends on the free layer's energy barrier for magnetization switching and is normalized to unity at the energy level $E = k_BT$ [26], where k_B is the Boltzmann constant. It is represented as follows:

$$TSF=\frac{E_b}{k_BT}=\left[\frac{\left(k_i-2\pi\left|\vec{M}_s\right|^2t_f\right)A}{k_B\times T}\right]. \quad (2.5)$$

In the above equation, E_b, A, and t_F, respectively, denote the voltage-dependent energy barrier between two stable states, the free layer's cross-sectional area, and the difference in switching fall time between fixed and free magnetic layers. The magnetization factor (k_i) is the ratio of the free layer magnetization (M_F) to the fixed layer magnetization (M_p) in MTJ.

The theoretical field that can align the magnetization perpendicular to the easy axis is called an anisotropic field. This field is dependent on the PMA and current density in the magnetic layer. The variation in TSF is inversely proportional to the anisotropic field, expressed as:

$$(TSF)_v\propto\frac{1}{H_a}, \quad (2.6)$$

where $(TSF)_v$, and H_a, represent the variation in TSF and the magnitude of the anisotropic field, respectively.

The TSF of the MRAM cell is influenced by the VCMA effect due to variation in the magnetization's initial angle. According to the Fokker-Plank distribution [27], the probability density function (*S*) for magnetic switching induced by the VCMA effect is represented as:

$$S(\theta)\Big|_{t=0} = \frac{e^{-TSF\cdot sin^2\theta}}{\int_0^{\pi} \sin\theta \cdot e^{-TSF\cdot sin^2\theta} d\theta}, \tag{2.7}$$

where $S(\theta)\Big|_{t=0}$ and TSF are the switching probability density function for initial magnetization aligned at an angle θ to the easy axis and thermal stability factor, respectively.

According to the Jullier's formula [28], the voltage and temperature dependence of TMR is expressed as:

$$TMR(T,V) = \frac{2P_0^2\left(1-\alpha_{sp}T^{3/2}\right)^2}{1-P_0^2\left(1-\alpha_{sp}T^{3/2}\right)^2} \cdot \frac{1}{1+\left(V/V_0\right)^2}, \tag{2.8}$$

where P_0, α_{sp}, and V_0 represent the polarization factor, a material-dependent empirical constant ($= 2e^{-5}$), and the bias voltage for $0.5TMR$, respectively.

The block diagram of the TST operation by the VCMA mechanism is represented in Figure 2.2. The magnetic anisotropy saturated at critical temperature t_c with applied MTJ voltage variation. This process was controlled using TSF mode activation and a high *K*-factor. H_{efx}, H_{efy}, and H_{efz} represent the magnetic field parameters along the x-axis, y-axis, and z-axis, respectively. The ambient temperature of

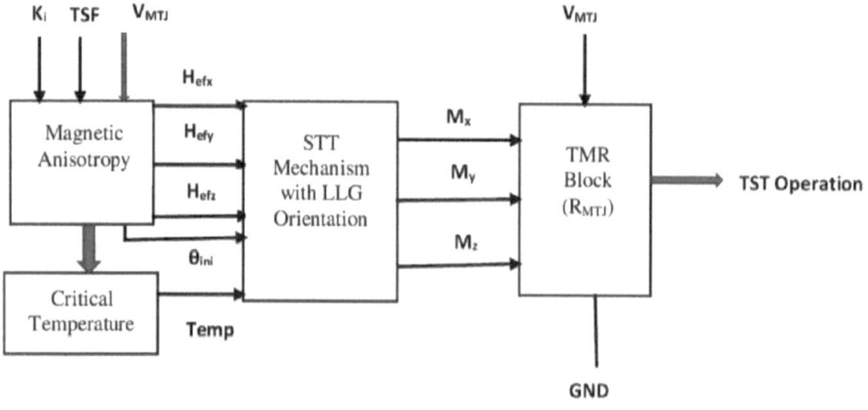

FIGURE 2.2 Block diagram of the VCMA dependent TST Mechanism.

MTJ, initial angular momentum, and an equivalent field coefficient are applied to the STT block using the LLG equation. The magnetization factor orientation with MTJ voltage is shifted in TMR and AMR blocks. This process generates a TST effect. The ground terminal is used to nullify the magnetic disturbances in the free layer.

Table 2.1 shows the TST-MRAM cell's design simulation based on the VCMA model. The saturated temperature, damping factor, and anisotropic coefficient of this model are fixed at 357, 0.024, and 105 K, respectively.

Figure 2.3(a) represents the different layers and material composition of a TST-MTJ cell, i.e., Co/Pt with 2.2 nm thickness, T_a with 0.3 nm and 2 nm thickness, CoFeB with 1.3 nm thickness with t nm boundary, and MgO with 1.5 nm thickness. A layer with the composition of different materials like CoFeB, MgO, and Fe is inserted at fixed terminal temperatures T_1 and T_2. This process is shown in Figure 2.3(b). The pinned FM layer with parasitic capacitances [29] is introduced by magnetic anisotropy at node voltage V_X.

TABLE 2.1
Design Specifications of a VCMA-Based TST-MRAM Cell

S. No.	Design Parameters	Numerical Values
1.	Saturated magnetization at 0 K (KA/M)	945
2.	Polarization factor	0.52
3.	Material	CoFeB/MgO
4.	Saturated temperature (K)	357
5.	Damping factor	0.024
6.	Anisotropic coefficient	105

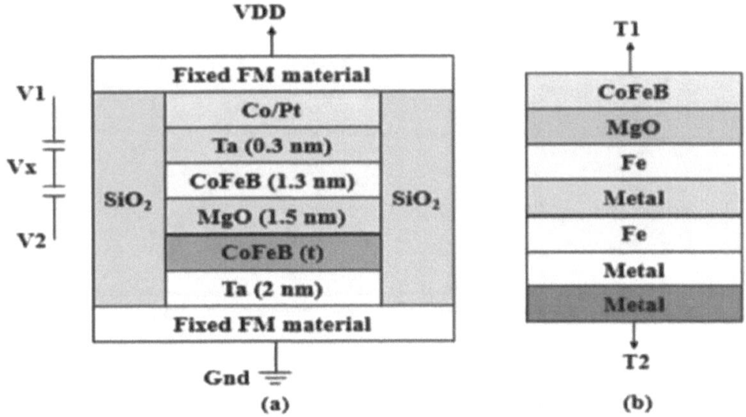

FIGURE 2.3 (a) MTJ structure under the VCMA condition. (b) VCMA structure with various material variations by the current and voltage biasing conditions.

Figure 2.4(a) demonstrates the spin current variation with time simulated via a SPICE hole density node in the *XY* direction. The spin current fluctuates during magnetic switching due to the resistor nodes. The spin current $I_s = 16$ nA for parallel to antiparallel switching at 0.45 ns and $I_s = 40$ nA for antiparallel to parallel switching at 2.2 ns. The spin current reaches a steady-state at $t = 3.8$ ns, and the value of

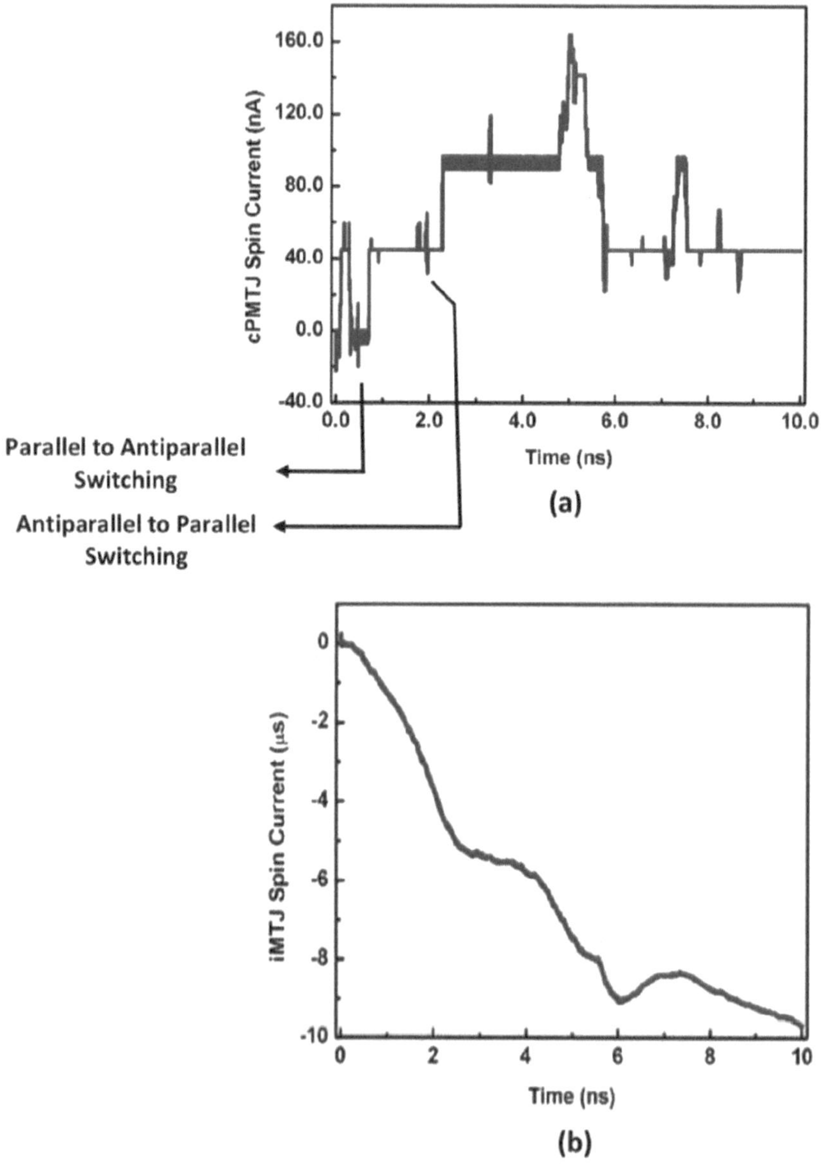

FIGURE 2.4 Spin current variation with the switching time in the (a) CPMTJ model; (b) IMTJ model.

the spin current is 98 nA in the CPMTJ model. The spin current fluctuations in the interface perpendicular magnetic tunnel junction (iPMTJ) model are greater than in the CPMTJ model, as represented in Figure 2.4(b). The spin current falls instantaneously at $t = 3$ ns and achieves the minimum value of−10 μA in the iPMTJ model. Although, the spin current fluctuates at $t = 4.42$ and 8.65 ns, respectively, for an in-plane orientation of the free layer.

Figure 2.5 demonstrates the variations in conductance with the sensing voltage. The total conductance achieves a maximum value of 4.55 m℧ and a minimum value of 2.98 m℧ at 0 mV, respectively.

2.2.3 The Rashba SOI Model

The Rashba effect instills a momentum-dependent splitting of spin bands [30] in bulk crystal and low-dimensional condensed matter systems. This process is similar to the splitting of particles and anti-particles in the systems described by the Dirac Hamiltonian. The combined effect of SOI and the asymmetry of crystal potential is responsible for spin band splitting with the splitting direction perpendicular to the two-dimensional plane [31]. The Rashba effect is mathematically described by the Rashba–Hamiltonian model [32, 33], expressed as:

$$\vec{H}_R = \alpha\left(\vec{\sigma} \times \vec{\rho}\right) \times \hat{z}, \tag{2.9}$$

where α , $\vec{\rho}$, and $\vec{\sigma}$ represent the Rashba coupling constant, the electron momentum, and Pauli matrix-vector describing the electronic spin, respectively. The two-dimensional version of the Dirac Hamiltonian follows the 90^0 rotation of spin along the z-axis.

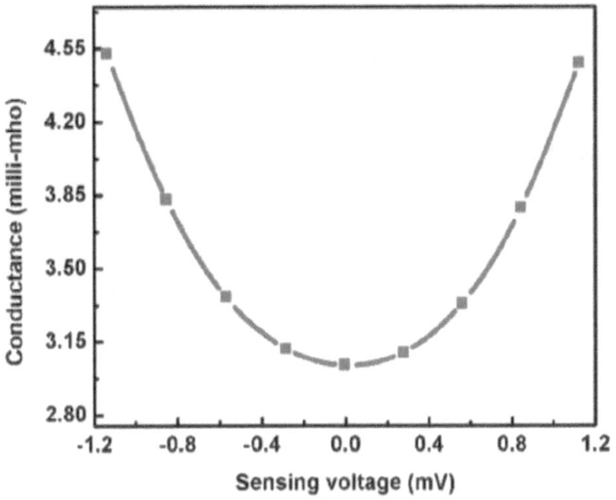

FIGURE 2.5 Conductance curve regarding MTJ sensing voltage.

The Rashba effect is a direct consequence of the inversion symmetry related to spin band splitting in the direction perpendicular to the two-dimensional plane. It is given as follows:

$$H_e = -\vec{E} \cdot \hat{z}, \tag{2.10}$$

H_e, and $\vec{E}$ are the Hamiltonian terms associated with the electric field and total electric field. Due to electronic transport in an electric field $(\vec{E})$, a resultant magnetic field $\vec{B}$ is induced in the relativistic electronic frame of reference, given by:

$$\vec{B} = -\frac{\vec{v} \times \vec{E}}{c^2}, \tag{2.11}$$

where c and $\vec{v}$ denote the speed of light and the absolute velocity of injected electrons, respectively. The effective magnetic field $\vec{B}$ interacts with the electron spin momentum of the dipole to generate the spin-orbit splitting, with the Hamiltonian given by:

$$H_{SO} = \frac{g\mu_b}{2c^2}\left(\vec{v} \times \vec{E}\right) \cdot \vec{\sigma}, \tag{2.12}$$

where $g\mu_b\left(\vec{\sigma}/2\right)$ is the electron magnetic moment [34].

The Rashba constant is expressed as:

$$\alpha = \frac{g\mu_b \left|\vec{E}_0\right|}{2mc^2}, \tag{2.13}$$

where m is the absolute mass of the electron, and the standard value of α is 0.13 for the Rashba SOI model. This model is derived from the Ehrenest theorem, which considers electron motion in the $\hat{z}$ direction. The Rashba constant contains the unity degree of freedom that extends to the two-dimensional space. In this model, the contacts are assumed to be in quasi-equilibrium with the zero time/mean of the electric fields [35].

The Zeeman energy due to the spin-orbit field is represented as:

$$H = \frac{\mu_b}{mc\left\{\vec{\sigma} \cdot \left(\vec{E} \times \vec{p}\right)\right\}} = -\frac{i\hbar^2}{2m^2c^2\left\{\vec{\sigma} \cdot \left(\nabla \times \vec{V}\right)\right\}}, \tag{2.14}$$

where $\hbar$, i, $\vec{B}$, and $\vec{\sigma}$ represent Plank's constant, a spin current, and the Pauli spin operator, respectively.

The two-dimensional (2D) Rashba field is expressed as below:

$$E_{2D} = K^2 + \alpha \times K, \tag{2.15}$$

where α and $K = \left(k_x^2 + k_y^2\right)^{1/2}$ are the Rashba effect constant and a scattering process coefficient, respectively.

The Rashba–Hamiltonian is dominated by diffused magnetic field H_0, which is represented as:

$$H_0 = \frac{m\rho^2}{2} + h_k \vec{\sigma} \cdot \hat{a}_z, \tag{16}$$

where $h_k = \alpha\left(\frac{Kl_z}{t}\right)$ is the Rashba SOI, $\vec{\sigma}$ is the Pauli spin operator, t is the thickness of FM layers separated by the materials, l_z is the perpendicular distance of the magnetic layer from the z-axis, and $\hat{a}_z$ is the unit vector in the z-direction.

The SOI Hamiltonian function $\vec{H}_\alpha$ [36] is expressed as below:

$$\vec{H}_\alpha = \alpha\left(\vec{\rho} \times \vec{z}\right) \times \vec{z}, \tag{2.17}$$

where α is the Rashba coupling constant.

The Rashba SOI model uses the magnetic layer saturation condition and follows the Dzyaloshinskii–Moriya interaction (DMI) equation [35]. DMI is represented as follows:

$$H_{DM} = D_{ij}\left(S_i \times S_j\right), \tag{2.18}$$

where S_i, S_j and H_{DM} represent the neighboring magnetic spins and the Hamiltonian function generated by the Dzyaloshinskii–Moriya interaction (DMI), respectively.

Although not discussed here, all the above equations related to the Rashba SOI can also be derived from the Dirac equation. The Dirac equation is a relativistic wave equation developed by British physicist Paul Dirac in 1928. This equation describes all spin particles, i.e., electrons and quarks, for symmetrical parity with electromagnetic interaction.

Mathematically, this equation is represented as:

$$\left(\beta mc^2 + c\sum_{n=1}^{3} \alpha_n p_n\right)\psi\left(x,t\right) = i\hbar\frac{\partial\psi\left(x,t\right)}{\partial t}, \tag{2.19}$$

where $\psi = \psi\left(x,t\right)$ is the wave function for the electrons of mass m with spatial coordinates x, t; p_1, p_2, and p_3 are the momentum components, c is the speed of light, and $\hbar$ is the reduced Plank's constant. These physics constants follow the special relativity and quantum mechanics principles.

The block diagram of the Rashba SOI model is represented in Figure 2.6. The input MTJ voltage is applied to the "Dependent current source with initial voltage" (Block 1) and "Rashba conductance before magnetization" (Block 2). Then the resultant "total Rashba conductance" (Block 3) depends on both Block 1 and Block 2. Block 3 is operated with the Rashba coefficient $\alpha = 0.13$ and it is applied to the "dependent current source with induced voltage" (Block 4) and "Rashba conductance after magnetization" (Block 5). The output MTJ voltage depends on Block 4 and Block 5. This model evaluates the Hamiltonian field vector along the x-axis.

The two-dimensional Rashba field increases exponentially with scattering coefficient *K*, as shown in Figure 2.7. After the spin-orbit interaction, the two-dimensional Rashba field achieves the maximum value of 0.11. The scattering coefficient is the activation factor to generate the Rashba spin-orbital process along the z-axis. It initiates the magnetic interaction along the axis perpendicular to the magnetic layer with reduced spin polarization.

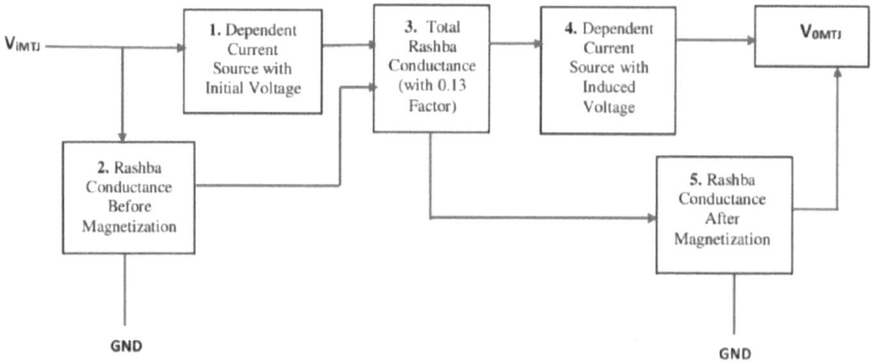

FIGURE 2.6 Block diagram of the Rashba SOI mechanism.

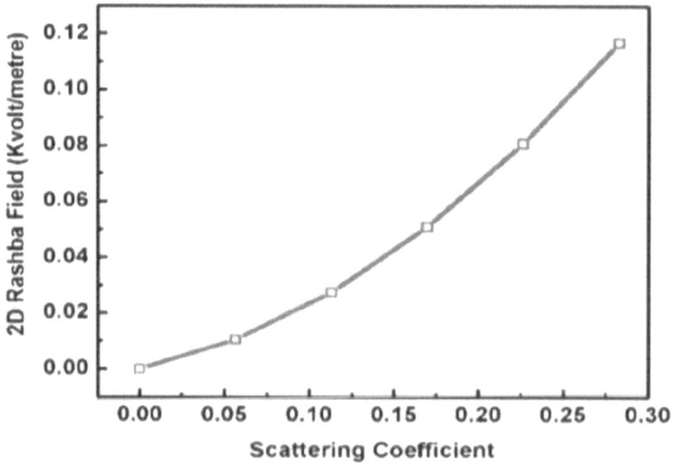

FIGURE 2.7 Two-dimensional Rashba field variation about scattering process coefficient *K*.

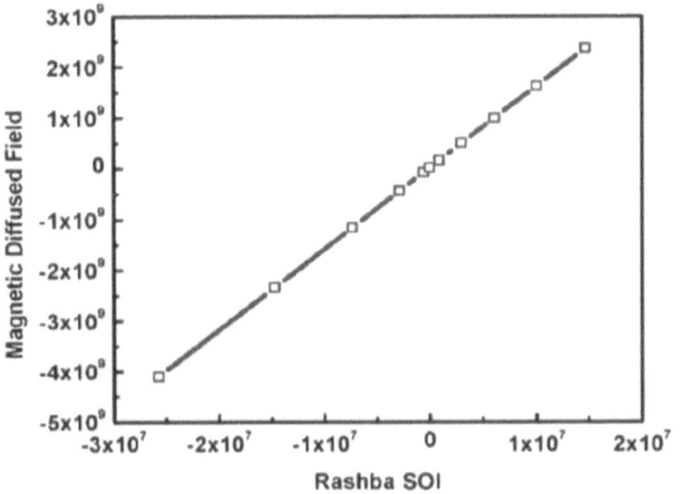

FIGURE 2.8 Magnetic diffused field variation about the increase in Rashba SOI.

If the Rashba effect increases the MRAM cell's magnetic anisotropy, it enhances the energy efficiency of SOT-MRAM device operation. Figure 2.8 shows the diffused magnetic field variation with Rashba SOI. The Rashba SOI increases monotonically with the thickness of the magnetic layer. The magnetic diffused field is 1.6 nH/m^2 with $h_k = 0.03$. This simulation is performed using the SPICE static simulation method of the Rashba field, and the magnetic field is applied at the magnetic layer interface along the spin-orbital axis.

Figure 2.9 demonstrates the variation of the Rashba SOI with magnetic layer thickness t. The Rashba SOI increases with a magnetic layer thickness (t), in the range between $t = 1$ nm to $t = 9$ nm.

So, to conclude, the Rashba effect occurs due to the variations in the Hamiltonian field vectors. The standard value of the Rashba coefficient is $\alpha = 0.13$. The overall scattering coefficient is produced from the Rashba effect and occurs in the domain wall for materials bracing the Rashba effect. The Rashba effect enhances the performance of MRAM cells by inducing the highest magnetization and minimum anisotropic coefficient.

2.3 MAGNETIC LAYERED INTERFACES FOR TST-MRAM

The spin interface in TST-MRAM is at the junction between a ferromagnetic layer and a non-magnetic layer in the presence of a heavy metal (HM) layer. The hybridization between the ferromagnetic and non-magnetic materials can be controlled by spin-orbital interaction. These interfaces contain two-dimensional layers lying between the ferromagnetic free and fixed layers. The spin-valve effect alters the composition and properties of these interfaces and can be used for hybrid MTJs. Different material interfaces like the FM/NM layer interface, FM/HM/NM layer

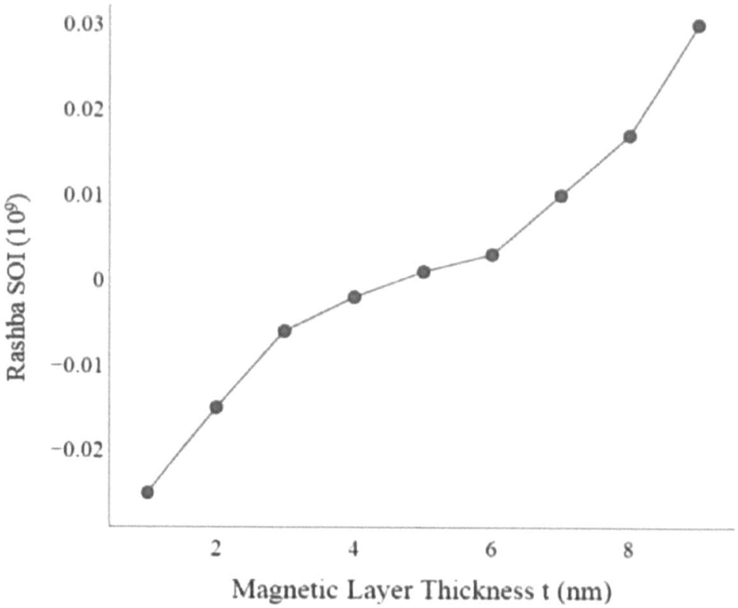

FIGURE 2.9 Rashba SOI variation in reference to magnetic layer thickness.

interface, and NM/FM interfaces are employed for designing the MRAM cells (for example, Ru, Ta, Mo, etc.). In the following part, we discuss the design specifics and modeling of the TST-MRAM cell that exploit interface properties to generate the spin-valve effect. In addition, we discuss heavy metal miniaturization in MRAM cells.

2.3.1 HM/FM Interfacial Modulation for Strong PMA

STT-MRAM suffers from some drawbacks, such as low adaptability and very low STT switching efficiency. In this section, we discuss perpendicular magnetized MTJs (p-MTJ) for applications in planer chips. The magnetization switching HM/FM multilayers is induced by SOT. These multilayers follow the Dzyaloshinskii–Moriya interaction (DMI) between molecules. The chiral domain nucleation supervises this process at the interfaces. The HM/FM interface's stiffness plays an important role in the DMI, and the outcomes of this process are directly reflected in the spin-orbit torque of the MRAM cell. Hence, the surface of the HM/FM multilayer becomes smooth due to the absence of impurities.

2.3.2 PMA in Cobalt-Dependent Multilayers

In-plane perpendicular magnetic anisotropy (IPMA) occurs at Pd/Co interfaces or junctions in MTJ [37] due to the presence of a Co layer. The combination of noble metal Pt and Co is employed to generate interfacial perpendicular magnetic anisotropy (iPMA) with the (111) texture [38, 39]. The Co/Pt multilayer consists of a free layer and reference layer in the p-MTJ structure. Due to the lower spin polarization coefficient, the TMR ratio is lower in Co/Pt multilayers [40, 41]. However, Co/Pd multilayers require a high threshold current for magnetization switching.

The Co/Pt multilayers with robust iPMA are used as a bottom reference layer in p-MTJ, as shown in Figure 2.10. The TMR ratio of a thermally robust p-MTJ is 1.5 when the annealing process [42] is applied at the interface for 30 minutes. The critical temperature of this process is 4000°C. The p-type synthetic anti-ferromagnetic (p-SAF) coupled reference structure is used at the spin transistor. The spin transistor is developed by a magnetic semiconductor and is a magnetically sensitive device.

Yakushiji [43] stated that the p-SAF system with interlayer exchange coupling (IEC) and Ru layer can replace the Co/Pt multilayer structure. This p-SAF reference layer is embedded in the p-MTJ structure and shows a TMR ratio of 1.45 or 145%. The bottom Ru layer is replaced with an Ir layer when IEC energy density variation occurs.

The MTJ conductance can be represented as a 2×2 matrix whose diagonal elements can be represented as:

$$G_{11} = G_{22} = G_0\left(1 + P_1P_2 \cdot \cos\theta\right), \tag{2.20}$$

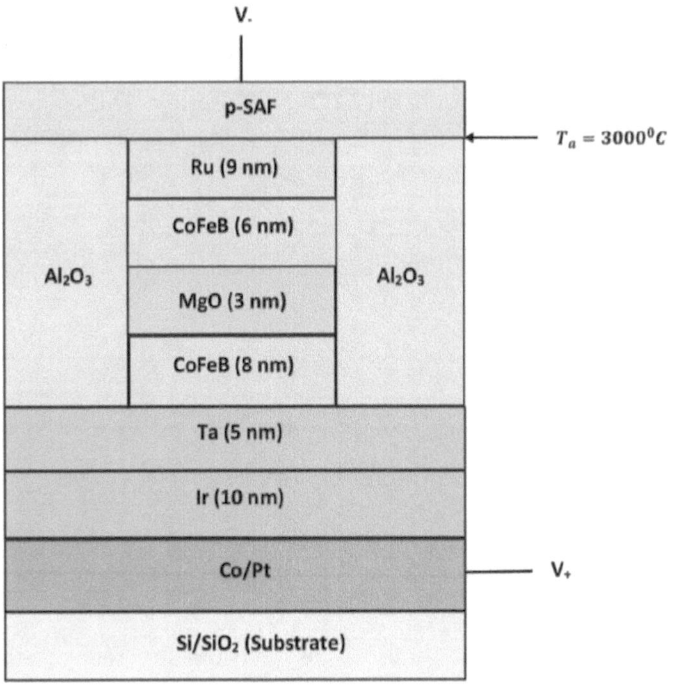

FIGURE 2.10 Schematic of the iPMTJ structure for CoFeB/MgO domains with Co/Pt multilayers.

G_{11} and G_{22} represent the MTJ conductance matrix's diagonal elements, while G_0 is the average of the MTJ conductance of parallel and antiparallel configurations. P_1, P_2, and θ represent the spin polarization coefficients of MTJ and spin polarization angle, respectively.

The non-diagonal elements of the MTJ conductance matrix are related to the diagonal elements, expressed as:

$$\begin{aligned} G_{12} &= -G_{11}, \\ G_{21} &= -G_{22}. \end{aligned} \tag{2.21}$$

The Co/Ni multilayers give rise to a tunable iPMA with large spin polarization [44] and a lower damping ratio [45] for the free layer. Such multilayers are bounded by considerable magnetic anisotropic energy (MAE) at the location of iPMA. The spin current is generated from the interface between thin Co and Ni layers. The iPMA occurs at a magnetic interface with spin-orbit coupling (SOC) [46] and gives rise to magnetoelastic anisotropy [47] for the free layer.

Due to the strong magnetic anisotropy, these multilayers are used as reference layers in p-MTJs. The TMR ratio of 1.45 or 145% is obtained in a Co/Ni-based p-MTJ stack with a channel thickness of 10.6 nm [48]. In the interfaces that support iPMA, the Hamiltonian field remains zero for the Rashba SOI in the range of between +300 and –300 MT/m^2, as demonstrated in Figure 2.11.

2.3.3 PMA in FM/Co-Fe-B/MgO Structures

The FM/CoFeB/MgO interface-based MTJ gives rise to a TMR ratio of 124%, a thermal stability of 43, and a switching current of 49 μA at the magnetic interface.

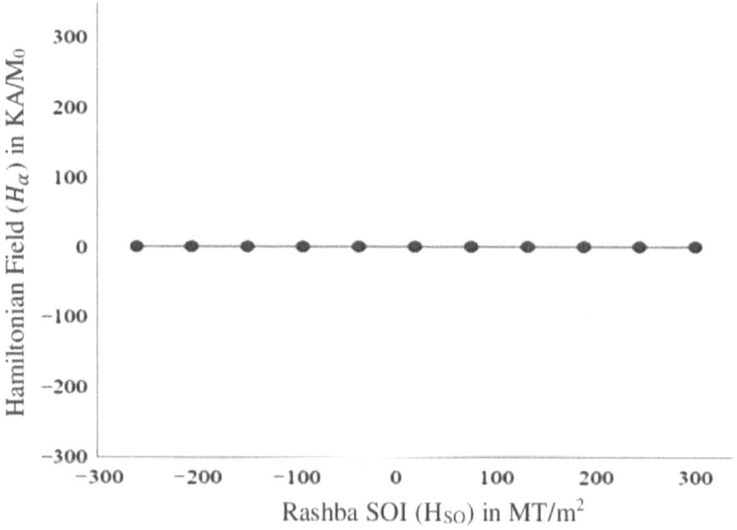

FIGURE 2.11 Hamiltonian field representation of Rashba spin-orbit interaction.

The channel length of the CoFeB composite is 20 nm. The thermal stability coefficient Δ [49] is expressed as:

$$\Delta = \frac{E_B}{k_B T} = \frac{KV}{k_B T}, \tag{2.22}$$

where E_B is the energy barrier between the parallel and antiparallel states for the magnetization in the free and reference layers. k_B, T, K, and V represent Boltzmann's constant, room temperature, the anisotropy energy density, and the free layer's volume, respectively. The anisotropy energy density of the free layer of CoFeB composite [50] is represented as follows:

$$K = M_s\left(\frac{\left|\vec{H}_a\right|}{2}\right) = \frac{K_i}{t_{CoFeB}} + K_b - 2\pi M_s^2, \tag{2.23}$$

where $K_i, M_s, \left|\vec{H}_a\right|$ are the interfacial magnetic anisotropy, the saturation magnetization, and the magnitude of the anisotropic field, respectively. K_b is the bulk magnetic anisotropy and t_{CoFeB} is the thickness of the CoFeB-free layer. The Ta/CoFeB/MgO structure demonstrates in-plane magnetic anisotropy when the CoFeB layer is very thin. The enhancement of magnetic anisotropy is not possible when the CoFeB layer thickness is 1.3nm.

The origin of the surface iPMA is generally believed to be the hybridization between the 3d orbitals of Fe atoms and 2p orbitals of O atoms, in conjugation with spin-orbit coupling. The process of over oxidation or under oxidation at the interface reduces the effective hybridization between 3d orbitals of Fe and 2p orbitals of O atoms. SOC causes a rapid decrease of iPMA at the interface. The aspect ratio is defined as the ratio of width and height of the magnetic layer, and this ratio of the Fe/MgO structure [51] is determined for the free layer.

It is found that the PMA occurs at the interfacial Fe monolayer in CoFeB composite and transports interfacial magnetic anisotropy into the Fe/Ru monolayer. These interfacial conditions result in the occurrence of iPMA at the Fe/MgO interface [52].

The p-MTJ with Ta/CoFeB/MgO structure is represented in Figure 2.12. This MTJ is formed in conjunction with the ferromagnetic layer CoFeB and non-magnetic layer MgO, along with Al_2O_3 spacer layers. The bulk material is SiO_2 with the Ta and Ru material layer interface and the FM/NM layer interface. This structure forms the Stoner particle domain and follows the Stoner principle.

2.3.4 PMA in HM/CoFeB/NM/MgO Structures

Strong PMA and considerable thermal stability are required to store the information in MTJ. When the storage limit of memory increases, the strong PMA ensures a lower chip deterioration rate and higher data retentivity. A thermal stability factor of 75 is required for a 128 GB memory chip. This factor is essential to accomplish a data confinement time of at least 10 years and chip deterioration rate [53–56] of 10^{-4}

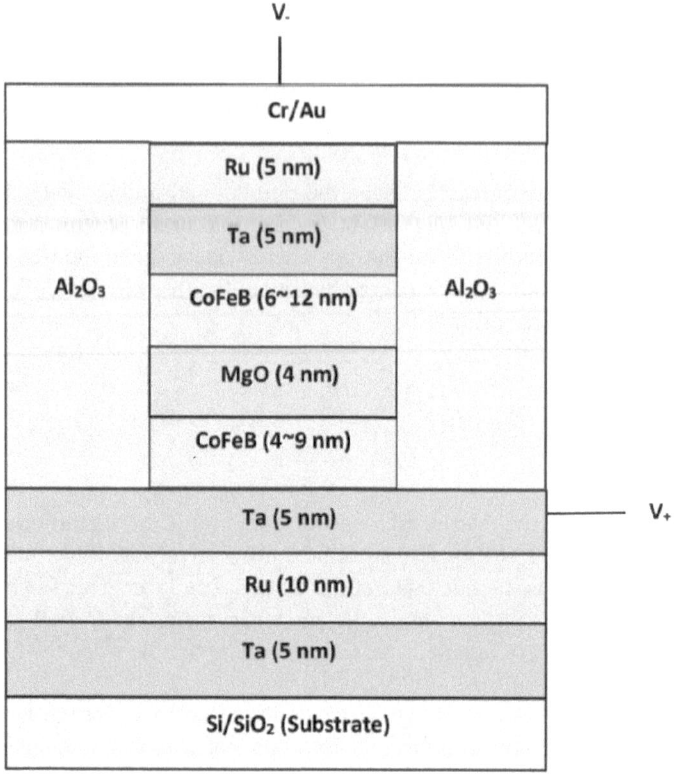

FIGURE 2.12 Schematic of the MTJ structure for in-plane and out-of-plane magnetization for CoFeB/MgO domains.

in particular MTJ. A desirable MRAM performance demands a higher anisotropic field. Considering this factor, the free layer's surface PMA should be lower than 4.7 mJm^{-2}, when the channel length of MTJ is less than 10 nm.

The interfacial PMA occurs at a Ta/CoFeB/MgO interface [57]. Generally, the applied flux density is less than 2 mJm^{-2}. The required flux density of the interfacial PMA system is higher than 3.1 mJm^{-2}, resulting in a data confinement time of less than 1 ms for cache memory. Therefore, it is essential to improve PMA for the large density of MRAM cells. Some researchers have investigated the impact of annealing temperature on the magnetic anisotropy, and PMA is increased at an annealing temperature of 3000°C [56–58]. When the annealing temperature is less than 350°C, the Ta/CoFeB/MgO composite doesn't develop the required strength of PMA. Implanting the Ta layer along the z-axis results in thermal saturation of magnetic flux at 400°C, and this phenomenon interferes with the coordination with CMOS circuits.

Figure 2.13 represents the p-MTJ stack with HM/CoFeB/NM/MgO composite. The ferromagnetic composite (CoFeB) channels are 1.3, 1, and 0.5 nm long with a

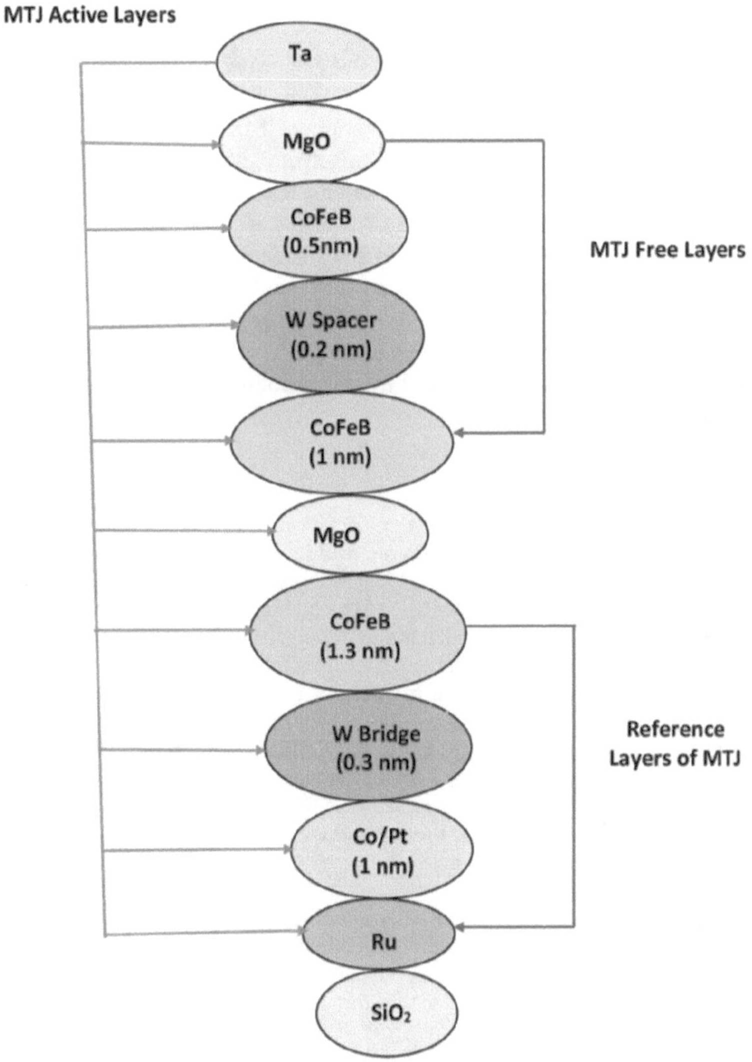

FIGURE 2.13 Structure of the p-MTJ stack through the magnetic anisotropy field at HM/FM/NM domains.

non-magnetic composite layer (MgO), which provides the interlayer interfacing. The interlayer (FM/HM/NM) interface contains a W spacer of 0.2 nm and a W bridge of 0.3 nm with Ta metal for a higher saturation magnetization and lowers read and write access time in the MRAM cell. The free layer of MTJ lies at the interface between MgO and 1nm CoFeB layer and the reference layer of MTJ lies at the interface between the 1.3 nm CoFeB layer and Co/Pt multilayer. The LLG equation governs these interfacial free and reference layer magnetizations.

2.4 FUTURE TRENDS IN SPINTRONICS MEMORY

In this chapter, we have mainly discussed the performance analysis of an advanced and modern MRAM cell. The VCMA and Rashba SOI effect are used to improve RW access time and saturation magnetization of MRAM cells. The toggle spin torque is generated due to large magnetic anisotropy in the spin channel. Compared to the traditional MRAM cells, these phenomena lead to lower read and write access times and high magnetization. The TST-MRAM cell relies on the VCMA effect. Currently, a lot of effort is being geared toward developing and commercializing these MRAM cells for applications in the near future. Although not discussed here, another direction toward achieving high-performance memory cells in a spin-based computer is the deployment of domain wall logic. The future generation of computing units would be heavily empowered by the upcoming generations of spin-based memory and its applications.

REFERENCES

1. J. Akerman, Toward a universal memory, *Science*, 308 (5721), 508–510, April 2005.
2. Y. Jin, M. Shihab, M. Jung, *Area, Power, And Latency Considerations of STT-MRAM to Substitute for Main Memory*, The Memory Forum, 2014.
3. S. Tehrani, Status and prospect for MRAM technology, IEEE Hot Chips 22 Symposium (HCS), 1–23, 2010, Chennai, India. https://doi.org/10.1109/HOTCHIPS.2010.7480057.
4. S. Li, C. Xu, Q. Zou, J. Zhao, Y. Lu, Y. Xie, Pinatubo: A processing-in-memory architecture for bulk bitwise operations in emerging non-volatile memories, Proceedings of the Design Automation Conference (DAC), 1–6, June 2016, Washington, DC.
5. H. Zhang, W. Kang, L. Wang, K. L. Wang, W. Zhao, Stateful reconfigurable logic via a single-voltage-gated spin hall-effect driven magnetic tunnel junction in a Spintronic memory, *IEEE Transaction Electron Devices*, 10 (64), 4295–4301, October 2017.
6. G.C. Adam, B.D. Hoskins, M. Prezioso, D.B. Strukov, Optimized stateful material implication logic for three-dimensional data manipulation, *Nano Research*, 12 (9), 3914–3923, December 2016.
7. Z. Chowdhury, J.D. Harms, S.K. Khatamifard, A.P. Lyle, Efficient in-memory processing using spintronics. *IEEE Computing Architecture Letter*, 17 (1), 42–46, January 2018.
8. M. Weisheit, S. Fähler, A. Marty, Y. Souche, C. Poinsignon, D. Givord, Electric field-induced modification of magnetism thin-film ferromagnets, *Science*, 315 (11), 349–351, January 2007.
9. T. Maruyama, Y. Shiota, T. Nozaki, K. Ohta, M. Shiraishi, Large voltage-induced magnetic anisotropy changes in a few atomic layers of iron, *Nature Nanotechnology*, 4, 158–161, January 2009.
10. M. Endo, S. Kanai, S. Ikeda, F. Matsukura, H. Ohno, Electric-field effects on thickness-dependent magnetic anisotropy of sputtered MgO/Co40Fe40B20/Ta structures, *Applied Physics Letter*, 21 (96), 212–503, May 2010.
11. W. G. Wang, M. Li, S. Hageman, C.L. Chien, Electric-field-assisted switching in magnetic tunnel junctions, *Nature Materials*, 11, 64–68, November 2011.
12. Z. Wang, H. Zhou, M. Wang, W. Cai, D. Zhu, W. Zhao, Proposal of toggle spin torques magnetic RAM for ultrafast computing, *IEEE Electron Devices Letters*, 4 (5), 726–729, May 2019.
13. J. Zhu, J.A. Katine, G.E. Rowlands, Y. J. Chen, Z. Duan. G. Alzate, P. Upadhyaya, J. Langer, K.L. Wang, I.N. Krivorotov, Voltage-induced ferromagnetic resonance in magnetic tunnel junctions, *Physics Review Letter*, 108 (12), 197–203, December 2012.

14. Y. Shiota, T. Nozaki, F. Bonell, S. Murakami, T. Shinjo, Y. Suzuki, Induction of coherent magnetization switching in a few atomic layers of FeCo using voltage pulses, *Nature Materials*, 39 (11), January 2012.
15. S. Kanai, Y. Nakatani, M. Yamanouchi, S. Ikeda, Magnetization switching in a CoFeB/MgO magnetic tunnel junction by combining spin-transfer torque and electric field-effect, *Applied Physics Letter*, 21 (104), 212406.1–212406.3, September 2014.
16. S. Yuasa, A. Fukushima, K. Yakushiji, T. Nozaki, M. Konoto, H. Maehara, H. Kubota, H. Arai, H. Imamura, K. Ando, H. Yoda, Future prospects of MRAM technologies, in electron devices meeting (IEDM), IEEE International Conference, Washington DC, 3.1.1., February 2013.
17. C.H. Lambert, A. Rajanikanth, T. Hauet, S. Mangin, E.E. Fullerton, S. Andrieu, quantifying perpendicular magnetic anisotropy at the Fe–MgO (001) interface, *Applied Physics Letter*, 102, 122–410, March 2013.
18. J.W. Koo, S. Mitani, T.T. Sasaki, H. Sukegawa, Z.C. Wen, T. Ohkubo, T. Niizeki, K. Inomata, K. Hono, Large perpendicular magnetic anisotropy at Fe = MgO interface, *Applied Physics Letter*, 103, 192–401, July 2013.
19. I. Ahmed, Z. Zhao, M. G. Mankalale, S. S. Sapatnekar, J. P. Wang, C. H. Kim, Comparative study between spin-transfer-torque and spin-hall-effect switching mechanisms in PMTJ using SPICE, IEEE *Journal Explorer Solid-State Computation Devices and Circuits*, 11 (3), 74–82, October 2017.
20. C. Grezes, A. Rojas Rozas, F. Ebrahimi, J.G. Alzate, X. Cai, J.A. Katine, J. Langer, B. Ocker, P. K. Amiri, K.L. Wang, In-plane magnetic field effect on switching voltage and thermal stability in electric-field-controlled perpendicular magnetic tunnel junctions, *AIP Advances*, 7 (6), 14–75, July 2016.
21. D.V. Berkov, *Handbook of Magnetism and Advanced Magnetic Materials, Volume Micromagnetism, Chapter Magnetization Dynamics Including Thermal Fluctuations: Basic Phenomenology, Fast Remagnetisation Processes and Transitions over High-energy Barriers*, John Wiley Ltd, n.d.
22. T.L. Gilbert, A phenomenological theory of damping in ferromagnetic materials, *IEEE Transactions on Magnetics*, 40 (6), 3443–3449, November 2004.
23. S. Yuasa, A. Fukushima, H. Kubota, Y. Suzuki, K. Ando, Giant tunneling magnetoresistance up to 410% at room temperature in fully epitaxial Co/MgO/Co magnetic tunnel junctions with BCC Co (001) electrodes, *Applied Physics Letters*, 4 (89), 042505, May 2006.
24. R. Heindl, A. Chaudhry, S.E. Russek, Estimation of thermal stability factor and intrinsic switching current from switching distributions in spin transfer-torque devices with out-of-plane magnetic anisotropy, *AIP Advances*, 1 (8), 015011, January 2018.
25. W. Kang, Y. Ran, Y. Zhang, W. Lv, W. Zhao, Modeling and exploration of the voltage-controlled magnetic anisotropy effect for the next generation low-power and high-speed MRAM applications, *IEEE Transaction of Nanotechnology*, 16(3), 387–395, May 2017.
26. G. Bihlmayer, O. Rader, R. Winkler, Focus on the Rashba effect, *New Journal Physics* 17, 050202, 2015.
27. W.H. Butler, T. Mewas, P.B. Visscher, S.E. Russek, Switching distributions for perpendicular spin-torque devices within the macrospin approximation, *IEEE Transactions on Magnetism*, 48(12), 4684–4700, December 2012.
28. A. Vatankhahghadim, S. Huda, A. Sheikholeslami, A survey on circuit modeling of spin-transfer-torque magnetic tunnel junctions, *IEEE Trans. Circuits Syst. I, Reg. Papers*, 61(9), 2634–2643, September 2014.
29. J. Song, I. Ahmed, Z. Zhao, D. Zhang, S.S. Sapatnekar, J. Wang, C.H. Kim, Evaluation of operating margin and switching probability of voltage-controlled magnetic anisotropy

magnetic tunnel junctions, *IEEE Journal on Exploratory Solid-State Computational Devices and Circuits*, 14, 37–50, November 2018.
30. G. Dresselhaus, Spin-Orbit coupling effects in zinc blende structures, *Physical Review Journal*, 580 (100), 50–67, 1955.
31. T. McGuire, R. Potter, Anisotropic magneto-resistance in ferromagnetic 3D alloys, *IEEE Transactions on Magnetics*, 4 (11), 34–48, July 1975.
32. L. Petersen, P. Hedegard, A simple tight-binding model of spin-orbit splitting of sp-derived surface states, *Surface Science Journal*, 449 (1), 49–56, July 2000.
33. P. Pfeffer, W. Zawadzki, Spin splitting of conduction subbands in III-V heterostructures due to inversion asymmetry, *Physical Review B*, 59 (8), 5312–5315, 1998.
34. J. Fabian, A. Matos-Abiague, C. Ertler, P. Stano, I. Zutic, Semiconductor spintronics, *Condensed Matter Material Science Journal*, 342, 61–72, November 2007.
35. Y. Gao, Z. Wang, X. Lin, W. Kang, Y. Zhang, W. Zhao, Scaling study of spin-hall-assisted spin transfer torque driven magnetization switching in the presence of Dzyaloshinskii–Moriya interaction, *IEEE Transactions on Nanotechnology*, 6 (16), 35–49, 2017.
36. P.F. Carcia, A.D. Meinhaldt, A. Suna, Perpendicular magnetic anisotropy in Pd/Co thin film layered structures, *Applied Physics Letters*, 47 (2), 178, April 1985.
37. P.F. Carcia, Perpendicular magnetic anisotropy in Pd/Co and Pt/Co thin film layered structures, *Applied Physics Letter*, 63 (10), 5066, January 1988.
38. S. Hashimoto, Y. Ochiai, K. Aso, Perpendicular magnetic anisotropy and magneto-striction of sputtered Co/Pd and Co/Pt multi-layered films, *Journal of Applied Physics*, 66 (10), 4909, August 1989.
39. J.H. Park, C. Park, T. Jeong, M.T. Moneck, N.T. Nufer, Co/Pt multilayer based magnetic tunnel junctions using perpendicular magnetic anisotropy, *Journal of Applied Physics*, 103 (7), 07A917, March 2008.
40. D. Lim, K. Kim, S. Kim, W.Y. Jeung, Study on exchange-biased perpendicular magnetic tunnel junction based on Pd/Co multilayers, IEEE *Transactions on Magnetism*, 45 (6), 32–41, June 2009.
41. M. Gottwald, J.J. Kan, K. Lee, X. Zhu, C. Park, Scalable and thermally robust perpendicular magnetic tunnel junctions for STT-MRAM, *Applied Physics Letter*, 106 (3), 032413, January 2015.
42. K. Yakushiji, A. Sugihara, A. Fukushima, H. Kubota, S. Yuasa, Very strong anti-ferromagnetic interlayer exchange coupling with iridium spacer layer for perpendicular magnetic tunnel junctions, *Applied Physics Letter*, 110 (9), 092406, February 2017.
43. S. Andrieu, T. Hauet, M. Gottwald, P. Ohresser, Co/Ni multilayers for spintronics: High spin polarization and tunable magnetic anisotropy, *Physical Review Materials*, 2, 064410, June 2018.
44. J.M.L. Beaujour, W. Chen, C.C. Kao, Ferromagnetic resonance study of sputtered Co/Ni multilayers, *European Physical Journal*, 59 (4), 475–483, October 2007.
45. G.H.O. Daalderop, P.J. Kelly, Prediction and conformation of perpendicular magnetic anisotropy in Co/Ni multilayers, *Physical Review Letter*, 68, 682, 1992.
46. M. Gottwald, S. Andrieu, F. Gimbert, L. Calmels, Co/Ni (111) superlattices studied by microscopy, x-ray absorption, and ab initio calculations, *Physical Review B*, 86, 014425, July 2012.
47. G.S. Kar, W. Kim, T. Tahmasebi, J. Swerts, S. Mertens, T. Min, Co/Ni-based p-MTJ stack for sub-20nm high density stand-alone and high-performance embedded memory application, IEEE International Electron Devices Meeting, December 2014, Chicago.
48. S. Ikeda, K. Miura, H. Yamamoto, M. Endo, A perpendicular anisotropy CoFeB–MgO magnetic tunnel junction, *Nature Materials*, 9, 721–724, July 2010.

49. M.T. Johnson, P.J.H. Bloemen, J.J. de Vries, Magnetic anisotropy in metallic multilayers, IOP Science *Reports on Progress in Physics*, 59(11), 54–60, 1996.
50. H.X. Yang, M. Chshiev, J.H. Lee, A. Manchon, First-principles investigation of the very large perpendicular magnetic anisotropy at Fe lMgO and Co lMgO interfaces, *Physical Review B*, 84, 054401, August 2011.
51. A. Hallal, H.X. Yang, B. Dieny, Anatomy of perpendicular magnetic anisotropy in Fe/MgO magnetic tunnel junctions: First principles insight, *Physical Review B*, 88, 184423, November 2013.
52. S. Nazir, S. Jiang, J. Cheng, K. Yang, Enhanced interfacial perpendicular magnetic anisotropy in Fe/MgO heterostructure via interfacial engineering, *Applied Physics Letters*, 114, 072407, February 2019.
53. N. Roschewsky, S. Schafer, F. Hellman, V. Nikitin, Perpendicular magnetic tunnel junction performance under mechanical strain, *Applied Physics Letters*, 112, 232401, 2018.
54. M. Cubukcu, O. Boulle, J. Langer, G. Gaudin, Spin-orbit torque magnetization switching of a three-terminal perpendicular magnetic tunnel junction, *Applied Physics Letters*, 104, 042406, January 2014.
55. K. Mizunuma, S. Ikeda, H. Ohno, K. Miura, Tunnel magnetoresistance properties and annealing stability in perpendicular anisotropy MgO-based magnetic tunnel junctions with different stack structures, *Journal of Applied Physics*, 109, 07C711, March 2011.
56. W.X. Wang, Y. Yang, H. Naganuma, X.F. Han, Y. Ando, The perpendicular anisotropy of Co40Fe40B20 sandwiched between Ta and MgO layers and its application in CoFeB/MgO/CoFeB tunnel junction, *Applied Physics Letters*, 99, 012502, July 2011.
57. N. Miyakawa, D.C. Worledge, K. Kita, Impact of Ta diffusion on the perpendicular magnetic anisotropy of Ta/CoFeB/MgO, *IEEE Magnetics Letters*, 4(1), February 2013.
58. H. Meng, R Sbiaa, C.C. Wang, S.Y.H. Lua, Annealing temperature window for tunneling magnetoresistance and spin torque switching in CoFeB/MgO/CoFeB perpendicular magnetic tunnel junctions, *Journal of Applied Physics*, 110(10), 103915, November 2011.

3 Evaluation of Magnetic Anisotropy via Intrinsic Spin Infusion

Divyansh Jain and Aniket Singha

CONTENTS

3.1 INTRODUCTION

In the recent past, exploiting non-volatile memory devices to facilitate in-memory computation benchmarks has demonstrated a considerable capability to address the von-Neumann drawbacks. This has resulted in a tremendous effort toward development of spintronic memory. The direct infusion of a spin-polarized current into a magnetic multilayer actuates a torque on the local magnetic moment. This in turn leads to a change in the magnetic orientation and, thus, magnetic switching occurs in the corresponding layer. These paradigms give rise to realistic strategies for storing data in small magnetic cells, in addition to paving the way toward energy-efficient nano-oscillators and rectifiers. In this chapter, we discuss spin-infusion and spin-transfer torque in magnetic multilayers, which leads to the giant magneto-resistance (GMR) effect in crystalline perpendicular to-the-plane (CPP) geometry, and the tunnel magneto-resistance (TMR) effect in magnetic tunnel junctions (MTJs). Further, spin-infusion magnetization switching (SIMS) and its impacts on toggle spin torque

DOI: 10.1201/9781003155751-3

(TST) based devices are discussed. The TST in the TST-MRAM cell is generated due to the interplay between spin-transfer torque (STT) and spin–orbit torque (SOT) mechanisms. The benefits of TST-MRAM include energy efficiency, a large TMR ratio, and superfast read-and-write operation. Also, the impacts of spin infusion, specifically spin-transfer oscillation (STO), anisotropic coefficient generation (ACG), and the spin torque diode effect on MRAM performance, are elaborated. We end this chapter with a brief note on the applications and performance enhancements of MRAM cells due to the spin-infusion mechanism.

3.2 SPIN INFUSION IN MTJ AND EFFECTIVE TORQUE

3.2.1 Spin Infusion in MTJ

An electron contains a charge (–e) and spin-angular momentum ($\hbar/2$). Hence, when an electron flows in a material, it transports both a charge and magnetic moment. This process gives rise to "spin infusion" and "spin current." The net spin-angular momentum is preserved when electrons are injected through the interface between a magnetic and non-magnetic material. Thus, the interchange of spins between particles is mediated by exchange interaction. The spin-current density characterizes the spin-angular momentum transport and can be represented as follows:

$$\vec{J}_s\left(\vec{x},t\right)=\sum_i \vec{v}_i(t)\cdot\vec{s}_i\left(\vec{x},t\right)+\left(\text{exchange mediated term}\right), \tag{3.1}$$

where $\vec{s}_i\left(\vec{x},t\right)$ and $\vec{v}_i(t)$ represent the spin and drift velocity, respectively, of the ith electron. The first term in Eq. (3.1) represents the spin-polarized current. In contrast, the second term considers the exchange of spin-angular momentum between electrons while preserving the total spin of the system. It should be noted that the spin-current density $\hat{j}^s$ is a tensor with directionality corresponding to both the current-density vector and spin-polarization vector. Under the assumption of the conservation of the total spin-angular momentum, continuity equations dictating the conservation of the charge and spin-angular momentum can be represented as:

$$\frac{\partial s}{\partial t}+\vec{\nabla}\cdot\vec{j}^{s}=0;\ \frac{\partial \rho}{\partial t}+\vec{\nabla}\cdot\vec{j}^{Q}=0, \tag{3.2}$$

where ρ and $\vec{j}^{Q}$, respectively, represent the charge and charge-current density, while s and $\vec{j}^{s}$ represent the spin and spin-current density, respectively.

Let us assume that an electric field is applied to a steady ferromagnetic material, which is magnetized along $-\vec{e}_{spin}\left(\left|\vec{e}_{spin}\right|=1\right)$. Since the density of the majority of the spin electrons n_+ is higher than that of the minority spins n_-, the spin-angular momentum density and the spin polarization are non-zero in ferromagnetic materials. This is expressed as:

$$\begin{cases} \vec{s} = \vec{s}_+ + \vec{s}_- = \frac{\hbar}{2}\vec{e}_{spin}\left(n_+ - n_-\right); \\ \rho = \rho_+ + \rho_- = \left(-e\right)\left(n_+ + n_-\right). \end{cases} \tag{3.3}$$

where $\vec{s}_+$ and $\vec{s}_-$ are vectors representing the density of the majority and minority spins, respectively, and ρ_+ and ρ_- represent the charge density of the majority and minority spins, respectively. The control of the angular momentum is opposed to the magnetization of the electrons.

Electrons encounter various scattering phenomena during transport through a non-magnetic channel when the channel length is greater than the mean free path. Such scattering processes can change the electron spin orientation and are known as spin-flip processes. Thus, the injected spin-polarized electrons gradually lose their net spin-angular momentum during the process of channel transport. In the simplest model for spin transport, the majority and minority spins are assumed to be flowing in their respective channels, which interact with each other via spin-flip processes. Such a model, which considers separate channels for the majority and minority spin flow, is known as the two-channel model.

A schematic of a two-channel representation of a parallel and antiparallel MTJ configuration is demonstrated in Figure 3.1. The left panel of Figure 3.1 represents the parallel (P) MTJ configuration. The up-spin (majority) electrons flow through the three layers with a minimum scattering coefficient. However, the down-spin (minority) electrons are strongly scattered in both magnetic layers. This results in a low resistance (R_L) and high resistance (R_H) for the up-spin and down-spin electrons, respectively, in both magnetic layers. Because the two spin channels are parallel, the total conductance is dominated by the highly conductive up-spin channel. Thus, this configuration demonstrates a low resistance. For the antiparallel (AP) MTJ configuration shown in Figure 3.1(b), both up-spin and down-spin electrons scatter weakly in one but firmly in the other ferromagnetic layer. In this case, a low-resistance path is not available for either majority or minority spin electrons. Thus, this configuration demonstrates a higher resistance.

The resistance of the parallel (P) and antiparallel (AP) configurations are represented as follows:

$$\begin{cases} R_p = \dfrac{2R_+R_-}{R_+ + R_-}, \\ R_{ap} = \dfrac{R_+ + R_-}{2}. \end{cases} \tag{3.4}$$

The resistances R_+ and R_- are developed due to the spin-up and spin-down electrons, respectively.

GMR is defined as the large change in the electrical resistance induced by a magnetic field in thin films composed of alternating ferromagnetic and non-magnetic layers [1]. The change in resistance occurs due to the re-orientation of the relative alignment magnetization vectors in adjacent ferromagnetic (FM) layers [2]. Thus,

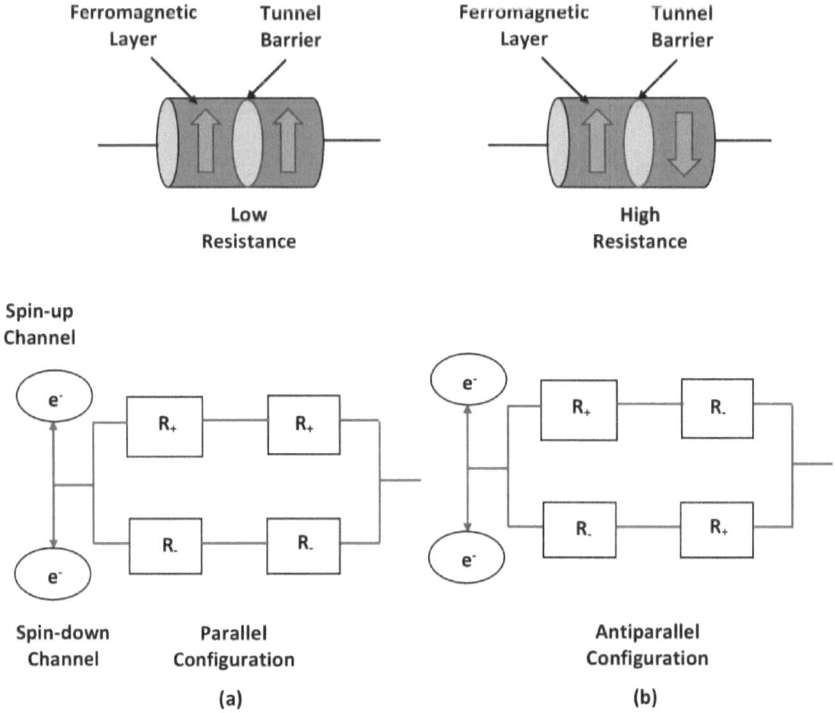

FIGURE 3.1 Schematic of the parallel (P) and antiparallel (AP) configuration in the two-channel model.

the total resistance is *lower* for the parallel configuration and *higher* for the antiparallel configuration [3]. Mathematically, GMR is expressed as:

$$\delta_h = \frac{R_H - R_0}{R_0}, \tag{3.5}$$

where R_H and R_0 are the sample's resistance when placed in a pre-defined non-zero magnetic field $(H = H_{std})$ and zero magnetic field, respectively.

TMR changes the resistance in MTJs when the relative magnetic alignment of the two ferromagnetic layers in an MTJ changes from a parallel to an antiparallel orientation. TMR is based on the effect of spin-dependent tunneling [4]. In an architecture by Jullier et al. [5], TMR can be expressed based on the two assumptions. First, the electronic spin-angular momentum is conserved during the tunneling process. Second, the tunneling rate for a selected spin orientation varies proportionally with the product of the spin-resolved effective density of states in the two magnetic electrodes [6]. Mathematically, TMR is represented as:

$$\delta_r = \frac{R_{ap} - R_p}{R_p} = \frac{2P_nP_s}{1 - P_nP_s}, \tag{3.6}$$

R_{ap} and R_p represent the resistances of the MTJ in the antiparallel and parallel configuration. $P_n = \left(N_L^U - N_L^D\right)/\left(N_L^U + N_L^D\right)$ and $P_s = \left(N_R^U - N_R^D\right)/\left(N_R^U + N_R^D\right)$, respectively, represent the spin-polarization coefficient of the left and right electrodes.

Figure 3.2(a) demonstrates electronic transport from a ferromagnetic to a non-magnetic material [7] through the interface. In a ferromagnetic material, since electron spins are polarized, a net spin flow accompanies electronic transport through the interface giving rise to spin current and spin accumulation in a non-magnetic material. The ferromagnet spins are then injected ("spin infusion") into the non-magnetic material through the interface. The infused electronic spins undergo spin-flip scattering due to the spin–orbit interaction, resulting in the loss of net spin polarization as they flow through the non-magnetic material. Spin-injection phenomena have

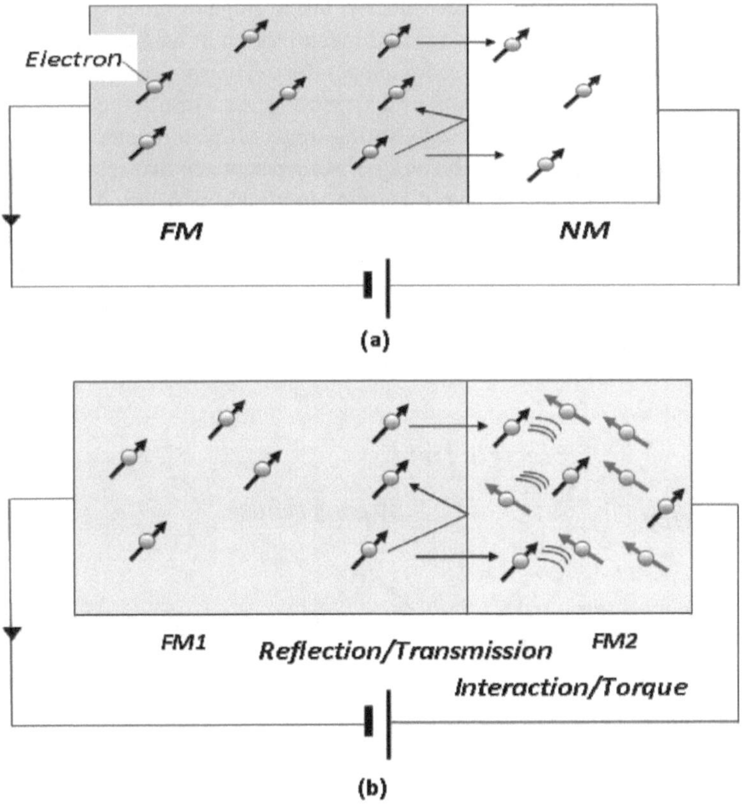

FIGURE 3.2 Spin infusion across an interface. (a) Spin infusion from a ferromagnetic material (FM) to a non-magnetic material (NM). The application of an electric charge current through the interface results in injecting a spin-polarized current into a NM. (b) Spin infusion from a ferromagnetic material (FM1) to another ferromagnetic material with a distinct magnetization direction (FM2). Spin-dependent reflection and transmission at the interface gives rise to the magneto-resistance effect.

been detected via the emission of circularly polarized light from injected electrons in semiconductors such as GaAs [8–11] and by measuring the difference between spin-dependent electrochemical potentials in so-called non-local magneto-resistance geometries [12–17] or using the magneto-resistive/filtering effect [18].

Figure 3.2(b) demonstrates the current flowing between two ferromagnetic blocks with different magnetic orientations. The spin-dependent scattering process at the interface results in tunnel magneto-resistance. The electron spins injected into the FM1 exchange the spins with the electrons in FM2 through exchange interaction, resulting in a net change in the magnetic orientation in FM2 [19–30].

Assuming the charge densities and drift velocities $\left(\vec{v}_{+}, \vec{v}_{-}\right)$ are different in the two spin subchannels, the electrical conductivities σ_{+} and σ_{-} are also different in the two spin subchannels as shown in Figure 3.3.

Under the influence of an electric field, an accelerating electron suffers repeated scattering events that randomize its velocity. The average distance between two successive scattering events is known as the mean free path., which is in the order of tens of nanometres for metals at room temperature. The spin orientation of an electron usually randomizes throughout its drift motion. As a result, the net polarization of drifting electrons gradually decays during transport phenomena. This process is known as spin-flip scattering. The average distance between two successive spin-flip scattering events is known as spin diffusion length and is denoted by the $\lambda_{+(-)}$ for majority (minority) spins. At zero electric field $\lambda_{+} = \lambda_{-}$.

Under the influence of an applied electric field, the drift charge-current density $\vec{j}^{Q,Drift}$ and drift spin-current density $\vec{j}^{S,Drift}$ in each spin co-channel may be represented as [24–26]:

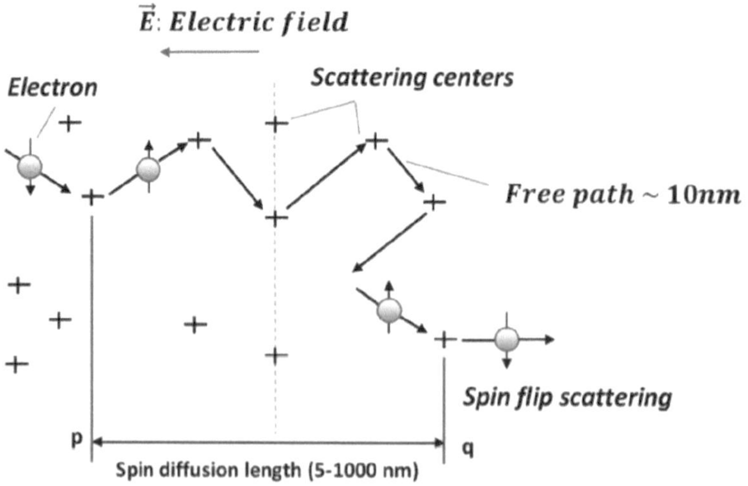

FIGURE 3.3 Drift flow of a free electron in a conductive material. Schematic concept adapted from [27].

$$\begin{cases} \vec{j}^{S,Drift} = \vec{v}_+ \vec{s}_+ + \vec{v}_- \vec{s}_- = -\frac{\hbar}{2}\frac{1}{e}(\sigma_+ - \sigma_-)\overrightarrow{Ee_{spin}}, \\ \vec{j}^{Q,Drift} = \vec{v}_+ \rho_+ + \vec{v}_- \rho_- = (\sigma_+ + \sigma_-)\vec{E}, \end{cases} \tag{3.7}$$

where ρ_+ and ρ_- are the charge density corresponding to the majority and minority spin electrons, respectively.

Figure 3.3(a) demonstrates scattering affected electronic transport at the interface between a ferromagnetic and a non-magnetic material. Two lines, p and q, embedded in the ferromagnetic and non-magnetic layers, are considered to lie at a distance greater than a few spin-flip scattering lengths from the interface. In such a case, the spin current is finite at p within the ferromagnetic material and negligibly small at q within the non-magnetic material. The law of conservation of spin-angular momentum [31] is given in Eq. (3.8). This equation predicts that the spins accumulate at the interface with time. The spin accumulation finally comes to thermal equilibrium (or quasi-equilibrium in the case of the non-zero applied voltage) as the accumulated spin polarization simultaneously decays through the spin–orbit coupling.

$$\frac{\partial \vec{s}}{\partial t} + \nabla \cdot \vec{j}^S = -\frac{\vec{s}_+ - \vec{s}_+^{eq}}{\tau_+^{sf}} - \frac{\vec{s}_- - \vec{s}_-^{eq}}{\tau_-^{sf}}, \tag{3.8}$$

where $\tau_\pm^{sf}$ and $\vec{s}_\pm^{eq}$ are the spin relaxation time and spin density, respectively, in the thermal equilibrium condition in each spin co-channel, respectively.

Assuming the density of states in metals $N_\pm$ is different from the carrier density in semiconductors $n_\pm$, for both majority and minority spin co-channels, the spin relaxation [32] times are also different for both these systems. This is represented as follows:

$$\begin{aligned} &\tau_+^{sf} N_- = \tau_-^{sf} N_+ \text{ for Metals,} \\ &\tau_+^{sf} n_- = \tau_-^{sf} n_+ \quad \text{for Non-degenerate semiconductors.} \end{aligned} \tag{3.9}$$

The diffusion spin-current density flows through the interface due to the difference in spin polarization inside the ferromagnetic and non-magnetic material and can be expressed as:

$$\begin{aligned} \vec{j}^{S,Diffusion} &= -D_+ \overrightarrow{\nabla s_+} - D_- \overrightarrow{\nabla s_-}, \\ \vec{j}^{Q,Diffusion} &= -D_+ \vec{\nabla} \rho_+ - D_- \vec{\nabla} \rho_-. \end{aligned} \tag{3.10}$$

where $D_{+(-)}$ is the diffusion constant of the majority (minority) spins in each spin co-channel.

The diffusion constant for the majority (minority) spin is related to its conductivity in Einstein's relation, and is represented as:

$$\begin{aligned} &\sigma_{+(-)} = N_{+(-)} e^2 D_{+(-)} \qquad \text{for Metals,} \\ &\sigma_{+(-)} = n_{+(-)} e^2 D_{+(-)} / kT \text{ for Non-degenerate semiconductors} \end{aligned} \tag{3.11}$$

where k and T denote the Boltzmann constant and particle temperature, respectively.

The spatial variation of majority (minority) density in each spin co-channel can be related to the spin-dependent chemical potential $\mu_{+(-)}$ as:

$$\begin{aligned} \vec{\nabla} n_{+(-)} &= N_{+(-)} \vec{\nabla} \mu_{+(-)} \quad \text{for Metals,} \\ \vec{\nabla} n_{+(-)} &\approx n_{+(-)} \vec{\nabla} \mu_{+(-)} / kT \quad \text{for Non-degenerate semiconductors.} \end{aligned} \tag{3.12}$$

The above equations are valid just for small spatial variations in $\mu_{+(-)}$. Under the application of an electric field, defining the spatial electrochemical potential $\bar{\mu}_{+(-)}(x) = \mu_{+(-)} - e\psi(x)$ (where $\psi(x)$ is the electric scalar potential), the total majority (minority) spin current may be expressed as:

$$\vec{j}^{\,Q}_{+(-)} = \vec{j}^{\,Q,Drift}_{+(-)} + \vec{j}^{\,Q,Diffusion}_{+(-)} = \frac{\sigma_{\pm} \vec{\nabla} \bar{\mu}_{+(-)}}{e}. \tag{3.13}$$

The net spin and charge densities are represented as follows:

$$\begin{cases} \vec{j}^{\,S} = -\dfrac{\hbar}{2} \dfrac{\vec{e}_{spin}\left(\vec{j}^{\,Q}_{+} - \vec{j}^{\,Q}_{-}\right)}{e}, \\ \vec{j}^{\,Q} = \vec{j}^{\,Q}_{+} + \vec{j}^{\,Q}_{-}. \end{cases} \tag{3.14}$$

where $\vec{j}^{\,Q}$ is the total charge-current density.

The charge and spin current, as well as the electrochemical potential, is continuous at the interface. The nature of spatial variation in the majority and minority spin electrochemical potentials is demonstrated in Figure 3.4. The gradient of electrochemical potentials is discontinuous at the interface. Under this condition, the spin polarization of the infused current can be represented as:

$$\pi = \frac{\left|\vec{j}^{\,Q}_{+}(0+) - \vec{j}^{\,Q}_{-}(0+)\right|}{\left|\vec{j}^{\,Q}_{+}(0+) + \vec{j}^{\,Q}_{-}(0+)\right|} = \frac{r^{FM}}{r^{FM} + r^{NM}} \beta^{FM}, \tag{3.15}$$

where the interface is assumed to lie at $x = 0$.

In Eq. (3.15), π represents the spin polarization of the injected current and $\beta^{FM} \cong \dfrac{\sigma_{+}^{FM} - \sigma_{-}^{FM}}{\sigma_{+}^{FM} + \sigma_{-}^{FM}}$; $r^{FM} \cong \left(\dfrac{1}{\sigma_{+}^{FM}} + \dfrac{1}{\sigma_{-}^{FM}}\right) \lambda_{spin}^{FM}$; $r^{NM} \cong \left(\dfrac{2}{\sigma^{NM}/2}\right) \lambda_{spin}^{NM}$.

The + and − represent the majority and minority spin in the magnetic layer. $\sigma_{\pm}^{FM}$ and σ^{NM} are the electrical conductivity of the majority/minority spin co-channel in the ferromagnetic layer and the net electric conductivity of the non-magnetic layer, respectively. β^{FM} represents the spin asymmetry of the electric conductivity in the ferromagnetic layer. A high value of β^{FM} gives rise to immense spin polarization

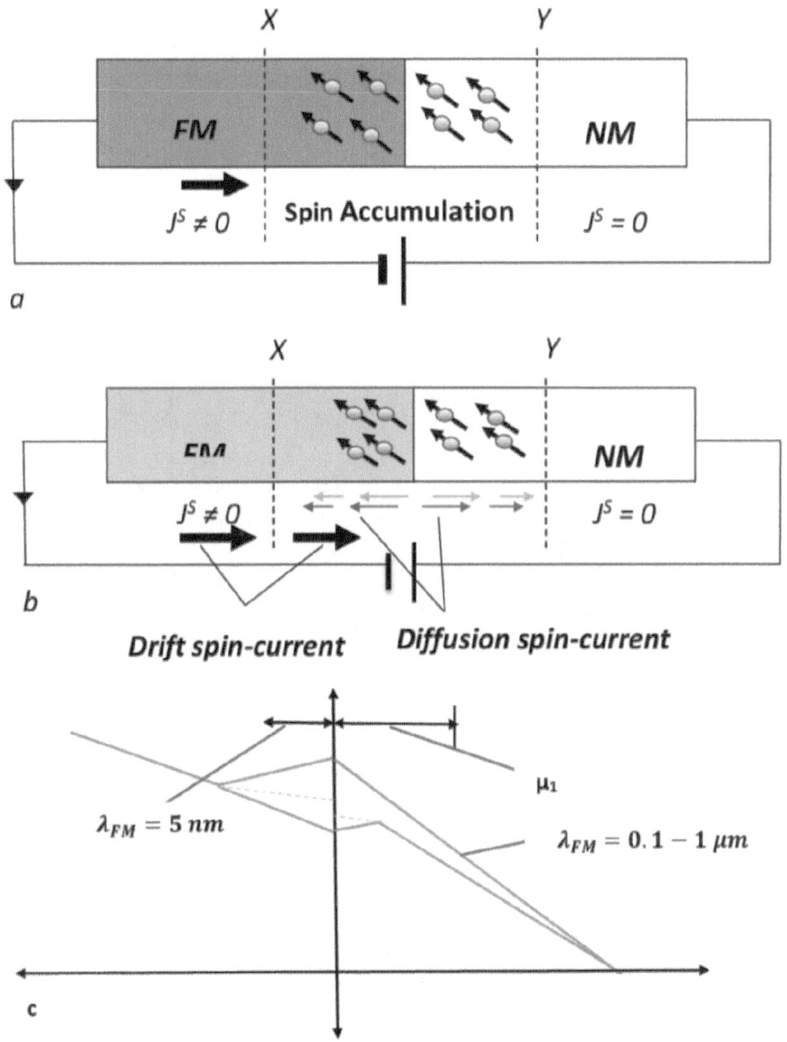

FIGURE 3.4 Spin accumulation, relaxation, and diffusion during spin injection from a ferromagnetic material (FM) to a non-magnetic material (NM). (a) The spin current at section P is finite, whereas that at Q is zero; that is, spins are injected into the interface region between P and Q but never leave it. As a result, the spins accumulate in the interface region. (b) The number of accumulated spins is decreased because of spin-flip scattering inside the interface region. Spin accumulation produces a gradient of spin density. It causes the spin diffusion current that moves spins away from the interface. (c) Associated electrochemical potential distribution. λ_{FM} and λ_{NM} are the average spin diffusion length of FM and NM. Schematic concept adapted from [27].

of the infused current. $\lambda_{spin}^{FM}\left(\lambda_{spin}^{NM}\right)$ is the spin diffusion length in the FM (NM) layer. $r^{FM}\left(r^{NM}\right)$ measures the spin-infusion resistance into the FM (NM) layer and is termed spin interface resistance.

3.2.2 Spin Torque

The magnetization vector's temporal dynamics in single-domain ferromagnets are governed by the Landau–Lifshitz [33] equation, which states that the change in magnetization vector per unit time is proportional to the magnetic torque. This can be expressed as:

$$\frac{\partial \vec{m}}{\partial t} \propto \vec{\tau}\ ;\ \vec{m} = \frac{\vec{M}}{M_s}. \tag{3.17}$$

where τ is the net magnetic torque and $\vec{m}$ is the normalized magnetization or unit vector along with the magnetic orientation. The concept of this magnetic torque is defined by the Stoner–Wohlfarth model [34]. According to this model, net torque is dependent on the magnetic field, damping the motion of spin electrons and the effect of spin-transfer torque induced by the spin current.

3.2.2.1 Magnetic Field Torque

This torque is generated due to a magnetic field applied to the Stoner particle. According to the Stoner–Wohlfarth theory, the Stoner particle is defined as a single-domain dipole in a magnetic field $\vec{H}$. The magnetic field acts on the magnetic dipole to exert a torque, which can be expressed as:

$$\vec{\tau}_{\vec{H}} = \gamma_0 \mu_0 \left(\vec{m} \times \vec{H}\right), \tag{3.18}$$

where $\vec{m}, \gamma_0$, and μ_0 are the magnetic dipole moment, the gyromagnetic ratio, and vacuum permeability, respectively. The Larmor frequency, $\omega_0 = \gamma_0 \left|H\right|$, refers to the rate of precession of the electron's magnetic moment around the external magnetic field. It is also called the precessional frequency and is dependent on the strength of the magnetic field, B_0.

The potential energy developed by a magnetic dipole under the influence of a magnetic field $\vec{H}$ can be expressed as:

$$E = -\mu_0 \int_{V_i}^{V_f} \vec{H} \cdot \partial \vec{M}. \tag{3.19}$$

The Stoner–Wohlfarth theory estimates energy by considering the complete single-domain ferromagnet as a macro-spin. Under the assumption that a single-domain ferromagnet behaves as a macro-spin, the energy E can be taken as the system's net potential energy. The total potential energy depends on the ferromagnet's form and composition. Often ferromagnets can be fabricated in a layered manner to assign a preferred magnetic orientation in the absence of a magnetic field. Such direction of preferred magnetic alignment is called the easy axis. In such cases, in the absence of an external magnetic field, the energy of a Stoner particle [35, 36] can be written as:

$$E = \frac{\mu_0 \left|\overrightarrow{M_s}\right|^2 V}{2}\left(h_X \sin^2\theta \cdot \cos^2 f + h_Y \sin^2\theta \cdot \sin^2 f + h_Z \cos^2\theta\right), \qquad (3.20)$$

where ϕ is the angle between the magnetization and the easy axis in the easy-plane, and θ is the incident angle between the magnetization vector and the fixed layer $(\hat{z})$. The shape and material thickness of the ferromagnetic layer are determined by the coefficients h_X, h_Y, and h_Z, respectively. The macro-spin energy is dependent on the Zeeman energy, the magneto-static energy, and the crystalline anisotropy energy [37] of the system. The Zeeman energy term is neglected in the circuit model for low magnetization operations. In such a case, Zeeman energy is created from an external magnetic field. The magneto-static energy is represented as follows:

$$\begin{aligned} E_d &= \frac{1}{2}\mu_0 V \left|\overrightarrow{M_s}\right| \vec{m} \cdot \overrightarrow{H_d}, \\ \overrightarrow{H_d} &= M_s \overline{N}_d \vec{m}. \end{aligned} \qquad (3.21)$$

where $\overline{N}_d$ and $\overrightarrow{H_d}$ are termed as the demagnetization parameter with diagonal factors N_x, N_y, and N_z, and the diffused magnetic field, respectively. The performance parameters of the spintronic memory cell are based on the formulation explained in [38]. The anisotropic crystalline energy of the system can be given as:

$$E_K = K_u V \sin^2 \varnothing - \theta, \qquad (3.22)$$

where K_u is the particle's energy density [36] and combines the effects of the magneto-static energy and crystalline anisotropy energy. The constants h_X, h_Y, and h_Z can be determined from K_u as:

$$\begin{cases} h_X = N_x - \dfrac{4K_u}{\mu_0 M_s^2 L_x}, \\ h_Y = N_y, \\ h_Z = N_z - \dfrac{H_k}{M_s}. \end{cases} \qquad (3.23)$$

In the above equation, H_k and L_x are the Stoner–Wohlfarth field and the ferromagnet's thickness perpendicular to the easy axis, respectively. From Eq. (3.18), the magnetic field magnitude can be represented as:

$$H = -\frac{1}{\mu_0 V}\frac{\partial E}{\partial M}, \qquad (3.24)$$

The different magnetization components of a Stoner particle can be expressed as:

$$\begin{cases} \vec{H}_x = -\left|\vec{M}_s\right| h_X \hat{m}_x, \\ \vec{H}_y = -\left|\vec{M}_s\right| h_Y \hat{m}_y, \\ \vec{H}_z = -\left|\vec{M}_s\right| h_Z \hat{m}_z. \end{cases} \tag{3.25}$$

The developed circuit model includes the rectification of the effective magnetic field along the X-axis. The effects of thermal noise can be incorporated through this model on a ferromagnetic body. The total magnetic field, including the effect of thermal noise can be given by:

$$H = -\left(\frac{1}{\mu_0}\right)\frac{\partial E}{\partial M} + \left|\vec{H}_T\right|. \tag{3.26}$$

where $\left|\vec{H}_T\right|$ is the magnitude of magnetic field generated by thermal noise. The thermal noise field is produced due to the electrons' thermal motion.

3.2.2.2 Damping Torque

According to the British physicist Gilbert [39], the electrons' damped motion produces a net damping torque that acts against the direction of magnetic motion. This torque represents a damped response of the system and generates a magnetization vector in the direction of the magnetic field. In Eq. (3.17), an additional torque term is required to account for the damped motion of magnetization. Gilbert added the following term to the Landau–Lifshitz equation to form the Landau–Lifshitz–Gilbert equation. The term accounting for the damping torque is given by

$$\vec{\tau}_\alpha = \frac{\alpha}{M_s}\left(\vec{M} \times \frac{\partial \vec{M}}{\partial t}\right). \tag{3.27}$$

α is known as the damping constant and is a system parameter [40].

3.2.2.3 Spin-Transfer Torque

An electron, propagating from a normal ferromagnet to a dominant ferromagnet, transfers spin-angular momentum. British scientist Zangwill [41] predicted that absorption of spin-angular momentum depends on spin reflection and transmission coefficients at the interface. The net ferromagnet is treated as a macro-spin for the standard torque equation. For modeling non-local spin torque, a ferromagnet is assumed to be single domain, thereby validating the treatment of the entire magnetization as a "macro-spin". The ratio of the net perpendicular spin current at the interface to the number of spins in the ferromagnet N_s is represented as follows:

$$N_s = \frac{2M_sV}{\gamma_0 \hbar} \tag{3.28}$$

The spin-transfer torque is proportional to the spin current component, with polarization transverse to the ferromagnet's magnetic orientation. This component can be written as:

$$\vec{I}_{s,-} = \vec{m} \times \left(\vec{I}_s \times \vec{m}\right), \tag{3.29}$$

where $\vec{I}_s$ is the total spin current. The net spin-transfer torque is defined as the average torque exerted per unit of electronic spin and can be written as:

$$\vec{\tau}_s = \frac{\vec{m} \times \left(\vec{I}_s \times \vec{m}\right)}{qN_s}. \tag{3.30}$$

where q is the electronic charge.

3.3 SPIN-INFUSION MAGNETIC SWITCHING

3.3.1 Expansion of the Precession

This section mostly concentrates on the two kinds of magnetic nano-pillars, represented in Figure 3.5. These architectures encompass all the essential features required to understand spin-transfer effects' physics and are also crucial for practical applications. We assume that the respective cross-section of the perpendicular

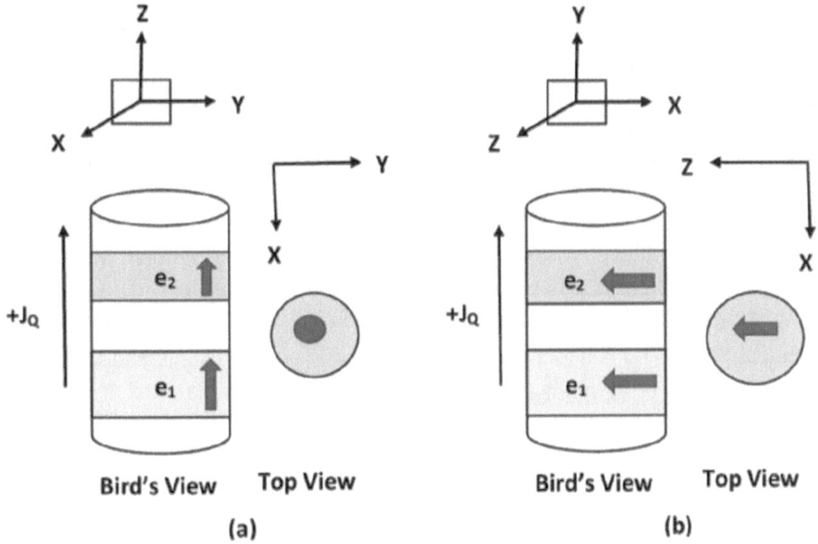

FIGURE 3.5 Standard structure magnetic nano-pillars developed for spin-injection magnetization switching (SIMS) processes. (A) A magnetic nano-pillar with perpendicular magnetization. (B) A magnetic nano-pillar with in-plane magnetization.

and in-plane magnetic nano-pillars are round and elliptical shaped. The magnetic orientation of the in-plane nano-pillars is parallel to the ellipse's long axis.

Figure 3.5 shows the electric current flowing through magnetic nano-pillars, which transports spin-angular momentum, thereby altering the free layer's magnetic orientation. The spin dynamics of the free layer are governed by the Landau–Lifshitz–Gilbert (LLG) equation, with an added spin-transfer torque term as given below:

$$\frac{\partial \vec{S}_2}{\partial t} = \gamma \vec{S}_2 \times \vec{H}_{eff} - \alpha \vec{e}_2 \times \frac{\partial \vec{S}_2}{\partial t} - g(\theta) \times \frac{\left|\vec{J}^{Q}\right|}{e} \frac{\hbar}{2} \vec{e}_2 \times \left(\vec{e}_1 \times \vec{e}_2\right), \tag{3.31}$$

The γ is the gyromagnetic ratio, where $\gamma < 0$ for electrons. The practical value of γ is -221000 m/A$-$s for free electrons. The first term represents the effective field torque in the above equation, which arises from the total effective magnetic field (both internal and external). Moreover, it describes the precession motion of $\vec{S}_2$. The second term represents the Gilbert damping, and the last term represents the efficiency of the spin-transfer torque. $g(\theta)$ represents the spin-transfer efficiency as a function of the inclination angle θ and is treated as constant. The $\vec{S}_2 = S_2\vec{e}_2$ is the net spin-angular momentum of the free layer and is aligned with its magnetic moment; $\vec{e}_2\left(\vec{e}_1\right)$ is the unit vector representing the direction of spin-angular momentum of the free (pinned) layer. It is assumed that the electrons ejected from the free (fixed) layer are aligned (in-phase) with its magnetic orientation. The exchange interaction inside the magnetic layers leads to in-phase precession of the electron spins, thereby generating a single-domain "macro-spin" [42–43]. The assumption of in-phase electron spin ejection and the macro-spin model are not valid under the influence of a demagnetizing field. The spin current or spin polarization, injected inside the non-magnetic channel, generates a faint magnetic field inside the channel itself, known as the Oersted field.

Inconsistencies occur in the domain wall model due to the priority of nearby electron spins. The device-level modeling of spintronics devices considers single domains, domain walls [44, 45], and macro-spin. It has been widely employed due to its transparency and validity in the quasi-equilibrium regime. The spin–orbital momentum is significantly less for 3D transition metals and is ignored in this model.

The effective field H_{eff} is the summation of the external field, demagnetization field, and anisotropic field. The demagnetization field and the anisotropic field are dependent on $\vec{e}_2$. H_{eff} is dependent on the magnetic energy (E_{mag}), and the net magnetic moment $\left(\vec{M}_2\right)$, of the free layer. It is expressed as:

$$H_{eff} = \frac{1}{\mu_0 M_2} \frac{\partial E_{mag}}{\partial e_2}, \tag{3.32}$$

where $\mu_0 = 4\pi \times 10^{-7}$ H/m is the magnetic susceptibility of the vacuum.

The spin dynamics in a perpendicular magnetic nano-pillar are described as cylindrical symmetry covered by an external magnetic field. According to Eq. (3.7), this field is parallel to the axis of symmetry.

3.3.2 Cylindrical Pillar With Uniaxial Anisotropy

In this case, the equation for magneto-static energy can be written as:

$$E_{mag} = -\frac{1}{2}\mu_0\left|\overrightarrow{M}_2\right|\left|\overrightarrow{H}_u\right|\cos^2\theta + \mu_0\left|\overrightarrow{M}_2\right|\left|\overrightarrow{H}_{ext}\right|\cos\theta + c_1, \tag{3.33}$$

where $\left|\overrightarrow{H}_u\right|$ and $\left|\overrightarrow{H}_{ext}\right|$ represent the magnitude of the effective uniaxial anisotropic and external fields, respectively. The magnetic uniaxial anisotropy is the summation of the crystalline anisotropy energy and the demagnetization energy. It is expressed as:

$$\frac{1}{2}\mu_0\left|\overrightarrow{M}_2\right|\left|\overrightarrow{H}_u\right| = K_u v + \frac{1}{2}\mu_0\left|\overrightarrow{M}_2\right|^2\left(\frac{N_{demag}}{v}\right), \tag{3.34}$$

where K_u, v, and N_{demag} are represented by a uniaxial anisotropic constant, the volume of the free layer, and a demagnetization factor, respectively. According to the Eq. (3.33), the effective field is represented as follows:

$$\overrightarrow{H}_{eff} = -\left|\overrightarrow{H}_u\right|\vec{e}_1\cos\theta + \left|\overrightarrow{H}_{ext}\right|\vec{e}_1. \tag{3.35}$$

Substituting the values in Eq. (3.32):

$$\begin{cases} \dfrac{\partial\vec{e}_2}{\partial t} \cong \gamma\left(-\left|\overrightarrow{H}_u\right|\cos\theta + \left|\overrightarrow{H}_{ext}\right|\right)\left(\vec{e}_2\times\vec{e}_1\right) - \alpha_{eff}(\theta)\gamma\left|\overrightarrow{H}_u\right|\cos\theta\,\vec{e}_2\times\left(\vec{e}_1\times\vec{e}_2\right), \\[2ex] \alpha_{eff}(\theta) \cong \alpha + \dfrac{1}{\gamma\left|\overrightarrow{H}_u\right|\cos\theta}g(\theta)\dfrac{\left|\vec{J}^Q\right|}{e}\dfrac{\hbar}{2\left|\vec{S}_2\right|}. \end{cases} \tag{3.36}$$

In Eq. (3.36), the degree of $0(\alpha^2)$ is ignored, as α is significantly less for 3D transition metals, and the practical value of α is 0.007 for FeNi composite [46]. $\alpha_{eff}(\theta)$ represents the effective Gilbert damping coefficient of the free layer under the STT mechanism.

For a cylindrical nano-pillar MTJ with uniaxial anisotropy, the current-density equation is given as:

$$\vec{J}_{c0}^{Q} = -e\alpha\frac{\gamma\left(\left|\overrightarrow{H}_u\right|\cos\theta - \left|\overrightarrow{H}_{ext}\right|\right)}{g(\theta)}\frac{2\vec{S}_2}{\hbar} \quad \text{for } \theta = 0 \text{ or } \pi. \tag{3.37}$$

The above equation represents the threshold current of SIMS, with $\theta = 0$ for parallel to antiparallel and $\theta = \pi$ for antiparallel to parallel switching. It contains an unstable steady cutoff cycle between the parallel (P) and antiparallel (AP) states.

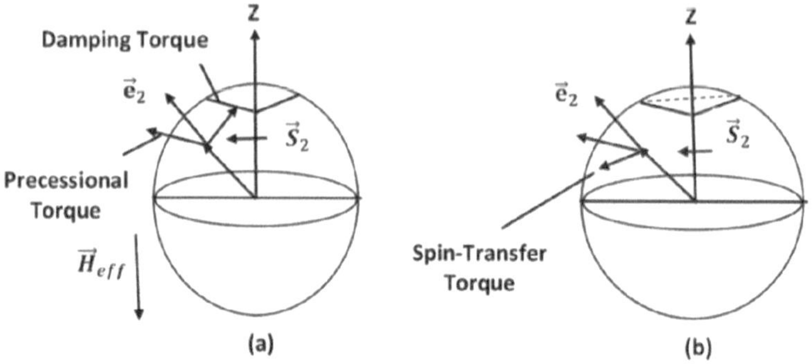

FIGURE 3.6 Illustration of each torque's direction and trajectory of the free layer spin momentum for a nano-pillar with perpendicular remnant magnetization. (a) In the absence of an electric current, the free layer spin's precession is damped. (b) Under an electric current, if the spin-transfer torque overcomes the damping torque, the free layer spin's precession is amplified.

Figure 3.6 shows the directions of the torques. An electric moment of $\vec{S}_2$ over $-\vec{H}_{eff}$ is increased due to the effective field torque. Moreover, the precession's inclination angle is decreased due to the damping torque. If a junction current, $\vec{J}^Q$, is not present in a flipped spin–orbit, the antiparallel transformation to the parallel magnetic field is denoted by $\vec{S}_2$ as in Figure 3.6(a). It is noticed that the direction of $\vec{S}_2$ is opposite to its magnetic moment. Figure 3.6(b) shows the STT mechanism's direction for both $g(\theta)$ and $\vec{J}^Q$. Suppose the induced current, $\vec{J}^Q$, is substantially large; the STT mechanism affects the damping torque. It results from the total torque applied to the ferromagnetic body in effective negative damping.

The negative damping factor sets the priority of the electron momentum. The increase in performance parameters of the STT mechanism is dependent on the inclination angle. From Eq. (3.36), the instability current is evaluated as $\alpha_{eff}(\theta) = 0$.

3.3.3 Linearized LLG Equation

The linearized LLG equation in cylindrical nano-pillars with uniaxial anisotropy is represented as:

$$\gamma \vec{S}_2^{(0)} \times \vec{H}_{eff}^{(0)} - g(\theta_0) \frac{\hbar}{2} \frac{\left|\vec{J}_0^Q\right|}{e} \vec{e}_2^{(0)} \times \left(\vec{e}_1 \times \vec{e}_2^{(0)}\right) = \vec{0}. \tag{3.38}$$

The linearized LLG equation is derived without a cylindrical structure. The primary condition is that the equivalent torque must be a null vector along the XY plane. The linearized form of the LLG equation generated a spin-logic node for the MRAM cell analysis.

3.4 OTHER TECHNIQUES IN SPIN-DEPENDENT COMPUTATION

This chapter has mainly discussed spin injection, transport, and spin switching phenomena in MTJs and spin valves. MTJs and spin valves are the basic building blocks of MRAM, where data is stored via magnetic domain polarization. The design parameters of spin valves and MTJs, i.e., GMR, TMR, spin current, etc., were discussed. MRAM technology is dependent on spin injection, transport, and manipulation. The magnetic switching in MTJ is dependent on spin injection and exploits the magnetic anisotropy in magnetic nano-pillars. Some of the other emerging technologies that can be employed for spin-dependent computation rely on voltage-controlled magnetic anisotropy and the Rashba effect in conjugation with quantum tunneling. A lot of global effort and investment has been geared toward the development and commercialization of MRAM and other spin-based technologies in consumer-level systems. Although the ultimate technological target in the current era is to achieve quantum computation in consumer-level systems, such techniques are still in their initial phase of research and demand a few decades of effort before commercialization. Spin-based classical computation in consumer-level systems, on the other hand, is not far away and can be achieved in the near future. Thus, such building units, such as MTJ, spin valves, MRAM, etc., constitute the "Beyond CMOS" industry research focus in the modern era.

REFERENCES

1. A.B. Granovsky, M. Liyn, A. Zhukov, V. Zhukova, Giant magneto-resistance of granular microwires: Spin-dependent scattering in intergranular spacers, *Physics of the Solid-State Journal*, 53, 320–322, 2011.
2. E. Hirota, H. Sakakima, K. Inomata, *Giant Magnetoresistance Devices*, Springer, 30 May 2020.
3. Pippard, A. Brian, Magnetoresistance in metals, *Cambridge Studies in Low Temperature Physics*, 8 (2), 31–42, 2006.
4. M. Bowen, A. Barthelemy, M. Bibes, E. Jacquet, J.P. Contour, A. Fert, Half-metallicity proven using fully spin-polarized tunneling, *Journal of Physics: Condensed Matter*, 41 (17), 41–51, September 2005.
5. M. Julliere, Tunneling between ferromagnetic films, *Physics Letter*, 54 (3), 225–226, August 1975.
6. F. Schleicher, U. Halisdemir, D. Lacour, M. Gallart, S. Boukari, G. Schmerber, V. Davesne, Localized states in advanced dielectrics from the vantage of spin and symmetry-polarized tunneling across MgO, *Nature Communication*, 5 (4), 45–47, August 2014.
7. M. Suzuki, N. Kawamura, Y. Yamamoto, Measurement of a Pauli and orbital paramagnetic state in bulk gold using x-ray magnetic circular dichroism spectroscopy, *Physical Review Letters*, 108, 047201, January 2012.
8. Y. Ohno, D.K. Young, B. Beschoten, Electrical spin injection in a ferromagnetic semiconductor heterostructure, *Nature Letters*, 402, 790–792, December 1999.
9. R. Fiederling, M. Keim, G. Schmidt, Injection and detection of a spin-polarized current in a light-emitting diode, *Nature Letters*, 402, 787–790, December 1999.
10. B.T. Jonker, Y.D. Park, B.R. Bennett, Robust electrical spin injection into a semiconductor heterostructure, *Physical Review B*, 62, 8180, September 2000.

11. B.T. Jonker, G. Kioseoglou, P.E. Thompson, Electrical spin-injection into silicon from a ferromagnetic metal/tunnel barrier contact, *Nature Physics Letters*, 3, 542–546, July 2007.
12. M. Johnson, Spin accumulation in gold films, *Physical Review Letters*, 70, 21–42, April 1993.
13. F.J. Jedema, M.V. Costache, H.B. Heersche, B.J. van Wees, Electrical detection of spin accumulation and spin precession at room temperature in metallic spin valves, *Applied Physics Letters*, 81, 5162, November 2002.
14. X. Lou, C. Adelmann, S.A. Crooker, P.A. Crowell, Electrical detection of spin transport in lateral ferromagnet-semiconductor devices, *Nature Physics*, 3, 197–202, March 2007.
15. N. Tombros, S.J. van der Molen, B.J. van Wees, Separating spin and charge transport in single-wall carbon nanotubes, *Physical Review B*, 73, 233403, June 2006.
16. N. Tombros, C. Jozsa, M. Popinciuc, Electronic spin transport and spin precession in single graphene layers at room temperature, *Nature Letters*, 448, 571–574, July 2007.
17. M. Ohishi, M. Shiraishi, R. Nouchi, T. Shinjo, Spin injection into a graphene thin film at room temperature, *Japanese Journal of Applied Physics*, 46 (7L), 605, June 2007.
18. I. Appelbaum, B. Huang, D.J. Monsma, Electronic measurement and control of spin transport in silicon, *Nature*, 447 (7142), May 2007.
19. J.Z. Sun, Batch-fabricated spin-injection magnetic switches, *Applied Physics Letters*, 81 (12), 2202, July 2002.
20. M. Tsoi, A.G.M. Jansen, J. Bass, Excitation of a magnetic multilayer by an electric current, *Physical Review Letters*, 80, 4281, May 1998.
21. S.I. Kiselev, J.C. Sankey, D.C. Ralph, Microwave oscillations of a nanomagnet driven by a spin-polarized current, *Nature Letters*, 425, 380–383, September 2003.
22. J.C. Slonczewski, Current-driven excitation of magnetic multilayers, *Journal of Magnetism and Magnetic Materials*, 159 (1), L1–L7, June 1996.
23. L. Berger, Emission of spin waves by a magnetic multilayer traversed by a current, *Physical Review B*, 54, 9353, October 1996.
24. E.B. Myers, D.C. Ralph, J.A. Katine, Current-induced switching of domains in magnetic multilayer devices, *Science*, 285 (5429), 867–870, June 1999.
25. J.Z. Sun, Current-driven magnetic switching in manganite trilayer junctions, *Journal of Magnetism and Magnetic Materials*, 202 (1), 157–162, July 1999.
26. J.A. Katine, F.J. Albert, E.B. Myers, Current-driven magnetization reversal and spin-wave excitations in Co/Cu /Co pillars, *Physical Review Letters*, 84, 3149, April 2000.
27. Y. Huai, F. Albert, P. Nguyen, Observation of spin-transfer switching in deep submicron-sized and low-resistance magnetic tunnel junctions, *Applied Physics Letters*, 84, 3118, April 2004.
28. S. Yakata, H. Kubota, T. Seki, Enhancement of thermal stability using ferromagnetically coupled synthetic free layers in MgO-based magnetic tunnel junctions, *IEEE Transactions on Magnetics*, 46 (6), 2232–2235, May 2010.
29. Z. Diao, D. Apalkov, M. Pakala, Spin transfer switching and spin polarization in magnetic tunnel junctions with MgO and AlO_x barriers, *Applied Physics Letters*, 87, 232502, October 2005.
30. I. Tudosa, J.A. Katine, S. Mangin, Perpendicular spin-torque switching with a synthetic antiferromagnetic reference layer, *Applied Physics Letters*, 96, 212504, May 2010.
31. R. Ouchida, T. Shiraishi, M. Ohta, Effects of Au/Cu ratio and gallium content on the low-temperature age-hardening in Au-Cu-Ga alloys, *Journal of Materials Science*, 30, 3863–3866, August 1995.
32. Y. Suzuki, A.A. Tulapurkar, C. Chappert, Spin-injection phenomena and applications, *Nanomagnetism and Spintronics*, 3, 93–153, September 2009.

33. L.D. Landau, E. Lifshitz, On the theory of the dispersion of magnetic permeability in ferromagnetic bodies, *Perspectives in Theoretical Physics*, 8 (2), 51–62, 1992.
34. E.C. Stoner, E.P. Wohlfarth, A mechanism of magnetic hysteresis in heterogeneous alloys, *Philosophical Transactions of the Royal Society A*, 240 (826), May 1948.
35. C. Kittel, Theory of the structure of ferromagnetic domains in films and small particles, *Physical Review*, 70 (11), 965, December 1946.
36. C. Tannous, J. Gieraltowski, The Stoner–Wohlfarth model of ferromagnetism, *European Journal of Physics*, 29 (3), 475, March 2008.
37. J. Xiao, A. Zangwill, M.D. Stiles, Macrospin models of spin transfer dynamics, *Physical Review B*, 72, 014446, July 2005.
38. M. Beleggia, M.D. Graef, Y.T. Millev, The equivalent ellipsoid of a magnetized body, *Journal of Physics D: Applied Physics*, 39 (5), 891, 2006.
39. T. Gilbert, J. Kelly, Anomalous rotational damping in ferromagnetic sheets, in Conference on Magnetism and Magnetic Materials, Pittsburgh, PA, 253–263,1955.
40. M.C. Hickey, J.S. Moodera, Origin of intrinsic Gilbert damping, *Physical Review Letters*, 102 (13), 137–601, March 2009.
41. M. Stiles, A. Zangwill, Anatomy of spin-transfer torque, *Physical Review B*, 66 (1), 014407, June 2002.
42. J.Z. Sun, Spin-current interaction with a monodomain magnetic body: A model study, *Physical Review B*, 62 (1), 570, July 2000.
43. Hillebrands, Burkard, Andre, *Spin Dynamics in Confined Magnetic Structures III*, Condensed Matter Physics Book, 2006.
44. J. Miltat, C. Vouille, Spin transfer into an inhomogeneous magnetization distribution, *Journal of Applied Physics*, 89 (11), 6982–6984, June 2001.
45. K.J. Lee, O. Redon, Analytical investigation of spin-transfer dynamics using a perpendicular-to-plane polarizer, *Applied Physics Letters*, 86 (2), 022505, January 2005.
46. S.M.S. Mizukami, T.M.T. Miyazaki, The study on ferromagnetic resonance linewidth for NM/80NiFe/NM (NM=Cu, Ta, Pd and Pt) films, *Japanese Journal of Applied Physics*, 40 (2), 580, November 2000.

4 Quantum-Dot Cellular Automata (QCA) Nanotechnology for Next-Generation Systems

Vijay Kumar Sharma

CONTENTS

4.1 INTRODUCTION

New emerging technologies can be used instead of complementary metal oxide semiconductor (CMOS) technology to improve the overall performance of digital systems [1, 2]. Many developments have been seen in the semiconductor industry during the last four decades, following the application of Moore's law. However, due to some technological limitations, its progress has slowed. The circuits have become more complex and more heat is being generated in the circuits as the density of devices increases day by day. The generated heat harms the circuit as it may cause damage because the existing heat removal techniques are not effective. A semiconductor material substrate is processed through different fabrication steps in order to make an integrated circuit (IC). The most important fabrication step is the lithography process, which is used to form three-dimensional (3D) patterns or images on the silicon substrate to which a certain silicon wafer is exposed for doping. A large number of devices are fabricated on a wafer for complex designs in the microelectronic field. All electronic devices, such as computers, mobile phones, television, remotes, digital versatile disk (DVD) and compact disc (CD) players, digital cameras, etc., need transistors, capacitors, inductors, resistors, and diodes to form their circuits. Due to the use of digital electronics, we can use binary numbers to represent the binary information mathematically as well as the binary '1' and '0' as 'ON' and 'OFF' states.

DOI: 10.1201/9781003155751-4

In the coming years, CMOS technology will reach its limit because of the high cost of lithography, physical scalability, and short channel effects (SCEs). Therefore, the semiconductor industry needs to start using new nanoelectronics devices [3]. Along these lines, arising advancements, such as a single-electron transistor (SET) [4], resonant tuning diode (RTD) [5], and quantum-dot cell automata (QCA) [6], could overcome the referenced difficulties. QCA has been presented as a viable solution. QCA is currently the best technology to design highly-dense digital circuits and offers fast switching, less size, and minimum power dissipation [7, 8]. QCA presents binary information on the cells. A QCA cell consists of four quantum dots or sites. Dots are localized at the opposite corners of the QCA cell. Two electrons reside in a QCA cell, which can tunnel between the other corner quantum dots. Electrons always reside diagonally at the cell's corner because of Coulombic repulsion.

In digital circuits, a full adder is used to perform addition and subtraction, and can be used in the construction of many applications. QCA technology presents a binary representation of data and a way to communicate this information. CMOS technology faces problems when using the nano-scale because of an increase in variations in nanometre design. The main concept of the QCA cell that it is transistor-free.

This chapter details the QCA fundamentals from the perspective of digital circuit design. A full adder is an important digital circuit widely used in many computation fields. Therefore, the various existing full adder circuits are overviewed in this chapter. Different performance metrics are estimated for full adder circuits, which helps the research community to identify promising ideas to investigate.

The rest of the chapter is divided into the following sections: the background of QCA nanotechnology is elaborated in Section 4.2. Section 4.3 discuss the existing QCA-based full adder circuits. A comparison analysis is provided in Section 4.4. The chapter concludes in Section 4.5.

4.2 BACKGROUND

The basic component of QCA technology is known as a quantum cell. QCA is one of the emerging technologies introduced by Lent et al. [6]. A QCA cell is the fundamental and simple component used in QCA technology. Each cell occupies a nanoscale area and represents a logical bit. Only two electrons are present in a cell and these electrons occupy two quantum dots. Boolean logic, such as logic zero or logic one, is denoted by the positions of free electrons in QCA technology. A square-shaped QCA cell includes four dots at the corners [9–12]. The two free electrons always reside diagonally and have the ability to tunnel between the dots. This diagonal arrangement of the electrons within a cell is due to the Coulombic repulsion force. There are two possible arrangements of a cell: 'P = +1' and 'P = –1', where P is stands for polarization. 'P = +1' signifies logic '1' and 'P = –1' denotes logic '0' [13–16]. Figure 4.1 shows both polarizations.

A QCA clock is needed for the operation of digital circuits. Clock pulses help drive the electrons. A QCA clock enables the flow of the electrons and permits the electrons to modify their orientation [17–20]. QCA clock pulses are given in Figure 4.2. A QCA clock comprises four phases: switch, hold, release, and relax.

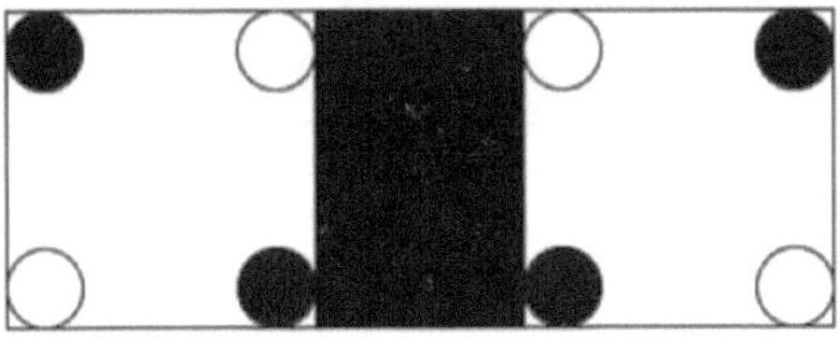

FIGURE 4.1 A QCA for (a) logic '0' (b) logic '1.'

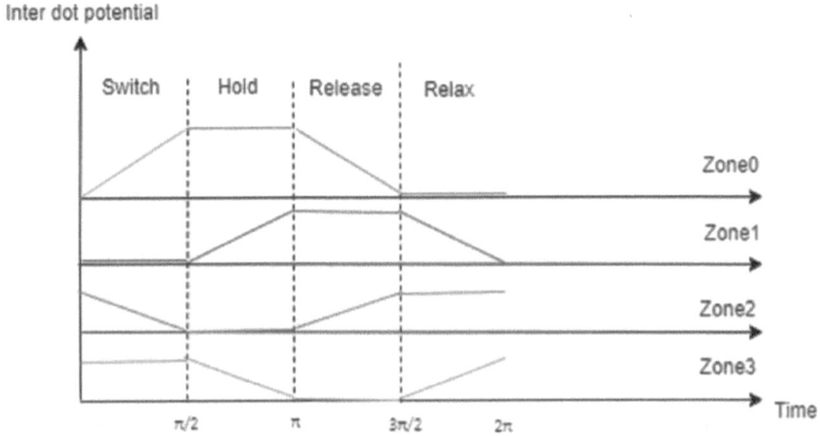

FIGURE 4.2 QCA clock phases with clock zones.

A QCA clock is also divided into clock zones and has four zones which are 90° in phase with the successive zone.

The inter-dot barrier rises in the switch phase, which causes polarization. QCA cells begin without polarization and become polarized. In the hold phase, the cell maintains its polarization as the polarization state remains preserved, whereas the potential barrier is high. The inter-dot barrier decreases when the clock pulse enters the release phase; the cell loses its polarization, and the cell becomes unpolarized. During the relax phase, polarization within the cell is completely lost, whereas the potential barrier remains lower [21–23]. No electric current flows from one cell to another for the information to flow from input to output, resulting in power dissipation in the circuit [24]. In CMOS technology, the clock is mostly used in sequential circuits, but in QCA technology the clock is used in combinational and sequential circuits. A QCA clock provides power as well as switch ability to the circuits.

In QCA technology, the signal is transferred from one end to another end through two types of QCA-based wires: binary and inverter chain wires. The electrons in the neighboring cell interact with each other, which results in the propagation of the information from one cell to another [25]. The arrangement of cells in a series, whether in a horizontal or vertical position, results in the formation of QCA wires [26]. The input cell is also known as the driver cell. A binary wire is shown in Figure 4.3.

Normal QCA wire   Diagonal QCA wire

FIGURE 4.3 Two ways of QCA wires.

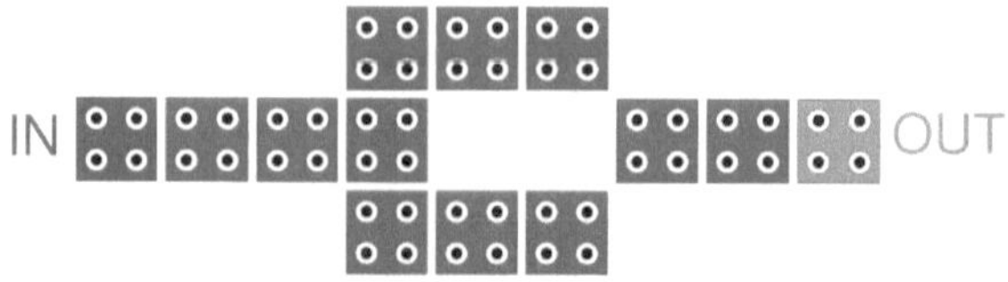

FIGURE 4.4 QCA inverter gate.

In the case of binary wire, the number of QCA cells are connected in a consecutive manner and transfer the input cell value to the output cell. In the inverter chain, the dots in the QCA cell are rotated by 45° as compared to the normal QCA cell [27]. The arrangement of QCA cells is such that each cell has the opposite polarization of its neighboring cell. The clock provides the power for accurate computing in QCA technology. Synchronization is required for the computation of the circuit.

An important gate in QCA is the NOT or inverter gate. An inverter gate inverts the input signal that is applied. If logic '0' is applied at input, then logic '1' will be available at output, and if logic '1' is applied at input, then logic '0' will be present at output. Figure 4.4 shows a QCA-based inverter gate [28, 29]. We can implement many logical functions using basic gates.

Wire crossings play an important role in QCA-based circuits [30] as they are required to cross the QCA wires in the circuit. The two types of wire crossing techniques are coplanar and multilayer. Only a single layer throughout the circuit is used in the coplanar technique, while many layers are applied to build a circuit in multilayer crossing [31–33]. Multilayer wire crossing design has proven to be more robust. Previously the majority of designs are based on coplanar because of its simplicity. In coplanar-based designs, the total area required by a QCA-based circuit layout is too vast, making it practically impossible to implement a complex Boolean function-based circuit, as it is not suitable in the nano-scale arena. Therefore, we can use a multilayer approach to solve the problem of area requirements.

In multilayer cases, the design uses more than one layer, and two cells are close to each other where one cell is placed directly over another cell in the same location but on two different layers. Multilayer designs allow for the propagation of signals by reducing the area occupied by the circuit. A coplanar crossover uses a normal wire as well as a diagonal wire. The normal wire has two sections, and the diagonal wire passes through the normal wire. In a multilayer crossover, vertical cells are marked by circles and are placed in different layers. Figure 4.5 depicts coplanar and multilayer crossovers.

The majority gate is also another one of the basic elements of QCA technology. A total of five QCA cells are utilized in a three-input majority gate, where three cells

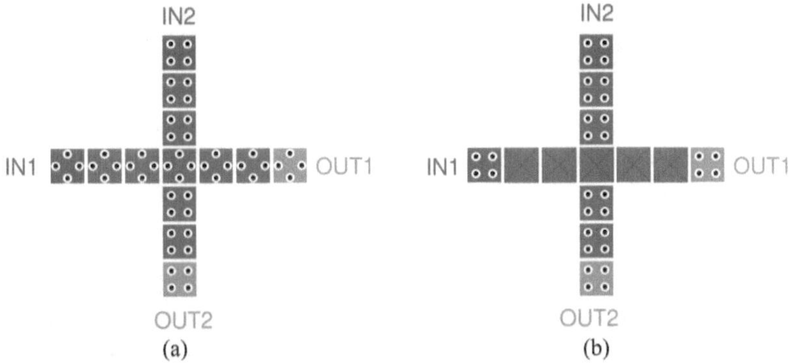

FIGURE 4.5 Crossover (a) coplanar (b) multilayer.

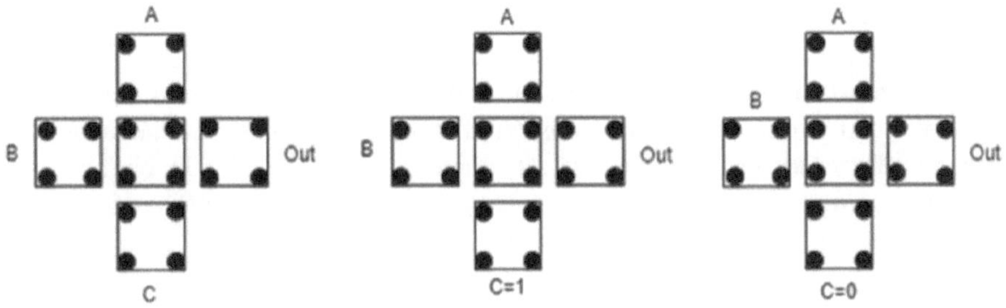

FIGURE 4.6 (a) Five-input majority gate (b) OR gate (c) AND gate.

are input cells, one cell is a driver cell, and one cell is an output cell [34, 35]. The output cell is easily fed into the next stage of the QCA circuit. The implementation of the logic function of the majority gate using CMOS logic requires 26 transistors, while QCA technology implementation needs only 5 QCA cells. Therefore, this provides a more compact logic. Both two-input OR and AND gates are designed using a three-input majority gate, and it depends on the fixing of the input at logic '1' or at logic '0'. If one input is fixed at logic '1' then it generates a two-input OR gate function; if one input is fixed at logic '0' then it generates a two-input AND gate function. Figure 4.6 presents a three-input majority gate.

The logical expression of the three-input majority gate is given in Eq. (4.1).

$$MAJ3(A,B,C) = AB + BC + CA \tag{4.1}$$

A five-input majority gate can also be used for performing complex QCA circuit implementations [36, 37]. Diagonal QCA cells can also form diagonal QCA wires and inverter gates. If the number of diagonal QCA cells is odd, it forms a diagonal QCA wire, whereas if the number of diagonal QCA cells is even it forms an inverter gate. In QCA technology, the clock is responsible for synchronization and controls the information flow. It also provides the power required to run the circuit [38].

4.3 FULL ADDER USING QCA NANOTECHNOLOGY

A full adder is a backbone circuit in the computational field. This section discusses the existing full adder circuits using QCA nanotechnology. Researchers have presented different ideas for implementing a full adder using QCA nanotechnology.

A one-bit full adder can be designed using a three-input XOR gate based on half distance. Output is obtained with the use of half space and interaction between the cells. A XOR gate consists of three inputs A, B, and C, and a five-input majority gate is used with 14 QCA cells, as shown in Figure 4.7 [39].

A 0.01 um^2 area is covered by this three-input XOR gate. A QCA full adder is designed using the QCA three-input XOR gate based on half distance. QCA full adder architecture is shown in Figure 4.8 [39] using the three-input XOR gate.

From Figure 4.9, we can observe that the three-input XOR gate produces the Sum, whereas the three-input majority gate produces the carry. The latency of this QCA three-input XOR gate circuit is 0.5 clock cycles.

The QCA full adder has a coplanar design and is made up of 29 cells. It offers a latency of 0.5 clock cycles and occupies an area of 0.02 um^2.

QCA mostly uses coplanar designs due to the simplicity of designing the circuits. However, multilayer designs are considered to be more robust. Coplanar designs require a larger area for complex circuits; therefore, they are not suitable for nanoscale technology. Complex circuits require various Boolean functions, which makes

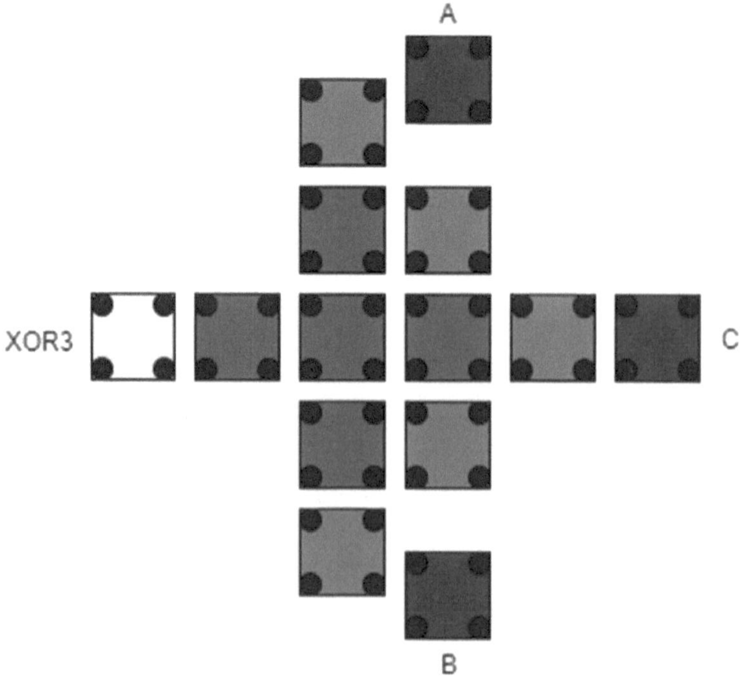

FIGURE 4.7 Three-input XOR gate designed in a single layer.

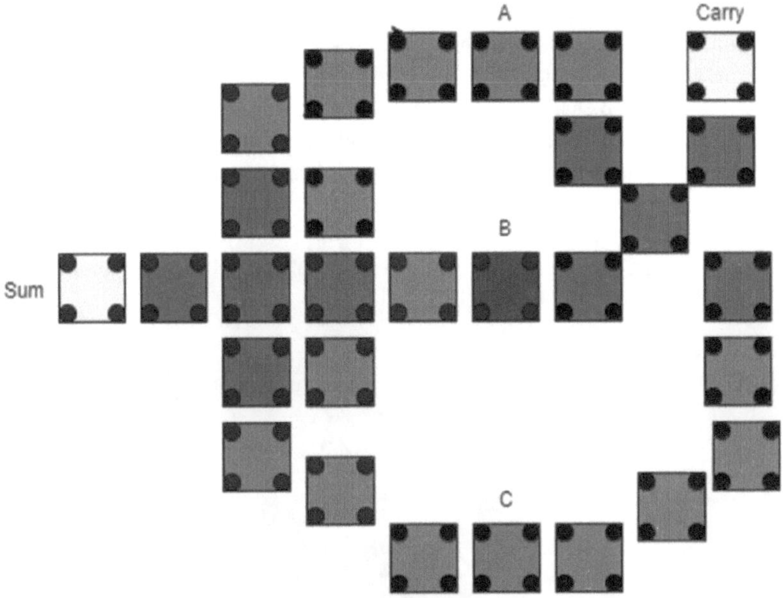

FIGURE 4.8 A QCA cells based full adder.

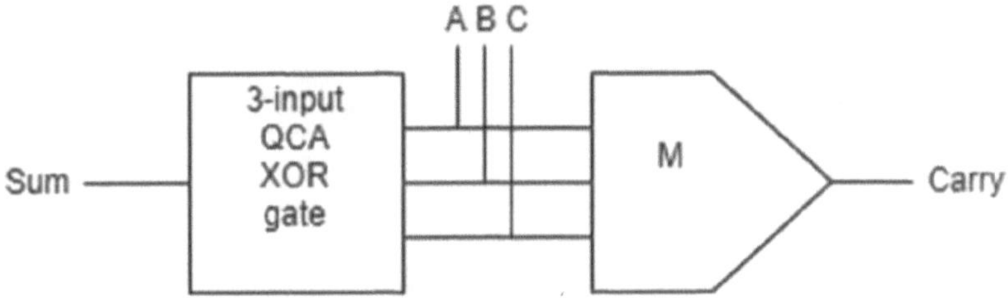

FIGURE 4.9 Logical way for implementation of QCA full adder.

it is practically impossible to implement a circuit layout using coplanar design, as this design is not acceptable in the nano-scale arena. This problem can be solved using multilayer design architecture. A multilayer design is preferred as two cells can be placed over one another in the same location, but both on different layers [40]. In coplanar wire crossing, there is the possibility of undesirable crosstalk between the crossing lines, as well as a discontinuity in the propagation of signals due to the loose binding of the coplanar design. Therefore, we can use multilayer wire crossings instead of coplanar wire crossings. A full adder is used to design arithmetic circuits, and it can be designed using a multilayer five-input majority gate. The Boolean function of the five-input majority gate is given in Eq. (4.2).

$$F(A,B,C,D,E) = ABC + ABD + ABE + ACD + ACE + ADE + BCD + BCE + BDE + CDE \tag{4.2}$$

A QCA layout of multilayer five-input majority gate is given in Figure 4.10 [41].

This multilayer five-input majority gate uses 13 cells, a 0.25 clock, and occupies 0.0096 um^2 of area. Output can be obtained from Layer 3, as it is not surrounded by QCA cells and can be easily assessed by other QCA-based circuits. Layer 1 consists of input E, Layer 3 consists of inputs A, B, and C, whereas Layer 5 consists of input D. No wire crossovers are required to transmit the output signal in this multilayer structure of a five-input majority gate. In multilayer design, as we move across layers, the input signals get inverted and instead of using five layers, we can implement the same design in three layers. This can be done by placing the upper layer cell diagonally opposite to the lower layer cell.

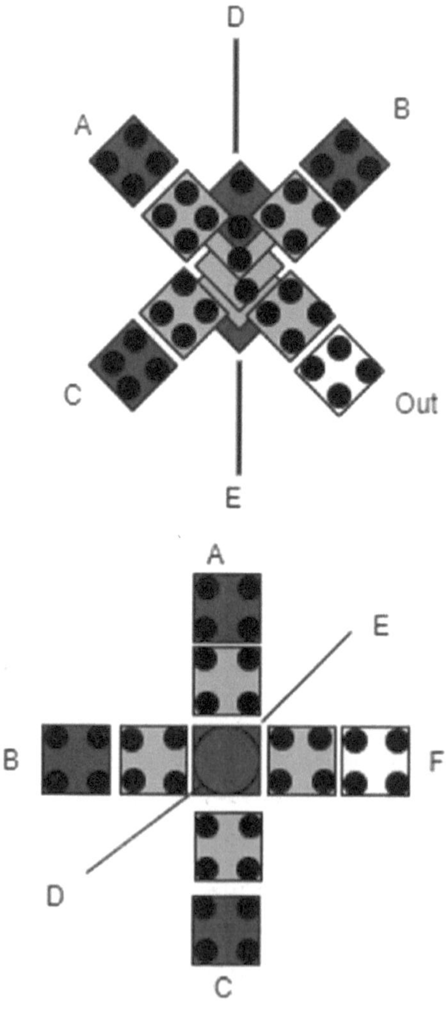

FIGURE 4.10 QCA layout of multilayer five-input majority gate.

The multilayer design of the full adder is shown in Figure 4.11, which uses two majority gates (a three-input majority gate and a five-input majority gate) [41].

This design of a full adder requires only three layers. The input gets inverted in the multilayer approach as we move across the layers. A five-input majority gate requires five layers, but the same design can be implemented using three layers by placing the upper layer cell diagonal to the lower layer cell (in the next layer), instead of placing it directly on the top of the lower layer cell. The carry bit is generated using a traditional three-input majority gate for a full adder in Layer 1. The carry bit is the output of the three-input majority gate. The generation of the carry bit as an output requires only one clock zone. As this carry bit propagates upwards toward Layer 2, the carry bit is inverted due to the multilayer approach. In Layer 2, the inverted carry bit (*CARRY*) acts as an input to the five-input majority gate in Layer 3. As *CARRY* enters Layer 3, it again gets inverted $(\overline{CARRY})$.

Some cells are also stacked on inputs A, B, and C to propagate them from Layer 1 to Layer 3. An additional clock zone (Clock Zone 1) is required for obtaining the output of a five-input majority gate (the Sum of the full adder). Inputs A, B, and C and $\overline{CARRY}$ present in Layer 3 are fed into the five-input majority gate using the Clock Zone 1. Therefore, only two clocks are needed for this 1-bit full adder. This proposed full adder uses 31 cells, 0.01 um^2 area, and a 0.50 clock.

Two variants of the five-input majority gate are represented in Figure 4.12, in which A, B, C, and D act as the inputs and the value of D is fed into two inputs [42].

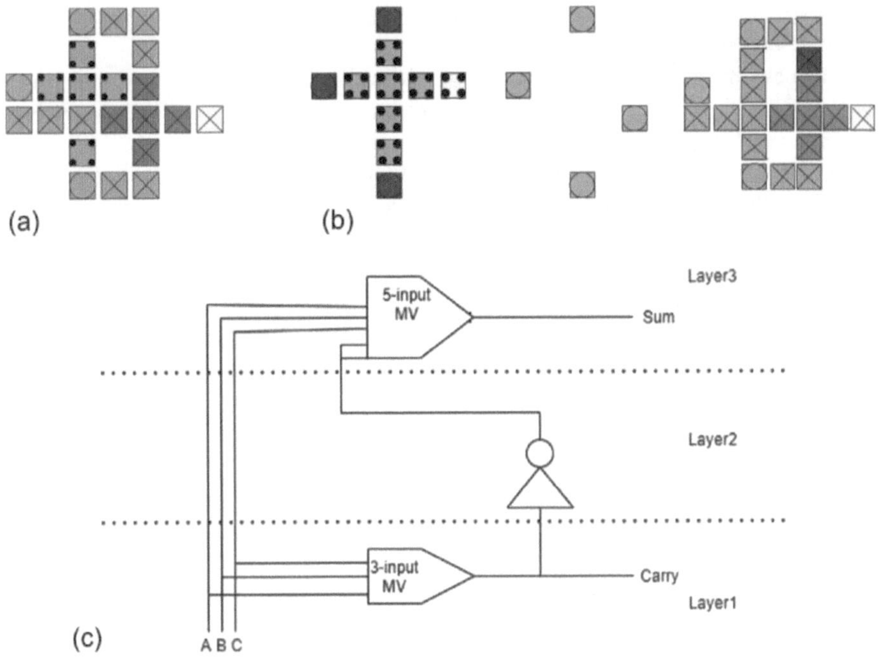

FIGURE 4.11 Full adder using multilayer (a) layout (b) three layers (c) schematic.

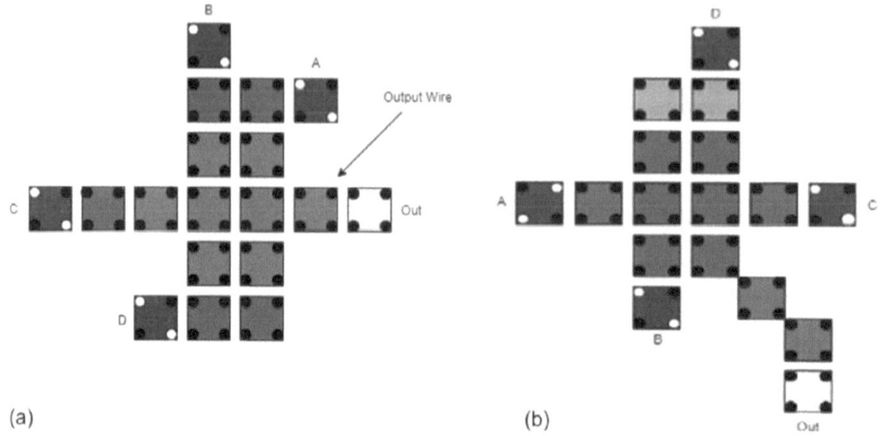

FIGURE 4.12 Two variants of a five-input majority gate.

Five-input majority gates and inverters can be used in the construction of full adders. A design is shown in Figure 4.12(a), in which the output value can be easily fed into the input of other QCA circuits because the output signal is not surrounded by other QCA cells. Input QCA cells are in the first clock zone, whereas the middle seven QCA cells (device cells) and output cells are in the second clock zone. In Figure 4.12(b), the inputs A, B, C, and D of the robust five-input majority gate are positioned at the first clock zone, whereas the eight device cells are used at the second clock zone.

The cells in the diagonal position are used in the output, and the output is transmitted using these cells that are positioned in the third clock zone. All these clock zones have a 90° phase delay between them. Using the first diagonal QCA cell maximizes the distance between input and output QCA cells, reducing the noise effect. However, this diagonal QCA cell produces the inversion of the five-input majority gate. The second diagonal QCA cell is used to produce the final output, which is then fed to the output cell. Efficient full adders can be implemented using these five-input majority gates.

A full adder designed using a five-input majority gate is shown in Figure 4.13 [42]. This full adder has a multilayer design and is simple in structure.

Figure 4.14 shows three layers of the full adder.

The first layer is shown in Figure 4.14(a), which consists of two diagonal positioned cells which act as an inverter, one three-input majority gate and one five-input majority gate. In the first layer, the output of the three-input majority gate gets inverted by the first diagonal cell $\left(\overline{CARRY}\right)$ and the second diagonal cell is used to produce C_{out}. The $\overline{CARRY}$ is used in the two inputs of the five-input majority gate. In Figure 4.14(b), the second layer consists of four cells, and these cells are used to transfer the carry values. As shown in Figure 4.14(c), the third layer consists of two cross wires, which complete the multilayer design.

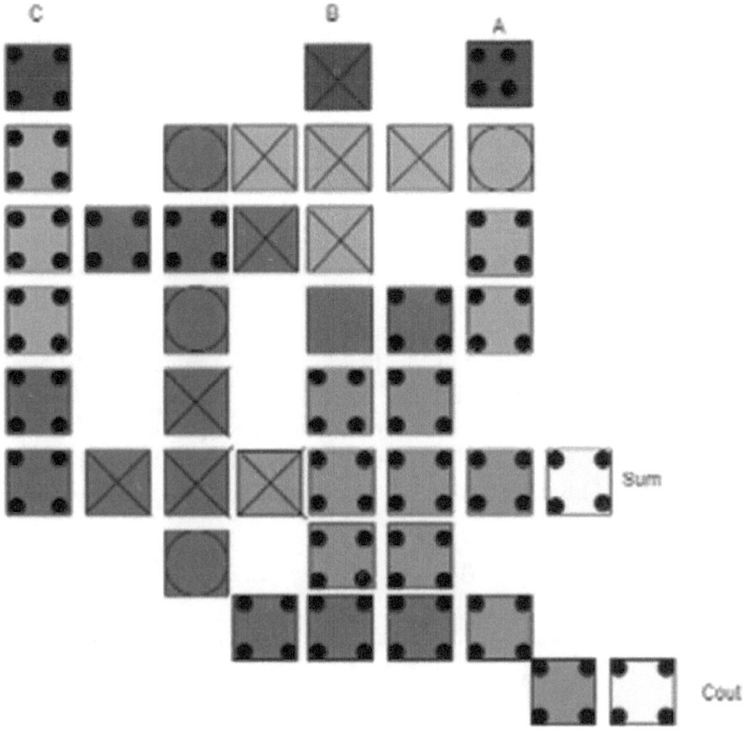

FIGURE 4.13 Layout of a QCA full adder.

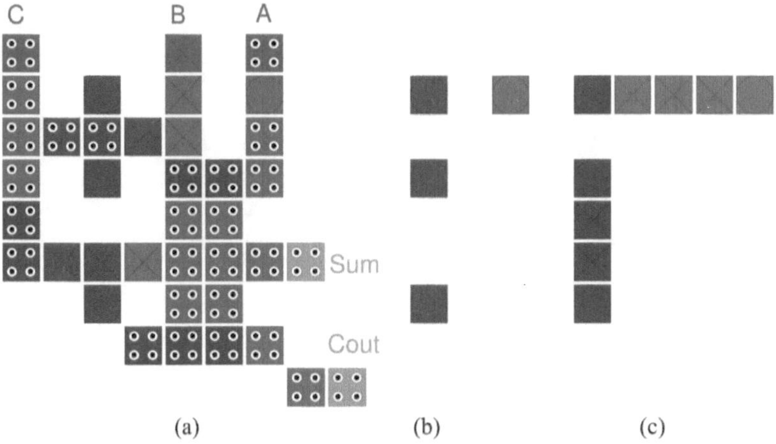

FIGURE 4.14 Three layers of the full adder (a) first, (b) second, and (c) third layer.

Three- and five input majority gates are used to construct robust full adders in QCA [43]. Figure 4.15 shows the schematic of a full adder consisting of a three-input majority gate (Maj1) and the output of a three-input majority gate producing the carry value.

The inversion of this carry value is used as the input for a five-input majority gate. The output of this five-input majority gate acts as Sum for full adder design. Figure 4.16 shows the QCA layout of the five-input majority gate.

This layout uses three clocking zones to produce output with a 90° phase difference between them. Clock Zone 0 is applied for the inputs B, C, and D, and inverted input A cells. In the middle of this layout, six QCA cells are positioned at Clock Zone 1, and the final output value is obtained from the QCA cells that are positioned at Clock Zone 2. Now the output of this five-input majority gate can be easily used as an input to other QCA-based circuits.

Figure 4.17 represents the layout of a robust one-layer QCA-based full adder [43]. This layout design consists of a three-input majority gate and uses three clock zones to produce output. This proposed full adder is based on a coplanar design and is

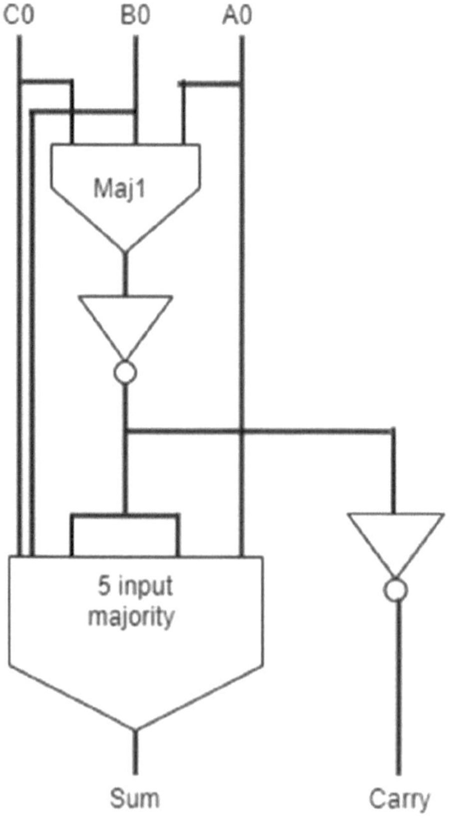

FIGURE 4.15 Schematic of a full adder.

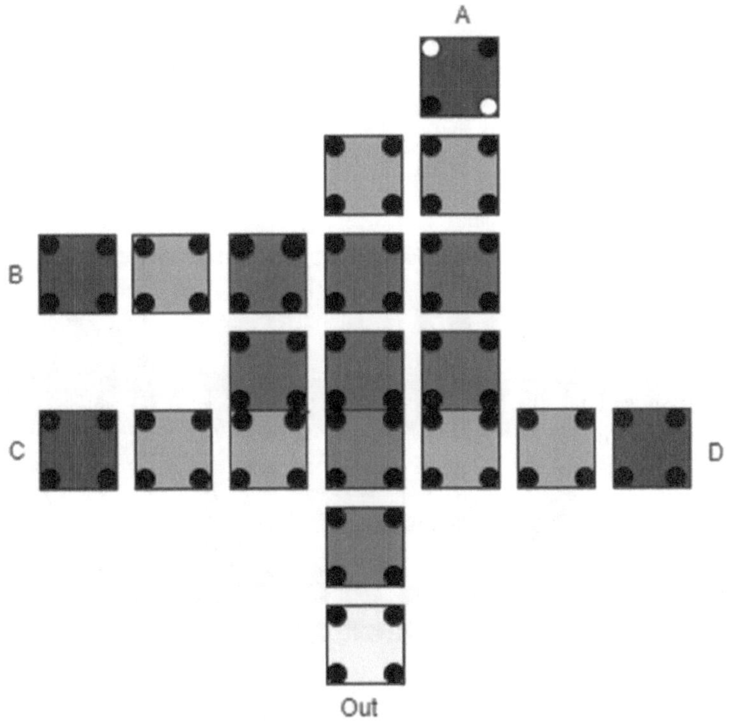

FIGURE 4.16 A layout for a five-input majority gate.

robust against sneak noise paths; therefore, it can be used to design many more QCA-based digital circuits.

Figure 4.18 contains the circuit diagram of the 1-bit full adder. It consists of three three-input majority gates, and the output is collected at the Sum and Carry pins [44].

Figure 4.19 presents the complete design of a 1-bit QCA full adder circuit. In this design, A, B, and C are the inputs, whereas Sum and Carry denote the outputs.

This design of a full adder uses 46 QCA cells and four clock zones. Design requires a 0.04 um^2 area and delay of four clock zones.

Figure 4.20 depicts that A, B, and C are the three inputs of a full adder and Sum and Carry are the two outputs where the output value Sum is the result of XOR inputs A, B, and C [45]. In contrast, the Carry value can be obtained using a three-input majority gate. This QCA full adder has a coplanar design, made up of 44 QCA cells and occupying an area of 0.06 um^2 with a latency of 1.25 clock cycles.

Figure 4.21 contains the circuit diagram of a full adder where two majority gates (one three-input majority gate and one five-input majority gate) are used for determining the outputs (Sum, Carry) [46].

The value of the output Sum can be obtained from the output of the five-input majority gate, and its equation can be written as given in Eq. 4.3.

$$Sum = Maj5\left(A, B, C, \overline{Carry}, \overline{Carry}\right) \tag{4.3}$$

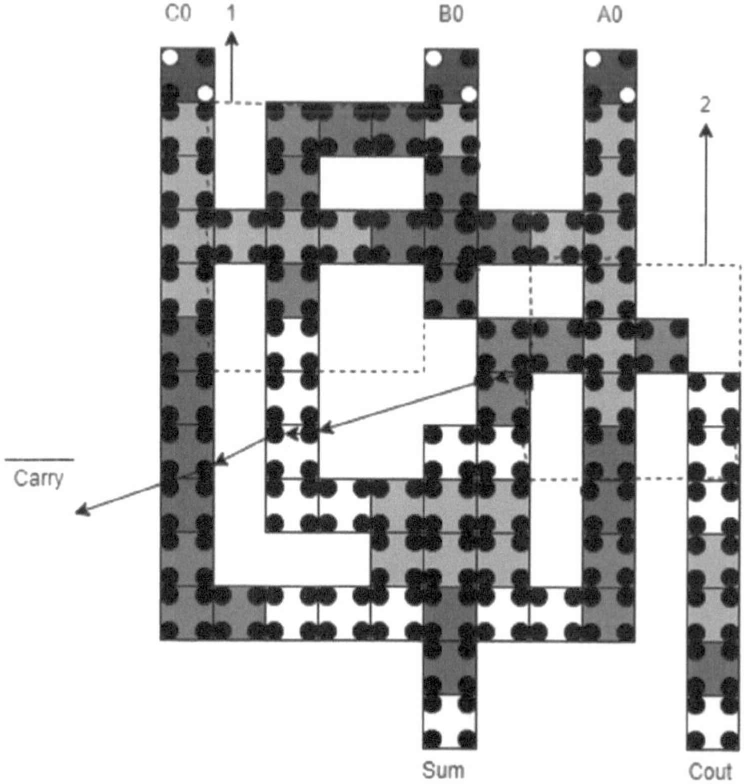

FIGURE 4.17 Layout for a single-layer full adder.

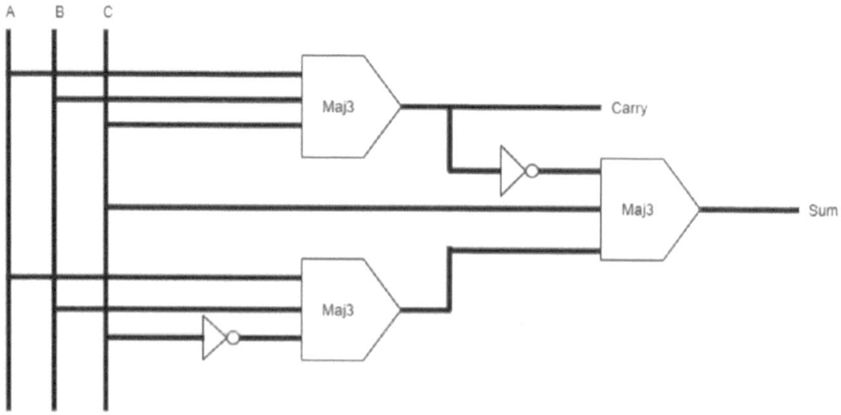

FIGURE 4.18 Circuit for a full adder.

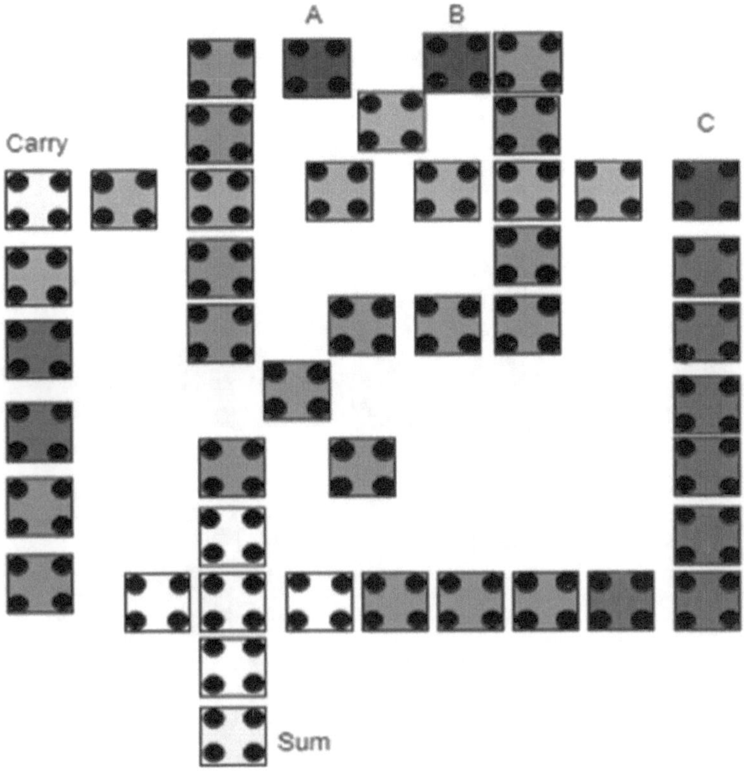

FIGURE 4.19 Circuit layout of a full adder.

The circuit diagram of the full adder shown in Figure 4.21 is much simpler as it contains only one three-input majority gate, one five-input majority gate and one inverter as compared to other full adder circuits, which uses only three-input majority gate and inverters. The QCA layout of the full adder using majority gates are given in Figure 4.22 [46].

The cost of the circuit can be determined as given in Eq. (4.4).

$$Cost = Area \times Latency^2 \tag{4.4}$$

Where *Area* is the space occupied by the design and *Latency* is the number of clock cycles. This coplanar design of a full adder uses 63 cells with a latency of 0.75, occupies an area of 0.05 um2, and has a cost of 0.028. This single-layer design is generally better than many multilayer designs.

Figure 4.23 represents the logical diagram of the full adder circuit, which consists of a three-input majority gate and a three-input XOR gate [47].

The output of the three-input majority gate generates Carry, whereas the output of the three-input XOR gate generates Sum. The layout of the QCA full adder is shown in Figure 4.24, which has only 26 cells implemented in a single layer [47].

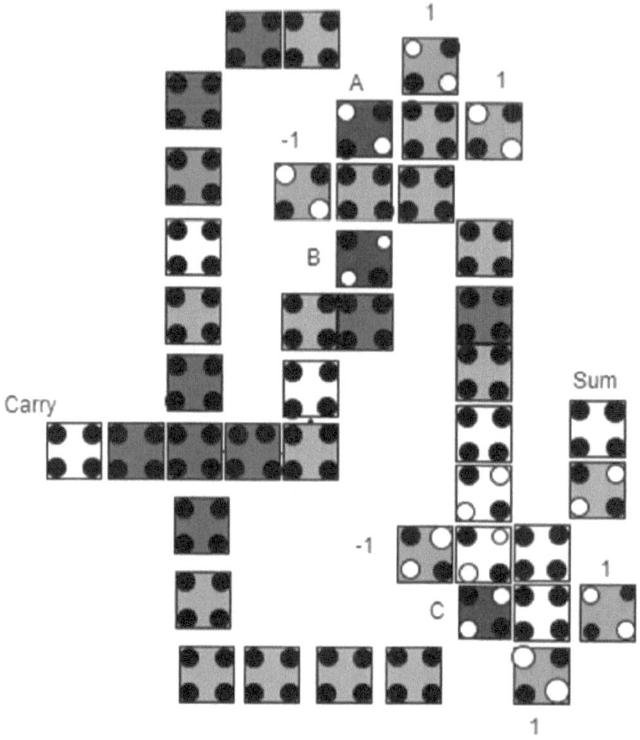

FIGURE 4.20 Implementation of a full adder in QCA technology.

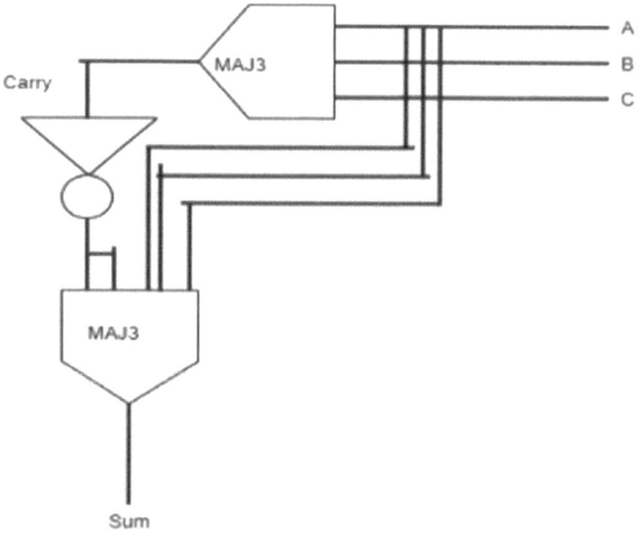

FIGURE 4.21 Circuit for a full adder.

FIGURE 4.22 Layout for a full adder.

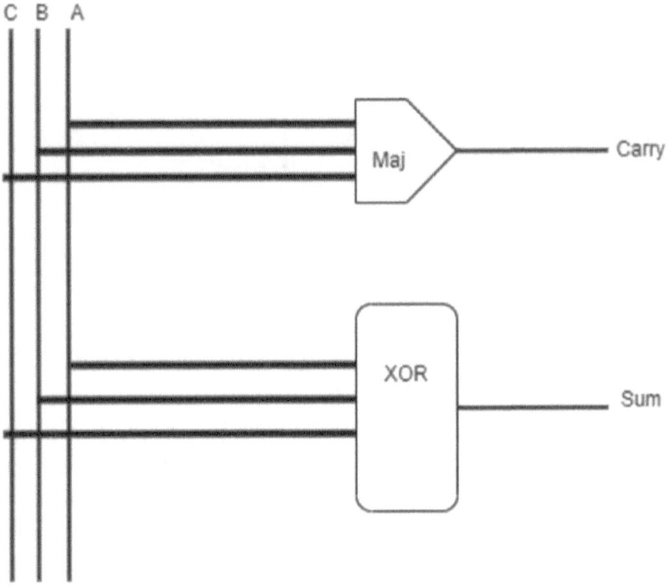

FIGURE 4.23 Logical representation of a full adder.

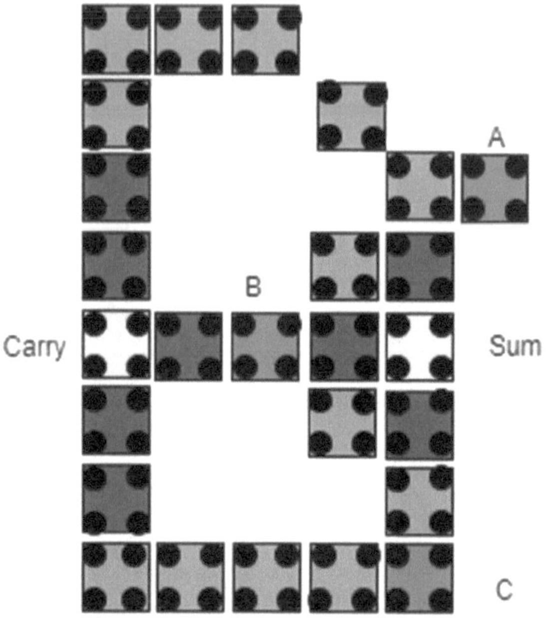

FIGURE 4.24 QCA layout of a full adder.

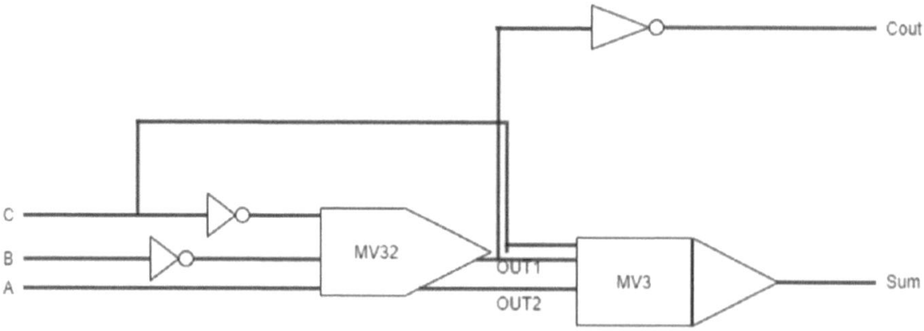

FIGURE 4.25 Logical circuit of a QCA full adder.

This full adder requires 0.03 um^2 of area and has a latency of 0.5 clock cycles. The different colors of the cells represent the different used clock zones.

A MV32 gate produces two outputs which include MV $(\bar{A}, \bar{B}, \overline{Cin})$ and MV (A, B, $\overline{Cin}$) functions [48]. Input *Cin* combined with these two outputs using a three-input majority gate. The Sum can be obtained from the inputs *A*, *B*, and *Cin*. MV32 can be used for implementing the full adder. The logical circuit of the full adder is shown in Figure 4.25.

Figure 4.26 presents the single-layer structure of the QCA full adder using the ideas in Figure 4.25 [48].

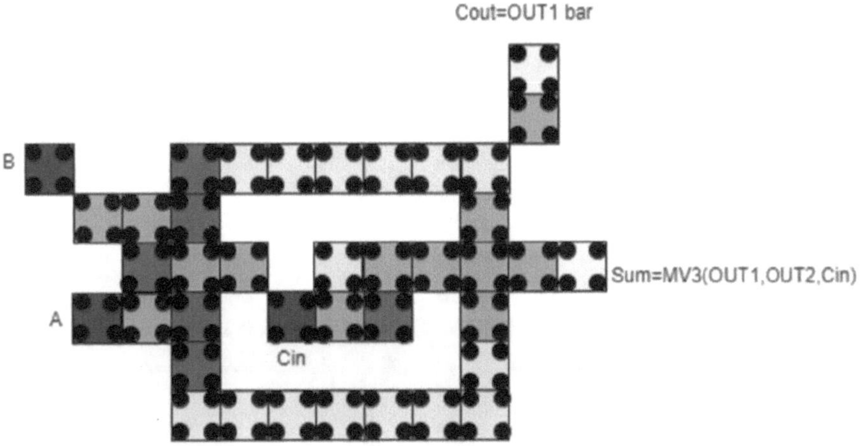

FIGURE 4.26 A QCA full adder with a single layer.

This full adder design has a delay of four clock phases, occupies an area of 0.07 um^2 and is made up of 38 cells. This full adder is designed using one MV32 gate, one MV3 gate and three NOT gates.

Figure 4.27 shows a 1-level layout of a QCA full adder [49]. It is made up of a total of 59 QCA cells with a latency of 1 clock cycle. The area covered by this design is 0.043 um^2.

4.4 COMPARISON ANALYSIS

This section compares the full adder circuits to provide an understanding of the more effective QCA-based full adder circuits. The important performance metrics for QCA circuits are the number of QCA cells, area, latency, layout cost, and type of layer. Table 4.1 shows the comparisons of different full adder circuits.

Cell count shows the total number of QCA cells used for a design. The cell dimensions are 18 nm × 18 nm, and the cell's spacing is 5 nm. Area requirement is calculated by counting the maximum number of QCA cells in horizontal and vertical directions and multiplying them with the QCA cell dimensions. It is always desired that the minimum number of clocks is used for a design. Latency is the clocks used in a circuit. The quantum cost is observed as per the product of the required area and square of the latency.

It can be observed from Table 4.1 that the full adder presented in [48] is made up of 23 QCA cells, which has the least QCA cells as compared to other full adder circuits. A minimum 0.50 clock latency is needed to design the full adder. The full adder design given in [41] occupies the smallest area and represents a space-efficient design compared to others. A Multilayer concept is utilized for design of a full adder, which reduces the area requirement. It is also cost-effective because it has less clock latency and area requirements.

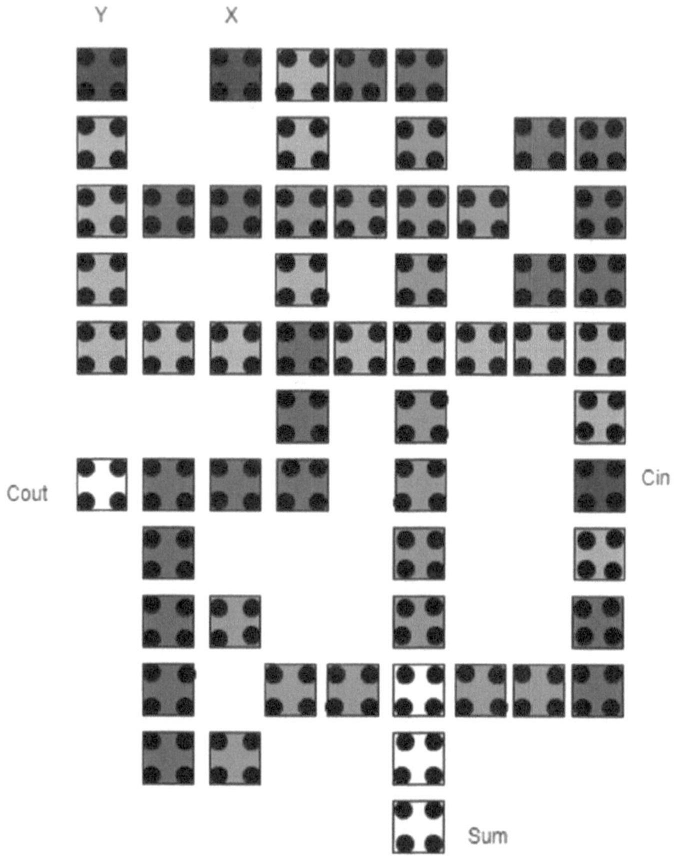

FIGURE 4.27 QCA Layout of a full adder.

TABLE 4.1
Comparisons of QCA-Based Full Adder Circuits

Design	Cells count	Area (um²)	Latency	Cost	Layer
[39]	29	0.02	0.50	0.005	Single
[41]	31	0.01	0.50	0.003	Multi
[42]	79	0.05	1.25	0.078	Multi
[43]	71	0.06	1.25	0.094	Single
[44]	46	0.04	4.00	0.640	Single
[45]	44	0.06	1.25	0.094	Single
[46]	63	0.05	0.75	0.028	Single
[47]	38	0.07	4.00	1.120	Single
[48]	23	0.03	0.50	0.008	Single
[49]	59	0.04	1.00	0.040	Single

4.5 CONCLUSION

QCA technology has various advantages in designing high-performance circuits as compared to CMOS technology. One of the most common circuits used to perform addition is the full adder. The full adder plays an important role in computer arithmetic fields. So, efficient implementation of a full adder can increase the efficiency of the arithmetic circuits. In this chapter, we compared the various available QCA-based full adders. The designed circuits were implemented using the QCA designer tool using QCA cells with dots of 20 nm in diameter. Few full adder designs are single-layered, and most are multi-layered designs. The implementation results confirmed that the designed circuits are modified in terms of complexity, latency, and required area. Few full adder designs use three-input majority gates, whereas some designs use five-input majority gates. These circuits can be used in processors with high operating speeds.

REFERENCES

1. Arden, Wolfgang M. "The international technology roadmap for semiconductors—perspectives and challenges for the next 15 years." *Current Opinion in Solid State and Materials Science* 6, no. 5 (2002): 371–377.
2. Compano, R., L. Molenkamp, and D. J. Paul. "Roadmap for nanoelectronics." *European Commission IST programme, Future and Emerging Technologies* 2000 (2000): 1–81.
3. Jeon, Jun-Cheol. "Low-complexity QCA universal shift register design using multiplexer and D flip-flop based on electronic correlations." *The Journal of Supercomputing* 76, no. 8 (2020): 6438–6452.
4. Likharev, Konstantin K. "Single-electron devices and their applications." *Proceedings of the IEEE* 87, no. 4 (1999): 606–632.
5. Chen, Kevin J., Koichi Maezawa, and Masafumi Yamamoto. "InP-based high-performance monostable-bistable transition logic elements (MOBILEs) using integrated multiple-input resonant-tunneling devices." *IEEE Electron Device Letters* 17, no. 3 (1996): 127–129.
6. Lent, C. S., P. D. Taugaw, W. Porod, and G. H. Berstein. "Quantum cellular automata." *Nanotechnology* 4, no. 1 (1993): 49.
7. Bahar, Ali Newaz, and Khan A. Wahid. "Design of an efficient N× N butterfly switching network in quantum-dot cellular automata (QCA)." *IEEE Transactions on Nanotechnology* 19 (2020): 147–155.
8. Timler, John, and Craig S. Lent. "Power gain and dissipation in quantum-dot cellular automata. " *Journal of Applied Physics* 91, no. 2 (2002): 823–831.
9. Gao, Mingming, Jinling Wang, Shaojun Fang, Jingchang Nan, and Li Daming. "A new nano design for implementation of a digital comparator based on quantum-dot cellular automata." *International Journal of Theoretical Physics* 60, no. 7 (2020): 2358–2367.
10. Riyaz, Sadat, Syed Farah Naz, and Vijay Kumar Sharma. "Multioperative reversible gate design with implementation of 1-bit full adder and subtractor along with energy dissipation analysis." *International Journal of Circuit Theory and Applications* 49, no. 4 (2020): 990–1012.
11. Senthilnathan, S., and S. Kumaravel. "Power-efficient implementation of pseudo-random number generator using quantum dot cellular automata-based D flip flop." *Computers & Electrical Engineering* 85 (2020): 106658.

12. Das, Jadav Chandra, and Debashis De. "Feynman gate based design of n-bit reversible inverter and its implementation on quantum-dot cellular automata." *Nano Communication Networks* (2020): 100298.
13. Sadhu, Arindam, Kunal Das, Debashis De, and Maitreyi Ray Kanjilal. "Area-Delay-Energy aware SRAM memory cell and M× N parallel read/write memory array design for quantum dot cellular automata." *Microprocessors and Microsystems* 72 (2020): 102944.
14. Wang, Lei, and Guangjun Xie. "A novel XOR/XNOR structure for modular design of QCA circuits." *IEEE Transactions on Circuits and Systems II: Express Briefs* 67, no. 12 (2020): 3327–3331.
15. Hani, Alamdar, Ardeshir Gholamreza, and Gholami Mohammad. "Phase-frequency detector in QCA nanotechnology using novel flip-flop with reset terminal." *International Nano Letters* 10, no. 2 (2020): 111–118.
16. Khakpour, Mahdi, Mohammad Gholami, and Shokoufeh Naghizadeh. "Parity generator and digital code converter in QCA nanotechnology." *International Nano Letters* 10, no. 1 (2020): 49–59.
17. Bahar, Ali Newaz, Sajjad Waheed, Nazir Hossain, and Md Asaduzzaman. "A novel 3-input XOR function implementation in quantum dot-cellular automata with energy dissipation analysis." *Alexandria Engineering Journal* 57, no. 2 (2018): 729–738.
18. Yang, Binfeng, and Sonia Afrooz. "A new coplanar design of multiplier based on nanoscale quantum-dot cellular automata." *International Journal of Theoretical Physics* 58, no. 10 (2019): 3364–3374.
19. Seyedi, Saeid, Mehdi Darbandi, and Nima Jafari Navimipour. "Designing an efficient fault tolerance D-latch based on quantum-dot cellular automata nanotechnology." *Optik* 185 (2019): 827–837.
20. Kandasamy, Nehru, Firdous Ahmad, D. Ajitha, Balwinder Raj, and Nagarjuna Telagam. "Quantum dot cellular automata-based scan flip-flop and boundary scan register." *IETE Journal of Research* (2020): 1–14.
21. Cocorullo, Giuseppe, Pasquale Corsonello, Fabio Frustaci, and Stefania Perri. "Design of efficient QCA multiplexers." *International Journal of Circuit Theory and Applications* 44, no. 3 (2016): 602–615.
22. Shiri, Ahmadreza, Abdalhossein Rezai, and Hamid Mahmoodian. "Design of efficient coplanar comprator circuit in QCA technology." *Facta Universitatis, Series: Electronics and Energetics* 32, no. 1 (2019): 119–128.
23. Vankamamidi, Vamsi, Marco Ottavi, and Fabrizio Lombardi. "Two-dimensional schemes for clocking/timing of QCA circuits." *IEEE Transactions on Computer-Aided Design of Integrated Circuits and Systems* 27, no. 1 (2007): 34–44.
24. Singh, Gurmohan, R. K. Sarin, and Balwinder Raj. "A novel robust exclusive-OR function implementation in QCA nanotechnology with energy dissipation analysis." *Journal of Computational Electronics* 15, no. 2 (2016): 455–465.
25. Chabi, Amir Mokhtar, Arman Roohi, Hossein Khademolhosseini, Shadi Sheikhfaal, Shaahin Angizi, Keivan Navi, and Ronald F. DeMara. "Towards ultra-efficient QCA reversible circuits." *Microprocessors and Microsystems* 49 (2017): 127–138.
26. Hashemi, Sara, and Keivan Navi. "New robust QCA D flip flop and memory structures." *Microelectronics Journal* 43, no. 12 (2012): 929–940.
27. Safoev, Nuriddin, and Jun-Cheol Jeon. "Design of high-performance QCA incrementer/decrementer circuit based on adder/subtractor methodology." *Microprocessors and Microsystems* 72 (2020): 102927.
28. Nath, Rajdeep Kumar, Bibhash Sen, and Biplab K. Sikdar. "Optimal synthesis of QCA logic circuit eliminating wire-crossings." *IET Circuits, Devices & Systems* 11, no. 3 (2017): 201–208.

29. Wang, Yuliang, and Marya Lieberman. "Thermodynamic behavior of molecular-scale quantum-dot cellular automata (QCA) wires and logic devices." *IEEE Transactions on Nanotechnology* 3, no. 3 (2004): 368–376.
30. Graunke, Christopher R., David I. Wheeler, Douglas Tougaw, and Jeffrey D. Will. "Implementation of a crossbar network using quantum-dot cellular automata." *IEEE Transactions on Nanotechnology* 4, no. 4 (2005): 435–440.
31. Devadoss, R., K. Paul, and M. Balakrishnan. "Coplanar QCA crossovers." *Electronics Letters* 45, no. 24 (2009): 1234–1235.
32. Sen, Bibhash, Anirban Nag, Asmit De, and Biplab K. Sikdar. "Towards the hierarchical design of multilayer QCA logic circuit." *Journal of Computational Science* 11 (2015): 233–244.
33. Bajec, Iztok Lebar, and Primož Pečar. "Two-layer synchronized ternary quantum-dot cellular automata wire crossings." *Nanoscale Research Letters* 7, no. 1 (2012): 221.
34. Sheikhfaal, Shadi, Shaahin Angizi, Soheil Sarmadi, Mohammad Hossein Moaiyeri, and Samira Sayedsalehi. "Designing efficient QCA logical circuits with power dissipation analysis." *Microelectronics Journal* 46, no. 6 (2015): 462–471.
35. Angizi, Shaahin, Soheil Sarmadi, Samira Sayedsalehi, and Keivan Navi. "Design and evaluation of new majority gate-based RAM cell in quantum-dot cellular automata." *Microelectronics Journal* 46, no. 1 (2015): 43–51.
36. Farazkish, Razieh, and Keivan Navi. "New efficient five-input majority gate for quantum-dot cellular automata." *Journal of Nanoparticle Research* 14, no. 11 (2012): 1252.
37. Majeed, Ali H., Esam AlKaldy, M. S. B. Zainal, and D. B. M. D. Nor. "A new 5-input majority gate without adjacent inputs crosstalk effect in QCA technology." *Indonesian Journal of Electrical Engineering and Computer Science* 14, no. 3 (2019): 1159–1164.
38. Rao, Nandini G., P. C. Srikanth, and Preeta Sharan. "A novel quantum dot cellular automata for 4-bit code converters." *Optik* 127, no. 10 (2016): 4246–4249.
39. Balali, Moslem, Abdalhossein Rezai, Haideh Balali, Faranak Rabiei, and Saeid Emadi. "Towards coplanar quantum-dot cellular automata adders based on efficient three-input XOR gate." *Results in Physics* 7 (2017): 1389–1395.
40. Singh, Rupali, and Devendra Kumar Sharma. "Design of efficient multilayer RAM cell in QCA framework." *Circuit World* 41 (2020): 31–41.
41. Sen, Bibhash, Ayush Rajoria, and Biplab K. Sikdar. "Design of efficient full adder in quantum-dot cellular automata." *The Scientific World Journal* 2013 (2013): 250802 (1–10).
42. Hashemi, Sara, Mohammad Tehrani, and Keivan Navi. "An efficient quantum-dot cellular automata full-adder." *Scientific Research and Essays* 7, no. 2 (2012): 177–189.
43. Hashemi, S., and K. Navi. "A novel robust QCA full-adder." *Procedia Materials Science* 11 (2015): 376–380.
44. Mokhtari, Dariush, Abdalhossein Rezai, Hamid Rashidi, Faranak Rabiei, Saeid Emadi, and Asghar Karimi. "Design of novel efficient full adder architecture for quantum-dot cellular automata technology." *Facta Universitatis, Series: Electronics and Energetics* 31, no. 2 (2018): 279–285.
45. Zoka, Saeid, and Mohammad Gholami. "A novel efficient full adder–subtractor in QCA nanotechnology." *International Nano Letters* 9, no. 1 (2019): 51–54.
46. Labrado, Carson, and Himanshu Thapliyal. "Design of adder and subtractor circuits in majority logic-based field-coupled QCA nanocomputing." *Electronics Letters* 52, no. 6 (2016): 464–466.
47. Babaie Shahram, Ali Sadoghifar, and Ali Newaz Bahar. "Design of an efficient multilayer arithmetic logic unit in quantum-dot cellular automata (QCA)." *IEEE Transactions on Circuits and Systems II: Express Briefs* 66, no. 6 (2018): 963–967.

48. Safavi, A., and M. Mosleh. "Presenting a new efficient QCA full adder based on suggested MV32 gate." *International Journal of Nanoscience and Nanotechnology* 12, no. 1 (2016): 55–69.
49. Abedi, Dariush, Ghassem Jaberipur, and Milad Sangsefidi. "Coplanar full adder in quantum-dot cellular automata via clock-zone-based crossover." *IEEE transactions on nanotechnology* 14, no. 3 (2015): 497–504.

5 An Overview of Nanowire Field-Effect Transistors for Future Nanoscale Integrated Circuits

J. Ajayan, D. Nirmal, P. Mohankumar, and Shubham Tayal

CONTENTS

5.1 ZnO/CuO/PbS NW FETs

Nanowires have gained huge interest for electronic, optic, and optoelectronic device applications. Therefore, it is useful and important to study the charge transport methods of nanowires in detail. ZnO is a n-type native semiconductor material, and the electron transport in undoped ZnO mainly arises from structural irregularities and native defects, such as unintentionally doped materials. The Mott VRH (variable range hopping) transport model, Efros–Shklovskii VRH transport model, and thermal activity transport model are the various charge transport methods used in ZnO nanowires [1]. The vapor–liquid–solid technique can be used for growing ZnO nanowires. ZnO nanowires can be synthesized by vaporizing a mixture of graphite powder (99%) and ZnO powder (99%) with a 1:1 mass ratio. The ZnO nanowire growth temperature is 920°C. A ZnO NW FET consists of a heavily doped p-Silicon

DOI: 10.1201/9781003155751-5

wafer, which acts as a back gate, SiO_2 layer, and ZnO nanowire channel. The Ti/Au or Al/Au are the metal electrodes used in ZnO NW FETs [2]. ZnO NW FETs exhibit a higher I_{ON}/I_{OFF} ratio of over 10^6.

The I_{DS} *vs* V_{DS} curves of ZnO NW FET for V_{GS}, ranging from –14 V to –8 V, are illustrated in Figure 5.1(a), and the I_{DS} *vs* V_{GS} curve for an ZnO NW FET is plotted in Figure 5.1(b). When the temperature increases from 77 K to 227 K, the threshold voltage (V_{TH}) of the ZnO NW FET shifts toward the left and, therefore, the ON-current (I_{ON}) increases (Figure 5.1(c)). When the temperature increases from 77 K to 227 K, the trapped charge density also increases in the ZnO NW FET (Figure 5.1(d)). Proton irradiation can affect the drain conductance in ZnO NW FETs. That is, proton irradiation can shift the V_{TH} of ZnO NW FETs toward a more negative regime that leads to the increase of I_{ON} [3]. The I_{ON} of ZnO NW FETs increases with an increase in dose of proton irradiation. The band gap of ZnO is 3.37 eV, and this wide and direct band gap makes ZnO suitable for electronic, optic, and optoelectronic applications. Another advantage of ZnO is its low-cost production [4]. ZnO NW FETs are promising transistors for chemical sensors and other transparent and flexible applications [5]. ZnO has a large number of dangling bonds and surface defects that act as absorption centers for gas molecules, especially O_2 molecules. Therefore, in order to improve the electrical performance of ZnO NW FETs, non-destructive surface cleaning techniques can be used [6]. Smooth ZnO

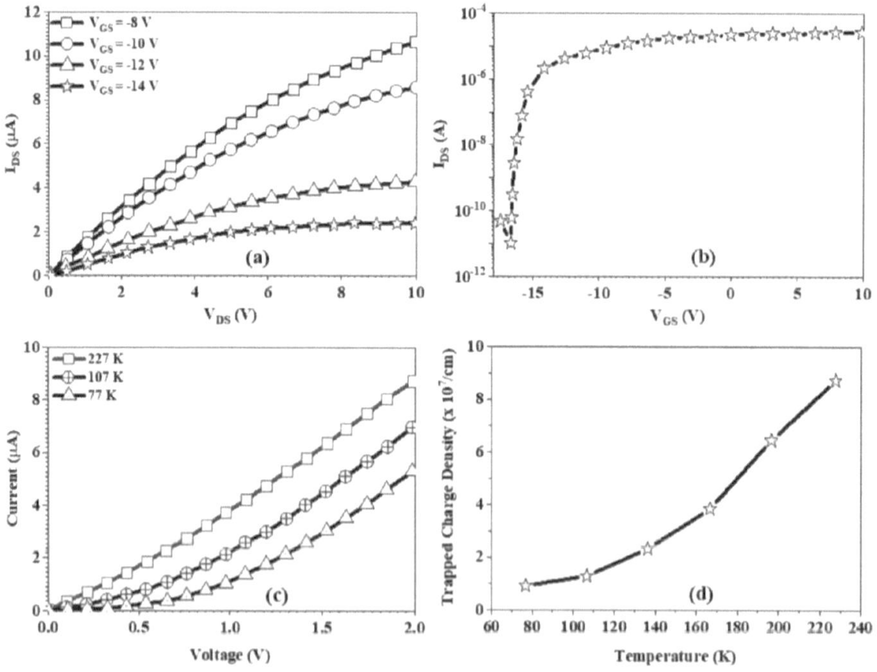

FIGURE 5.1 ZnO NW FET (a) I_{DS} *vs* V_{DS} curves, (b) I_{DS} *vs* V_{GS} curves, (c), thermal effects in I-V curves, and (d) the change in trapped charge density as a function of temperature [1].

nanowires exhibit a higher effective mobility and transconductance than rough ZnO nanowires [7]. The electrical performance of ZnO NW FETs are found to be improved after passivation [7]. Dopants like Mg, Sn, In, Ga, Al, etc., can be used as donor atoms in ZnO nanowires, whereas phosphorus (P) can be used as acceptor atoms in ZnO nanowires. Back-gated FETs, top-gated FETs, side-gated FETs and vertical-surrounding FETs are the popular device architectures for ZnO NW FETs [2]. ZnO NW FETs can be used for sensing different chemical gases (Figure 5.2) [8]. To improve the sensitivity of ZnO NW FETs, ZnO nanowires with smaller radii are preferable.

The mobility (μ) of a ZnO nanowire can be computed as [9]:

$$\mu = \frac{g_m L^2}{C V_{DS}} \tag{5.1}$$

$$n = \frac{C V_{TH}}{q \pi r^2 L} \tag{5.2}$$

L = length of the ZnO nanowire
C = ZnO nanowire gate capacitance
n = Carrier concentration in ZnO nanowire
r = radius of the ZnO nanowire

CuO nanowire FETs have recently emerged as an alternative choice for PH sensing and glucose sensing applications due to their low band gap of 1.2 eV and low-cost fabrication [10]. PbS nanowire semiconductors can also be used to making FETs for sensing applications due to their relatively lower band gap of 0.41 eV [11].

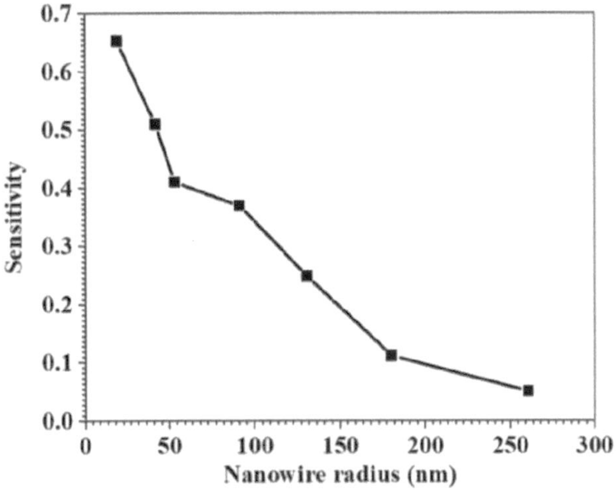

FIGURE 5.2 O_2 sensitivity of a ZnO NW FET [8].

5.2 III-V NANOWIRE FETs

Using III-V compounds as channel materials can help improve the I_{ON} using very low V_{DD} in future low-power, high-performance transistors. GaAs, InAs, GaN, InGaAs, InSb, etc., are considered suitable replacements for traditional silicon (Si) channels [12–16]. GaN is a wide band gap semiconductor considered highly suitable for RF, high-power, and logic applications due to their attractive material properties, such as large electron effective mass, low permittivity, and high thermal conductivity. Because of its outstanding gate controllability, GAA GaN NW FETs have shown improved performance in terms of good linearity, reduced trapping effects, high I_{ON}/I_{OFF} ratio, low V_{TH}, *SS* (subthreshold swing), *DIBL* (drain induced barrier lowering), and very low OFF-state leakage current (I_{OFF}). A combination of GAA technology and vertical channel design can effectively minimize the self-heating effects of the device. The rectangular cross-sectional view of a vertical GaN NW FET is shown in Figure 5.3. L_S, L_D, L_G, and d_{NW} represent source length, drain length, gate length, and diameter of the nanowire, respectively. Vertical GaN NW FETs can have a cylindrical or rectangular structure.

The interface traps (D_{it}) can significantly degrade the I_{ON}/I_{OFF} ratio, *SS*, and *DIBL* of vertical GaN NW FETs [17]. The shallow traps lead to an increase of I_{OFF}, and deep traps result in the reduction of I_{ON}. Self-heating in vertical GaN NW FETs leads to the degradation of g_m and I_{DS} (Figures 5.4(a) and 5.4(b)). When V_{GS} and V_{DS} are applied, the current begins to flow through the channel of NW FETs. At the edge of the drain channel, heat dissipation takes place. That is, the NW and substrate contacts and acts as heat source and heat sink, respectively [18–20]. Due to a high charge mobility, InAs nanowires have recently emerged as an alternative channel material for low-power, high-performance FETs. In 2020, Hironori Gamo et al. [21] reported

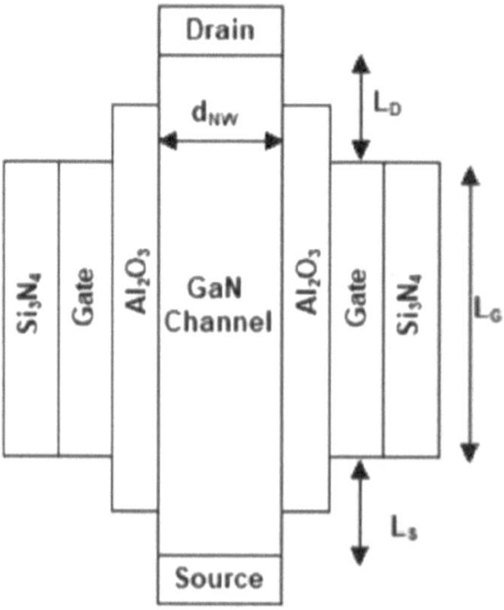

FIGURE 5.3 A rectangular cross-sectional view of a vertical GaN NW FET.

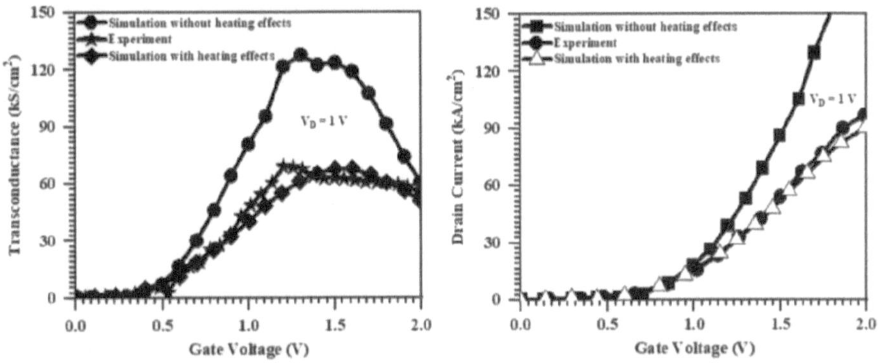

FIGURE 5.4 Self-heating effects in a vertical GaN NW FET [13].

a core–shell (CS) InAs/InP vertical NW GAA FET on Si. The *SS* of the CS-InAs/InP vertical GAA NW FET can be computed as [21]:

$$SS = \frac{KT}{q} \text{In}_{10}\left(1 + \frac{C_d + C_{it}}{\frac{C_{ox}C_{InP}}{C_{ox} + C_{InP}}}\right) \tag{5.3}$$

C_d = depletion capacitance
C_{it} = interface trap capacitance
C_{ox} = intrinsic gate oxide capacitance
C_{InP} = depletion capacitance of InP

The I_{ON}, *SS*, and I_{OFF} of GAA NW InAs FETs are found to be a strong dependent of L_G (Figures 5.5(a), 5.5(b), and 5.5(c)). In 2016, Cezar B. Zota et al. [23] reported an InGaAs lateral NW FET that exhibited an I_{ON} of 565 μA/μm at I_{OFF} = 100 nA/μm, *SS* of 77 mV/dec and g_m of 2.9 μS/μm. The variation of I_{ON} as a function of the width of the NW (W_{NW}) and gate length (L_G) of the InGaAs lateral NW FET is shown in Figure 5.5(d). The two factors that introduce 1/f noises in InAs NW FETs are [24]:

(1) current variation through the drain/source contacts
(2) interaction of charges with the oxide/nanowire interface through the channel

The resistance of the InAs NW (R_W) in a NW FET can be computed as [25]:

$$R_W = \frac{4t_{gs,gd}\rho_w}{d_{NW}^2 \pi n_x n_y} \tag{5.4}$$

$$\rho_w = \frac{1}{N_{D,w}\mu_i} \tag{5.5}$$

ρ_w = nanowire resistivity
n_x = number of nanowire columns

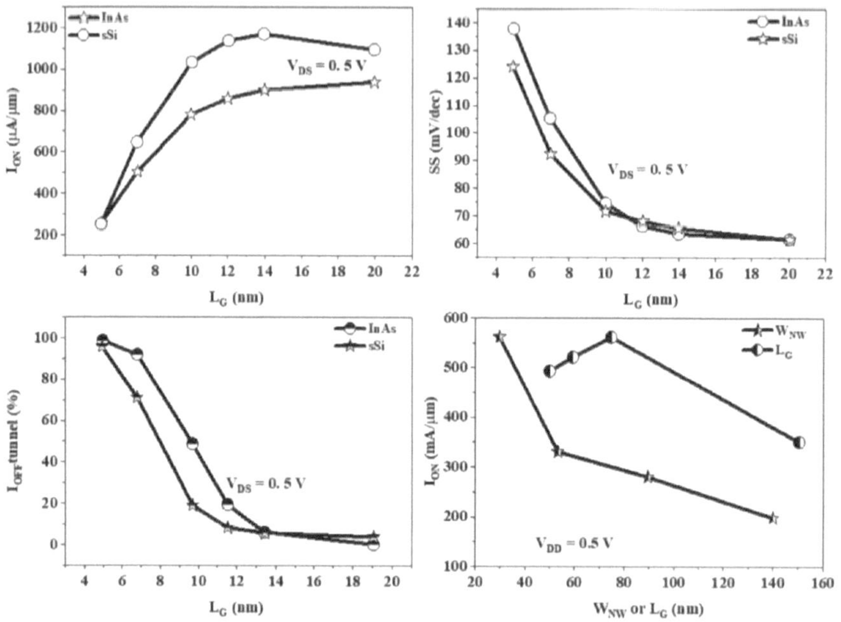

FIGURE 5.5 (a), (b), and (c) The variation of I_{ON}, SS, and I_{OFF} of a GAA NW InAs FET with L_G [22], and (d) the variation of I_{ON} as a function of the width of the NW (W_{NW}) and gate length (L_G) of a InGaAs lateral NW FET [23].

n_y = number of nanowire rows
$t_{gs,gd}$ = source or drain spacer thickness
μ_i = intrinsic mobility

The f_T (cut-off frequency) and f_{max} (maximum oscillation frequency) of NW FETs degrade with an increase in NW spacing and NW diameter. The band gap and effective electron mass of InAs and InSb NWs decreases with an increase in nanowire diameter (Figure 5.6(a) and (b)). For the same nanowire diameter (d_{NW}), InSb NW FETs offer a higher I_{ON} than InAs NW FETs. However, InAs NW FETs offer a higher I_{ON}/I_{OFF} ratio for the same d_{NW} [26, 27]. Core–shell NW FETs can also be developed using the GaP/GaAs material system [28]. The drain current of core–shell NW FETs decrease with an increase in core–shell thickness (T_{sh}) and an increase in EOT (Equivalent oxide thickness) (Figure 5.7).

5.3 III-V NANOWIRE TFETs

The high power consumption in electronic circuits is a critical issue for the microelectronics industry today. The P_{active} (active power consumption) of electronic circuits can be computed as:

$$P_{active} = \frac{1}{2} fCV_{DD}^2 \tag{5.6}$$

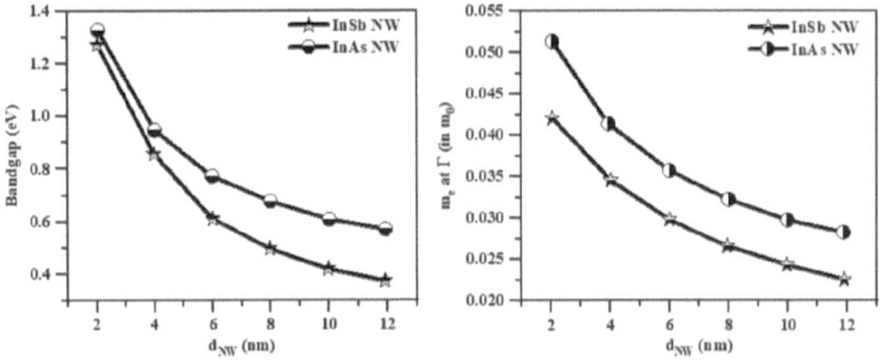

FIGURE 5.6 The influence of d_{NW} on the band gap and effective electron mass of InAs/InSb NWs [26].

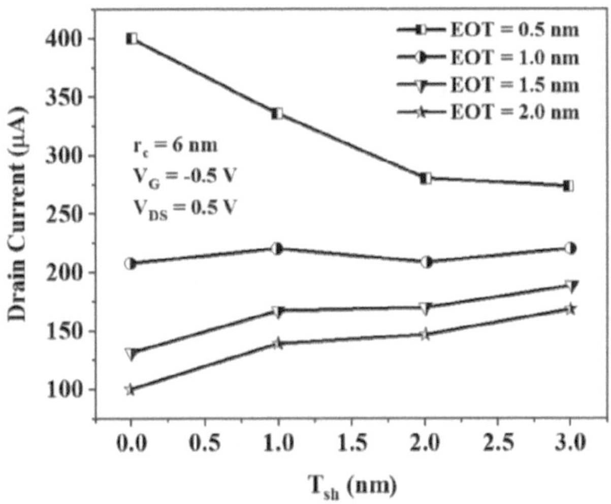

FIGURE 5.7 Influence of T_{sh} and E_{OT} on the drain current of a core–shell NW FET [28].

f = frequency
C = load capacitance
V_{DD} = operating voltage

Therefore, reducing V_{DD} is the most effective way of minimizing the P_{active} of electronic circuits. The main advantages of TFETs are that they are below 60 mV/dec SS and have a very low V_{DD}. The use of III-V materials in the development of TFETs helps to achieve a higher I_{ON}.

Tunnel field-effect transistors (TFETs) have three regions of operation: (1) OFF-state, (2) subthreshold region, and (3) ON-state. An energy band diagram of a TFET showing these three regions of operation is depicted in Figure 5.8. In the OFF-state,

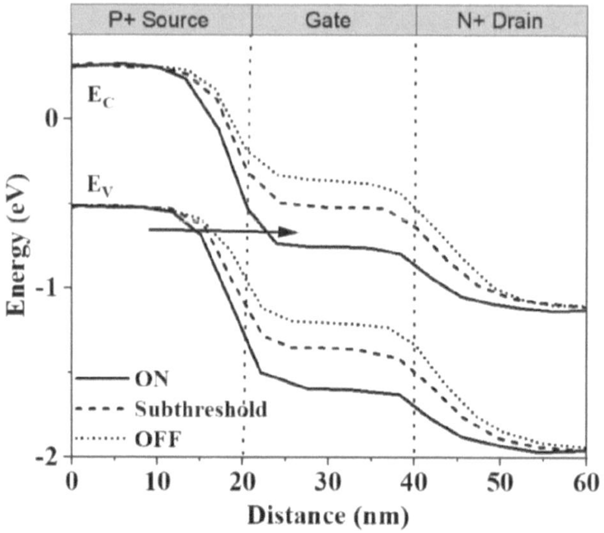

FIGURE 5.8 Energy band diagram of a TFET [29].

the conduction band energy of the channel is on top of the valence band energy of the source. Therefore, the tunneling barrier is large, and valence band electrons in the source cannot tunnel toward the channel and therefore the TFET remains OFF. In the subthreshold mode of operation, when the energy bands of the channel and source regions cross, a channel will be created for the electrons. Finally, in the ON state, the conduction band energy of the channel is below the valence band energy of the source and, therefore, a large amount of Zener tunneling by the electrons takes place through the tunneling barrier. To enhance the I_{ON} and I_{ON}/I_{OFF} ratio in III-V NW TFETs, a low nanowire diameter and higher source doping concentration is required [29]. InSb nanowire TFETs can provide a higher I_{ON} compared with InAs NW TFETs. In III-V TFETs, the subthreshold leakage current increases with an increase in temperature. Therefore, the *SS* of III-V NW TFETs also increases with an increase in temperature [30]. InAs, InSb, GaAs, GaN, InGaAs, etc., are the commonly used materials for realizing III-V NW TFETs [31, 32]. Therefore, traps in NW TFETs significantly degrade the I_{ON} and I_{ON}/I_{OFF} ratio of the device [33, 34]. A III-V heterostructure can be used to effectively improve III-V NW TFET performance [35]. GaSb/InAs, GaAsP/AlGaSb, and InAs/GaSb/InAs hetero-junctions are the most widely used hetero-junctions in the realization of III-V NW TFETs [35–39]. In 2017, Xin Zhao et al. [40] reported an InGaAs/InAs vertical NW hetero-junction TFET, the subthreshold current and *SS* of which were found to increase with a rise in temperature.

5.4 JL NW FETs

Junctionless (JL) technology has recently emerged as an attractive solution for eliminating critical issues like low thermal budget and high doping density gradients. A

JL transistor consists of a channel, drain, and source with a uniform doping density and homogeneous dopants. Therefore, the fabrication process becomes much simpler than conventional FETs with heterogeneous doping regions. JL FETs rely on bulk current rather than surface current. Since a cylindrical NW architecture exhibits excellent robustness against SCEs, cylindrical JL NW FETs are popular FET architectures [41, 42]. To suppress SCEs, shallow junctions are very important, but it is very difficult to precisely control the depths of the junctions in FETs. For this reason, research groups are now focusing on JL transistors to mitigate this issue, especially for developing circuits like non-volatile memory [43]. JL NW FETs can be used for piezoresistive sensing applications like blood pressure monitoring and fuel pressure monitoring in automobiles [44, 45]. A cross-sectional view of an n-type JL NW FET, which can be used for piezoresistive sensing, is shown in Figure 5.9.

A JL NW FET is built on an 8-inch p-type SOI (silicon-on-insulator) wafer with a p++surrounding gate, two n+nanowire channels, and a gate oxide. The temperature significantly affects the subthreshold leakage current of JL NW FETs, and it is observed that the subthreshold leakage current increases with a rise in temperature from 20°C to 80°C [46]. Strain and doping concentration are the other two factors that can change the drain currents of JL NW FETs, and it is found that the drain current increases with an increase in strain, and decreases with an increase in doping concentration [47–57]. To reduce V_{TH}, a low L_G is highly preferable (Figure 5.10(a). But when L_G reduces *SS* and *DIBL* increases, this shows severe SCEs (Figure 5.10(b) and 5.10(c)). It is also observed that JL NW FETs with source/drain (*S*/*D*) extensions suffer from SCEs (Figure 5.10(b) and 5.10(c)). The I_{ON} and I_{OFF} of JL NW FETs also increase with an increase in the number of nanowires [46].

In 2014, Po-Yi Kuo et al. [46] reported a GAA SW DNW (side-wall damascened nanowire) FET that exhibited an I_{ON}/I_{OFF} ratio of 8×10^7 with outstanding thermal stability. In 2016, Kian-Hui Goh et al. [59] reported a GAA $In_{0.53}Ga_{0.47}$ as a JL NW FET with an extrinsic transconductance of 820 μS/μm at V_{DS}=0.5 V and R_{SD} (*S*/*D* series resistance) of 275 Ω.μm. The use of III-V channel materials like

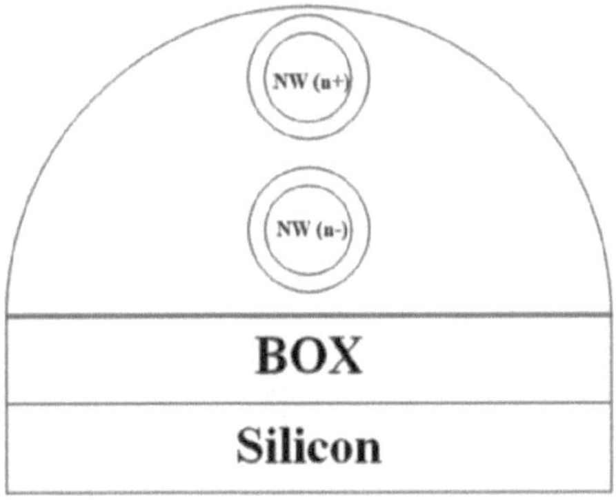

FIGURE 5.9 Cross-sectional view of the n-type JL NW FET, which can be used for piezoresistive sensing [44].

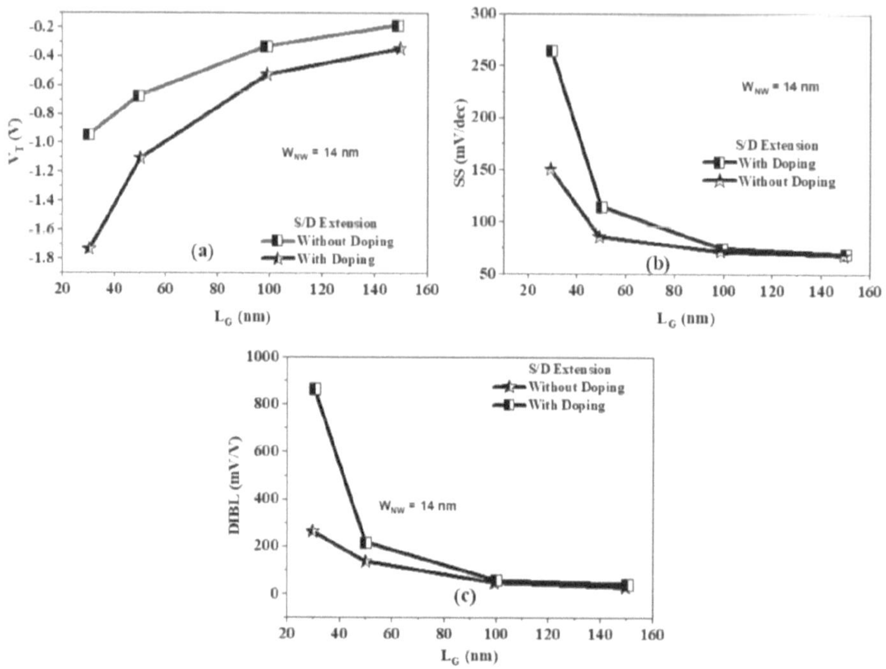

FIGURE 5.10 Variation of V_T, SS, and DIBL with gate length reduction [58].

InAs, InGaAs, GaN, InSb, etc., can improve I_{ON} at low V_{DD}. Gated raised *S/D* architecture can be used for effectively suppressing SCEs in JL NW FETs [60]. d_{NW} is an important parameter in the design of JL NW FETs because it can significantly affect the I_{ON}, I_{OFF} and I_{ON}/I_{OFF} ratio. It is observed that both I_{ON} and I_{OFF} of JL NW FETs increase with an increase in d_{NW} (Figure 5.11 (a) & (b)). A smaller d_{NW} is attractive for improving the I_{ON}/I_{OFF} ratio (Figure 5.11(c)). Lateral band-to-band tunneling (L-BTBT) of electrons from the NW channel to drain extension results in the increase of I_{OFF} in JL NW FETs [61, 62]. The cross-sectional view of a dual metal-gate (DMG) NW JL FET is shown in Figure 5.12. In a DMG-JL NW FET, the gate metal 1 (M1) acts as the control gate and gate metal 2 (M2) acts as the tunnel gate. DMG architecture helps reduce subthreshold conduction, thereby reducing I_{OFF} and improving the I_{ON}/I_{OFF} ratio. The work function of the control gate and tunnel gate materials significantly affects the drain current and cut-off frequency of JL NW FETs. The use of materials with large work functions for tunnel gates helps to reduce subthreshold currents. However, a low work-function material for the control gate is preferable for improving the cut-off frequency of JL NW FETs. BTBT can be controlled by using core–shell (CS) architecture. A CS-JL NW FET is illustrated in Figure 5.13. Shell thickness (T_{shell}), core diameter (d_{core}), S/D extension length (L_{EXT}), and thickness of the oxide (t_{ox}) are the key geometrical parameters that influence the ON and OFF characteristics of CS-JL NW FETs. The influence of core doping concentration on the transfer curves of conventional NW JL FETs and CS-JL NW

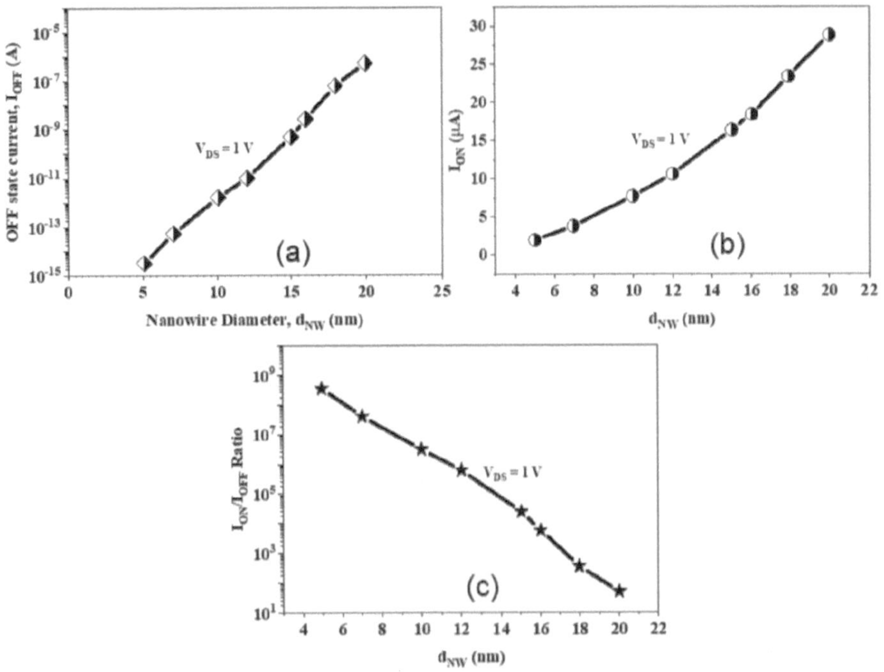

FIGURE 5.11 The influence of d_{NW} on I_{ON}, I_{OFF} and I_{ON}/I_{OFF} ratio of JL NW FETs [63].

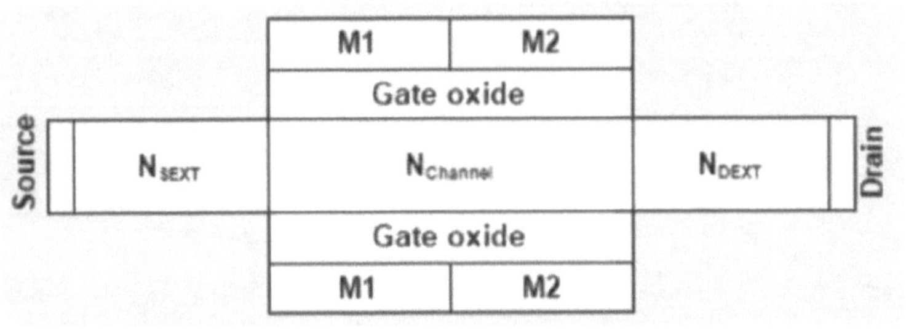

FIGURE 5.12 The cross-sectional view of a dual metal gate (DMG) NW JL FET [62].

FETs are demonstrated in Figure 5.14(a). I_{ON} decreases with an increase in core doping (Figure 5.14 (b)). However, the I_{ON}/I_{OFF} ratio improves with increased core doping (Figure 5.14(c)).

I_{ON} also decreases with an increase in core diameter (Figure 5.14 (d)). The I_{ON}/I_{OFF} ratio enhances with an increase in d_{core} (Figure 5.14(e)). When channel length is reduced from 20 nm to 7 nm, I_{OFF} is found to increase along with a reduction in the I_{ON}/I_{OFF} ratio (Figure 5.14(f)). This is due to the enhanced BJT action in CS-JL NW FETs. When core doping increases, I_{OFF} and I_{ON} decreases due to the increased depletion of the drain and source regions. This also results in an increase in series

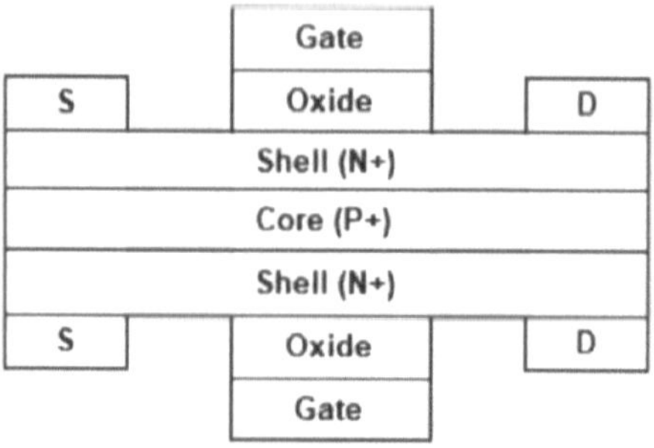

FIGURE 5.13 The cross-sectional view of Core–Shell (CS) JL NW FET [61].

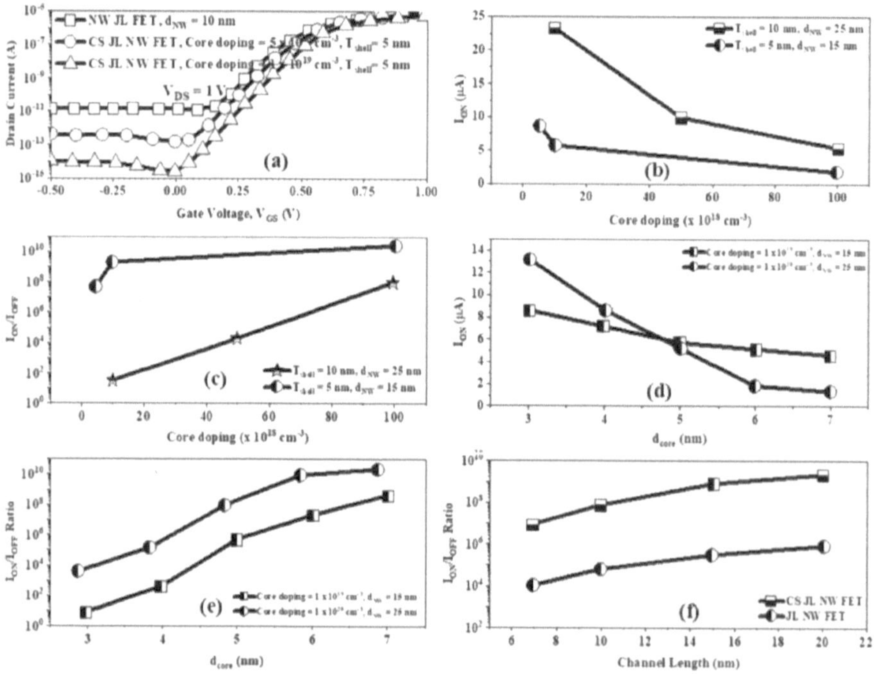

FIGURE 5.14 Performance of a CS-JL NW FET [61].

resistance. In 2018, A. K. Bansal et al. [57] demonstrated that the series resistance can be reduced using a vertically stacked JL accumulation-mode (AM) NW FET. The cross-sectional diagram of JL-DM (junctionless dual metal), CP-DM (charge plasma dual metal), JL GS-DM (junctionless gate stacked dual metal), and CP GS-DM (charge plasma gate stacked dual metal) NW FETs are depicted in Figure 5.15. CP-GS-DM NW FETs are considered as most promising for future analog circuit applications due to their low *SS* and higher I_{ON}/I_{OFF} ratio compared with other devices (Figure 5.16).

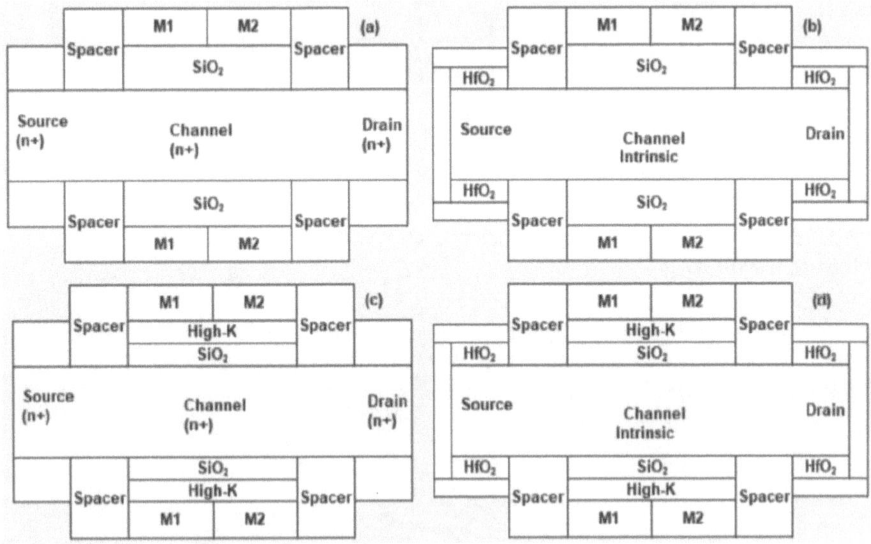

FIGURE 5.15 (a) JL-DM, (b) CP-DM, (c) JL-GS-DM, and (d) CP-GS-DM [64].

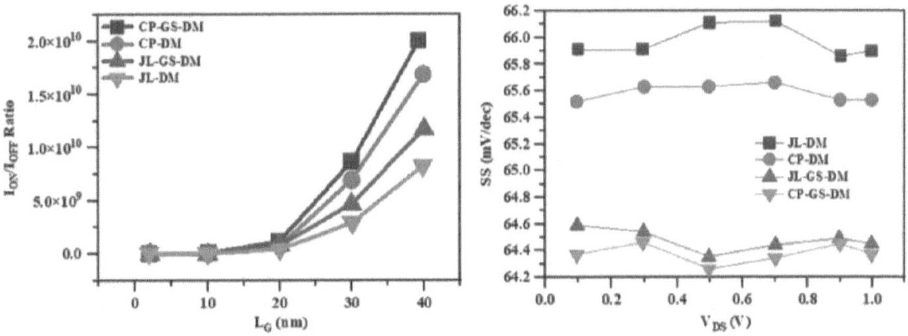

FIGURE 5.16 Performance comparison of CP-GS-DM, CP-DM, JL-DM, and JL-GS-DM FETs [64].

5.5 NW NCFETs

Negative-capacitance FETs (NC FETs) are considered as one of the most promising FET technologies for sub-3 nm nodes. A steep *SS* is the major advantage of NW NC FETs. The NC FETs consist of ferroelectric materials like HZO ($Hf_{1-x}Zr_xO_2$), PZT ($Pb_{0.2}Zr_{0.8}TiO_3$), and STO ($SrTiO_3$). There are two different types of NW NC FET architectures, and they are

1. MFIS (metal ferroelectric insulator semiconductor)
2. MFMIS (metal ferroelectric metal insulator semiconductor)

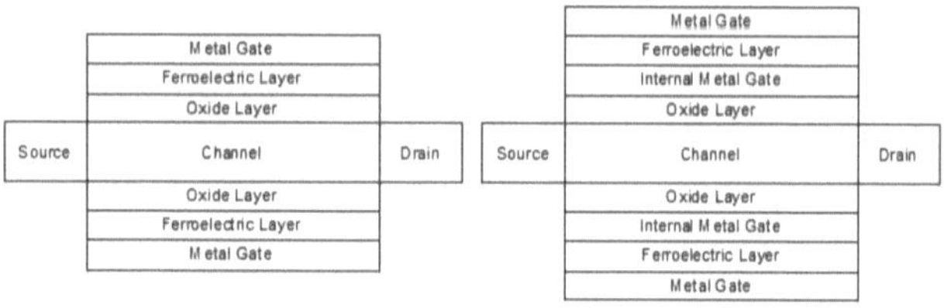

FIGURE 5.17 The cross-sectional schematics of NW GAA NC FETs with (a) MFIS and (b) MFMIS structures.

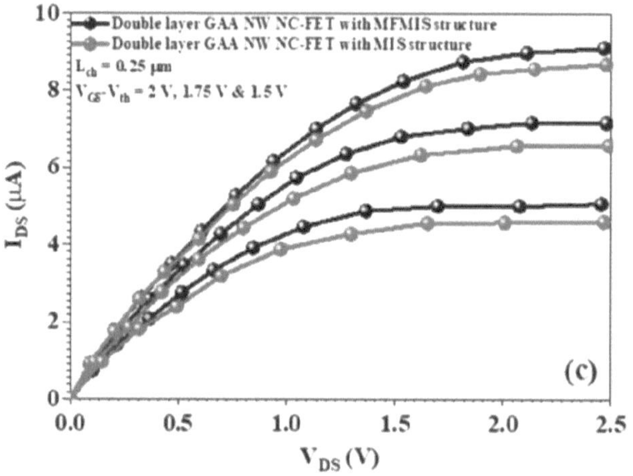

FIGURE 5.18 Output characteristics of GAA NW NC FETs with MFIS and MFMIS structures [83].

Depending on bias conditions, a negative-capacitance effect will be induced in the ferroelectric layer of NC FETs. S-shaped polarization in the ferroelectric layer is the major reason for the negative-capacitance found in the ferroelectric layer. The cross-sectional schematics of NW GAA NC FETs with MFIS and MFMIS structures are illustrated in Figure 5.17. The channel materials can be Si, Ge, SiGe, III-V materials like InAs, InSb, InGaAs, GaN, and 2D materials like MoS_2, carbon nano tube, graphene, etc. [65–82]. GAA NW NCFETs with MFMIS structures outperform GAA NW NCFETs with MFIS structures due to enhanced gate capacitance, improved transconductance, and reduced parasitic [83–85]. The output curves of GAA NW NC FETs with MFIS and MFMIS architectures are illustrated in Figure 5.18.

5.6 RECONFIGURABLE NW FETs

NW reconfigurable FETs are multifunctional transistors that provide both p-type and n-type electrical characteristics based on the polarity of an applied electrical signal. That is, a single device can act as both a n-type and p-type transistor depending on the electrical signal. Reconfigurable NW FETs are four-terminal devices with enhanced functionality that rely on the unique characteristics of Schottky junctions in NWs to control the concentration and polarity of charges. Another highlight of this device is that no doping is required in the drain and source regions; therefore, metals can be used to form drain and source contacts. An intrinsic channel is used in reconfigurable NW FETs to allow the conduction of both electrons and holes. $NiSi_2$ can be used for forming S/D contacts. The cross-sectional view of a reconfigurable NW FET is shown in Figure 5.19. Reconfigurable NW FETs consist of two gates: a control gate and a program gate. The two gates act like carrier injection valves connected at the ends of a nanowire channel. These two gates control the charge concentration and polarity of reconfigurable NW FETs. The metallic *S/D* regions form two Schottky contact junctions on either side of the NW channel. The diameter of the NW channel determines the area of the Schottky contact junction. The NW channel is usually undoped; therefore, the metallic *S/D* regions inject carriers into the channel. The injection of carriers can be controlled by changing the shape of the Schottky barrier [86–90]. The transfer curve of reconfigurable NW FET is shown in Figure 5.20. The *SS* and I_{ON} of reconfigurable NW FETs depend on the length of the NW (L_{NW}) and d_{NW}. The I_{ON} degrades severely with an increase in L_{NW} (Figure 5.21(a)).

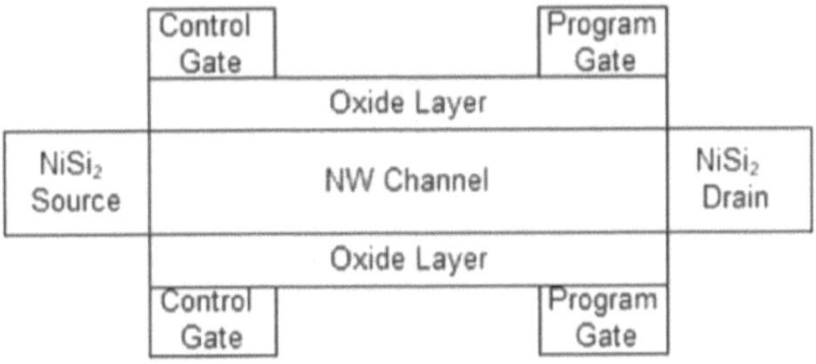

FIGURE 5.19 The cross-sectional view of a reconfigurable NW FET.

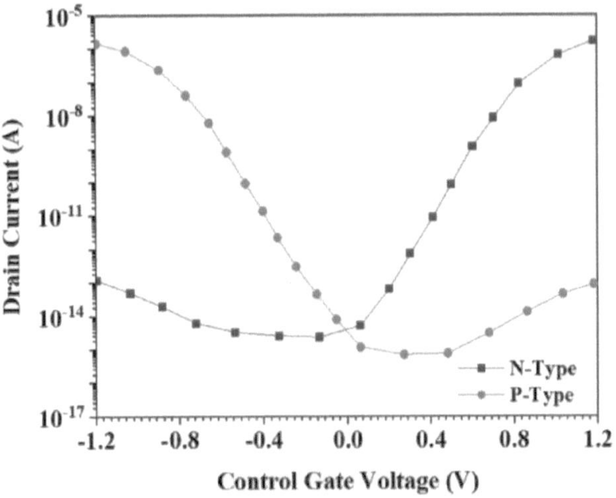

FIGURE 5.20 The transfer curve of a reconfigurable NW FET [88].

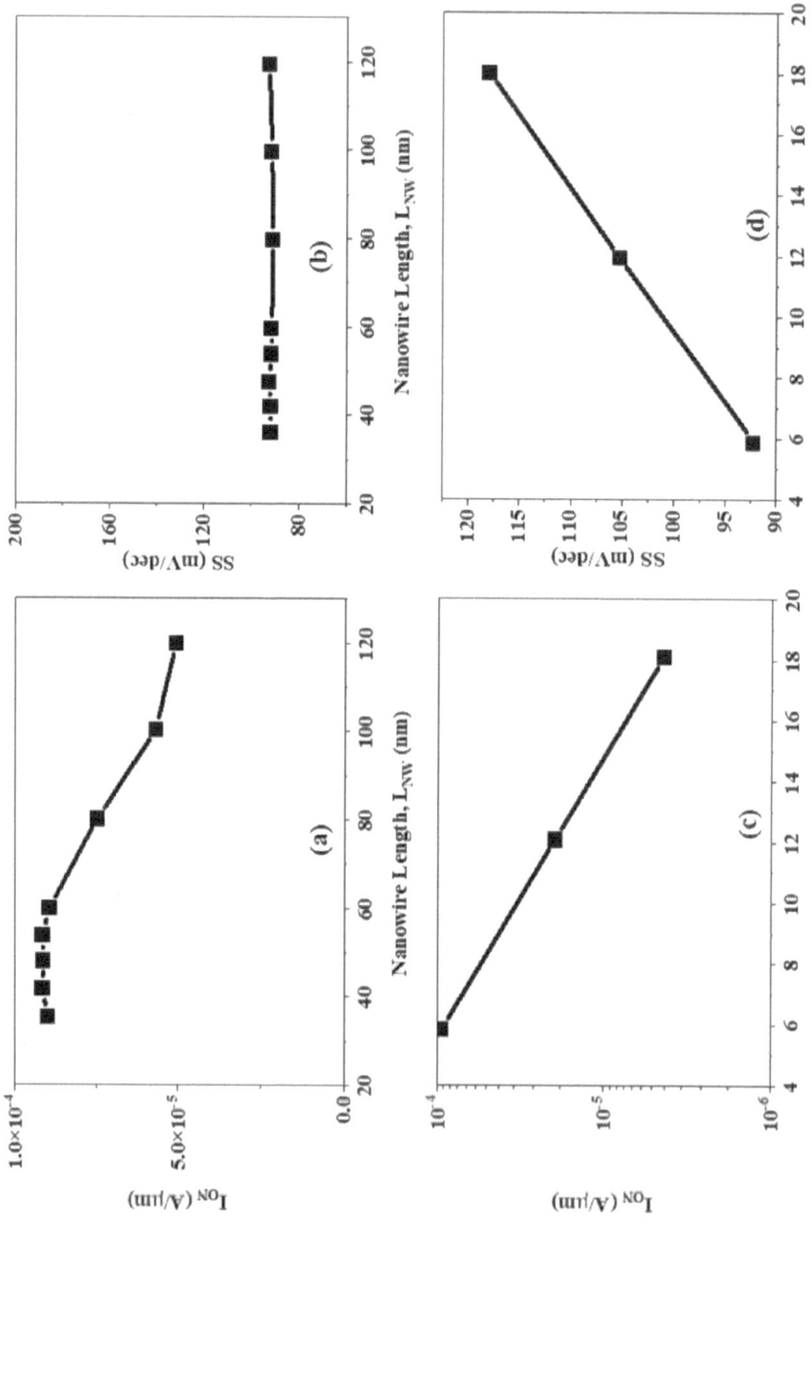

FIGURE 5.21 A reconfigurable NW FET (a) I_{ON} *vs* L_{NW}, (b) *SS vs* L_{NW}, (c) I_{ON} *vs* d_{NW}, and (d) *SS vs* d_{NW} characteristics [88].

However, the L_{NW} does not significantly affect the *SS* (Figure 5.21(b)). On the other hand, a large d_{NW} results in the reduction of the I_{ON} (Figure 5.21(c)) and an increase in *SS* (Figure 5.21 (d)).

5.7 RELIABILITY ISSUES IN NW FETs

Negative bias temperature instability (NBTI) reliability in a deep nanometer regime is a serious concern for p-type Si NW FETs because NBTI severely affects the life span of the NW FETs and may result in circuit failure. NBTI is a chemical and physical effect that occurs in p-type NW FETs under negative bias conditions at higher temperatures. NBTI leads to the formation of interface traps at the oxide/Si interface [91, 92]. The generation of interface traps increases the V_{TH} and reduces the I_{ON} of NW FETs. The NBTI-induced V_{TH} variation as a function of stress time and temperature is plotted in Figure 5.22. Temperature plays a key role in NBTI activation energy. At higher temperatures, NBTI-induced V_{TH} variations will be higher due to the generation of a large number of interface traps. Self-heating effects are another important concern in the reliability and performance of NW FETs, due to the use of low thermal conductivity materials [93]. NW FETs also suffer from lateral BTBT-induced current leakage due to parasitic bipolar junction transistor action [94, 95]. Hot carrier degradation is also an important reliability issue in GAA NW FETs. Hot carrier degradation leads to the reduction of I_{ON} due to the increase of V_{TH} [96].

5.8 SI NW FETs

FinFET is the leading transistor technology for current high-performance applications. But FinFET technology face critical issues when maintaining control of its electrostatics. NW FETs and nanosheet FETs have been considered as possible replacements for FinFETs [97]. Compared with nanosheet FETs and FinFETs, NW FETs provide relatively low *SS* due to better electrostatic control of the device. Due to the overlap between the drain conduction band and the channel valence band, electrons from the channel tunnel toward the drain using the L-BTBT process. This

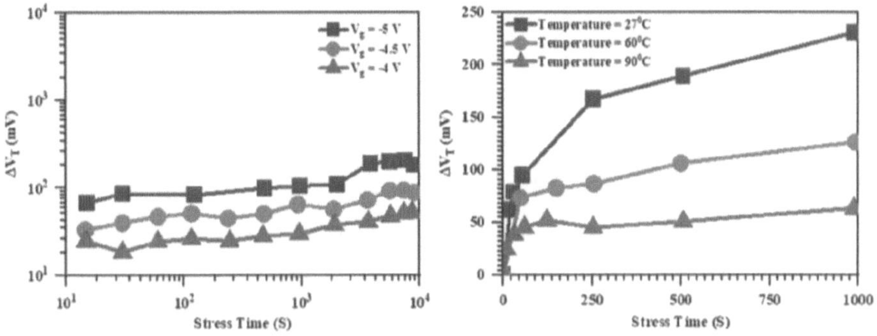

FIGURE 5.22 The NBTI-induced V_{TH} variation as a function of stress time and temperature [92].

results in an increase in the I_{OFF} [98, 99]. NW FETs can be used for chemical bio sensing [100–160]]. In NW FETs, dielectric materials with low thermal conductivity are used as an oxide covering [161–186]. Therefore, it is very difficult to dissipate the heat generated from the channel. This leads to the reduction of the I_{ON} and also creates reliability issues due to the self-heating effect [187]. In order to reduce the parasitic and improve energy efficiency and device performance, NW FETs can be designed with corner spacers. The 2D cross-sectional view of a GAA NW FET with a corner spacer is shown in Figure 5.23. The gate capacitance (C_{gg}) increases with an increase in corner spacer thickness (T_{cor}). Similarly, C_{gg} increases with a decrease in the length of the corner spacer (L_{cor}). The D_{NW}, length of the spacer (L_{SP}), and thickness of the *S/D* epi are the other factors that affect the C_{gg} of the NW FETs [188–191]. A 2D cross-sectional view of a NW FET with three vertically stacked Si NWs is illustrated in Figure 5.24. GAA NW FETs have emerged as an attractive transistor technology for future sub-5 nm technology nodes [193]. Variation of oxide thickness, random dopants, and work-function granularity of the gate metal are the different sources of variability that could severely degrade the yield and performance of NW FETs [194–210].

Channel doping concentration, S/D doping concentrations, gate work function, gate length, spacer length, width, and diameter of the nanowire are the key parameters that decide the performance of NW FETs [211, 212]. Insulator capacitances (C_{ins}) for NW FETs with circular and rectangular geometries are given in Eqs. (5.7) and (5.8), respectively [213].

$$C_{ins} = \frac{2\pi\varepsilon_{ins}}{\ln\left(1+\left(\frac{T_{ins}}{r}\right)\right)} \tag{5.7}$$

T_{ins} = insulator thickness
R = radius of the nanowire
ε_{ins} = permittivity of the insulator

The I_{DS} of the NW FET can be computed as [213]

$$I_{DS} = W_{ch}C_{ins}\upsilon_{eff}Q_{av} \tag{5.8}$$

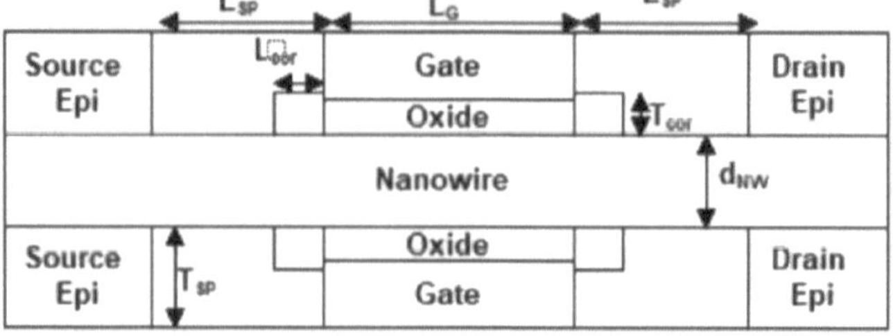

FIGURE 5.23 The 2-D cross-sectional view of a GAA NW FET with corner spacer [189].

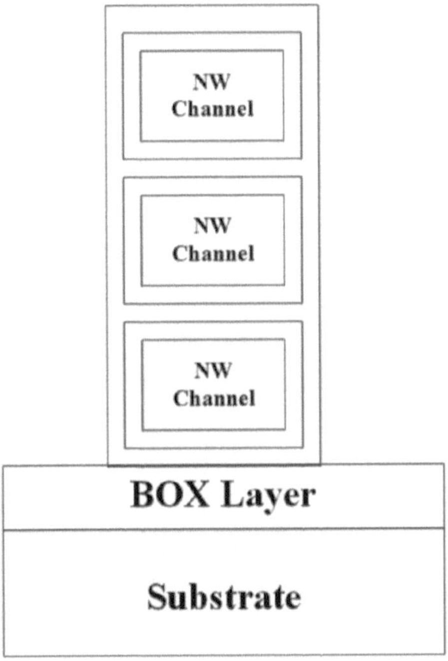

FIGURE 5.24 2-D cross-sectional view of a NW FET with three vertically stacked Si NWs [192].

Effective velocity,

$$\upsilon_{eff} = \frac{\upsilon_{dd}\upsilon_{inj}}{\upsilon_{dd} + \upsilon_{inj}} \tag{5.9}$$

υ_{dd}=drift-diffusive velocity
υ_{inj}=ballistic velocity

$$\upsilon_{dd} = \frac{\mu_{eff}\left(q_s - q_d\right)}{L_G} \tag{5.10}$$

μ_{eff}=effective mobility

$$\upsilon_{inj} = \sqrt{2kT\pi/m^*}\,\frac{\lambda}{\lambda + l} \tag{5.11}$$

l=distance along the channel over which the potential reduces by $\frac{KT}{q}$
λ=mean free path length

Effective charge density,

$$Q_{av} = \frac{q_s + q_d}{2} + \left(2 - \eta_{iv}\right)\upsilon_t \tag{5.12}$$

q_s=source charge density
q_d=drain charge density

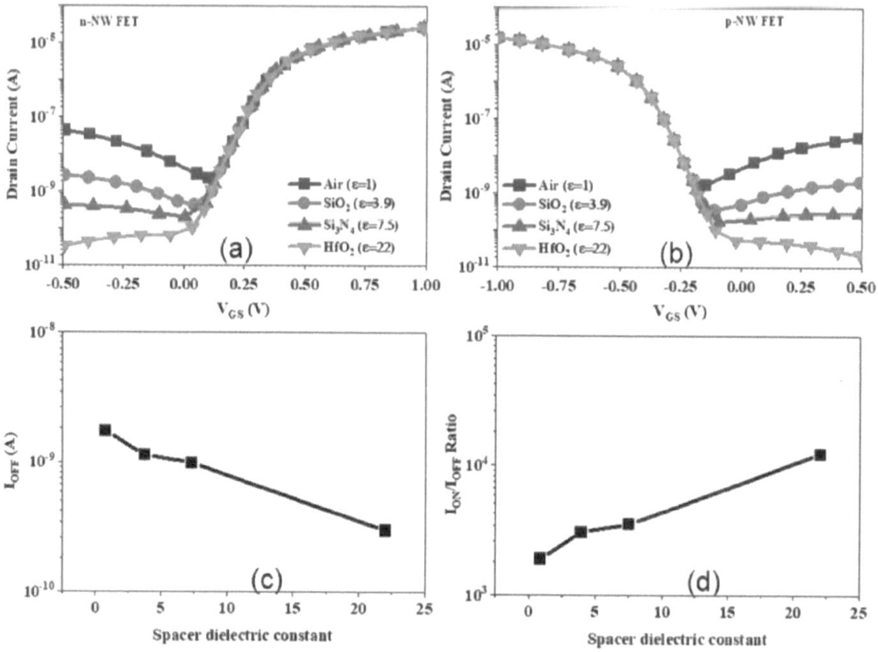

FIGURE 5.25 (a) and (b) The influence of different gate side-wall spacer materials on the transfer curves of NW FETs; (c) and (d) the impact of a dielectric permittivity spacer material on the I_{OFF} and I_{ON}/I_{OFF} ratio of NW FETs [214].

The influence of different gate side-wall spacer materials on the transfer curves of NW FETs is illustrated in Figure 5.25. The use of a high permittivity side-wall spacer reduces the I_{OFF} in NW FETs.

The impact of a dielectric permittivity spacer material on the I_{OFF} and I_{ON}/I_{OFF} ratio of NW FETs is demonstrated in Figures 5.25(c) and 5.25(d), respectively. Line edge roughness is a serious concern in aggressively scaled NW FETs [215–226].

5.9 SUMMARY

GAA NW architecture is considered the most promising solution for the aggressive scaling of CMOS transistors. This chapter covered the different types of NW FETs, such as ZnO/CuO/PbS NW FETs, III-V NW FETs, III-V NW TFETs, JL NW FETs, NW NC FETs, reconfigurable NW FETs, and Si NW FETs. The reliability issues in NW FETs were discussed in this chapter.

REFERENCES

1. Y. Shao, J. Yoon, H. Kim, T. Lee, W. Lu, Temperature dependence of electron transport in ZnO nanowire field effect transistors, *IEEE Transactions on Electron Devices*, 61 (2014) 625–630.

2. P.-C. Chang, J.G. Lu, ZnO nanowire field-effect transistors, *IEEE Transactions on Electron Devices*, 55 (2008) 2977–2987.
3. G. Jo, W.-K. Hong, M. Choe, W. Park, Y.H. Kahng, T. Lee, Proton irradiation-induced electrostatic modulation in ZnO nanowire field-effect transistors with bilayer gate dielectric, *IEEE Transactions on Nanotechnology*, 11 (2012) 918–923.
4. S.M. Sultan, K. Sun, O.D. Clark, T.B. Masaud, Q. Fang, R. Gunn, J. Partridge, M.W. Allen, P. Ashburn, H.M.H. Chong, Electrical characteristics of top-down ZnO nanowire transistors using remote plasma ALD, *IEEE Electron Device Letters*, 33 (2012) 203–205.
5. V.P. Verma, H. Jeon, S. Hwang, M. Jeon, W. Choi, Enhanced electrical conductance of ZnO nanowire FET by non-destructive surface cleaning, *IEEE Transactions on Nanotechnology*, 7 (2008) 782–786.
6. S.J. Pearton, D.P. Norton, L.-C. Tien, J. Guo, Modeling and fabrication of ZnO nanowire transistors, *IEEE Transactions on Electron Devices*, 55 (2008) 3012–3019.
7. W-Ki Hong, Gunho Jo, S-S. Kwon, S. Song, T. Lee, Electrical properties of surface-tailored ZnO nanowire field-effect transistors, *IEEE Transactions on Electron Devices*, 55 (2008) 3020–3029.
8. Z. Fan, J.G. Lu, Chemical sensing with ZnO nanowire field-effect transistor, *IEEE Transactions on Nanotechnology*, 5 (2006) 393–396.
9. S.-Y. Lee, J.-H. Hyung, S.-K. Lee, Dielectrophoresis (DEP)-prepared multiple-channel ZnO nanowire field-effect transistors, *Electronics Letters*, 44 (2008) 695–696.
10. A.K. Mishra, D.K. Jarwal, B. Mukherjee, A. Kumar, S. Ratan, S. Jit, CuO nanowire-based extended-gate field-effect-transistor (FET) for pH sensing and enzyme-free/receptor-free glucose sensing applications, *IEEE Sensors Journal*, 20 (2020) 5039–5047.
11. S. Lee, J.-S. Noh, J. Kim, M. Kim, S.Y. Jang, J. Park, W. Lee, The optoelectronic properties of PbS nanowire field-effect transistors, *IEEE Transactions on Nanotechnology*, 12 (2013) 1135–1138.
12. R.K. Kakkerla, C.-J. Hsiao, D. Anandan, S.K. Singh, S.-P. Chang, K.P. Pande, E.Y. Chang, Growth and crystal structure investigation of InAs/GaSb heterostructure nanowires on Si substrate, *IEEE Transactions on Nanotechnology*, 17 (2018) 1151–1158.
13. M.S. Ram, K.-M. Persson, M. Borg, L.-E. Wernersson, Low-power resistive memory integrated on III–V vertical nanowire MOSFETs on silicon, *IEEE Electron Device Letters*, 41 (2020) 1432–1435.
14. S.A. Fortuna, X. Li, GaAs MESFET with a high-mobility self-assembled planar nanowire channel, *IEEE Electron Device Letters*, 30 (2009) 593–595.
15. M.A. Khayer, R.K. Lake, The quantum and classical capacitance limits of InSb and InAs nanowire FETs, *IEEE Transactions on Electron Devices*, 56 (2009) 2215–2223.
16. K.-M. Persson, E. Lind, A.W. Dey, C. Thelander, H. Sjöland, L.-E. Wernersson, Low-frequency noise in vertical InAs nanowire FETs, *IEEE Electron Device Letters*, 31 (2010) 428–430.
17. T. Thingujam, D.-H. Son, J.-G. Kim, S. Cristoloveanu, J.-H. Lee, Effects of interface traps and self-heating on the performance of GAA GaN vertical nanowire MOSFET, *IEEE Transactions on Electron Devices*, 67 (2020) 816–821.
18. S. Shin, M.A. Wahab, M. Masuduzzaman, K. Maize, J. Gu, M. Si, A. Shakouri, D.Y., Peide, M.A. Alam, Direct observation of self-heating in III–V gate-all-around nanowire MOSFETs, *IEEE Transactions on Electron Devices*, 62 (2015) 3516–3523.
19. P.T. Blanchard, K.A. Bertness, T.E. Harvey, A.W. Sanders, N.A. Sanford, S.M. George, D. Seghete, MOSFETs made from GaN nanowires with fully conformal cylindrical gates, *IEEE Transactions on Nanotechnology*, 11 (2011) 479–482.

20. K. Tomioka, T. Tanaka, S. Hara, K. Hiruma, T. Fukui, III–V nanowires on Si substrate: Selective-area growth and device applications, *IEEE Journal of Selected Topics in Quantum Electronics*, 17 (2010) 1112–1129.
21. H. Gamo, K. Tomioka, Integration of indium arsenide/indium phosphide core-shell nanowire vertical gate-all-around field-effect transistors on Si, *IEEE Electron Device Letters*, 41 (2020) 1169–1172.
22. C. Grillet, D. Logoteta, A. Cresti, M.G. Pala, Assessment of the electrical performance of short channel InAs and strained Si nanowire FETs, *IEEE Transactions on Electron Devices*, 64 (2017) 2425–2431.
23. C.B. Zota, L.-E. Wernersson, E. Lind, High-performance lateral nanowire in GaAs MOSFETs with improved on-current, *IEEE Electron Device Letters*, 37 (2016) 1264–1267.
24. C.J. Delker, S. Kim, M. Borg, L.-E. Wernersson, D.B. Janes, 1/f noise sources in dual-gated indium arsenide nanowire transistors, *IEEE Transactions on Electron Devices*, 59 (2012) 1980–1987.
25. K. Jansson, E. Lind, L.-E. Wernersson, Performance evaluation of III–V nanowire transistors, *IEEE Transactions on Electron Devices*, 59 (2012) 2375–2382.
26. M.A. Khayer, R.K. Lake, Performance of $ n $-type InSb and InAs nanowire field-effect transistors, *IEEE Transactions on Electron Devices*, 55 (2008) 2939–2945.
27. E. Lind, M.P. Persson, Y.-M. Niquet, L.-E. Wernersson, Band structure effects on the scaling properties of InAs nanowire MOSFETs, *IEEE Transactions on Electron Devices*, 56 (2009) 201–205.
28. Y. He, Y. Zhao, C. Fan, J. Kang, R. Han, X. Liu, Performance evaluation of GaAs–gap core–shell-nanowire field-effect transistors, *IEEE Transactions on Electron Devices*, 56 (2009) 1199–1203.
29. S.S. Sylvia, M.A. Khayer, K. Alam, R.K. Lake, Doping, tunnel barriers, and cold carriers in InAs and InSb nanowire tunnel transistors, *IEEE Transactions on Electron Devices*, 59 (2012) 2996–3001.
30. K.E. Moselund, H. Schmid, C. Bessire, M. Bjork, H. Ghoneim, H. Riel, InAs–Si nanowire heterojunction tunnel FETs, *IEEE Electron Device Letters*, 33 (2012) 1453–1455.
31. F. Conzatti, M. Pala, D. Esseni, Surface-roughness-induced variability in nanowire InAs tunnel FETs, *IEEE Electron Device Letters*, 33 (2012) 806–808.
32. F. Conzatti, M. Pala, D. Esseni, E. Bano, L. Selmi, Strain-induced performance improvements in InAs nanowire tunnel FETs, *IEEE Transactions on Electron Devices*, 59 (2012) 2085–2092.
33. M.G. Pala, D. Esseni, Interface traps in InAs nanowire tunnel-FETs and MOSFETs—Part I: Model description and single trap analysis in tunnel-FETs, *IEEE Transactions on Electron Devices*, 60 (2013) 2795–2801.
34. D. Esseni, M.G. Pala, Interface traps in InAs nanowire tunnel FETs and MOSFETs—Part II: Comparative analysis and trap-induced variability, *IEEE Transactions on Electron Devices*, 60 (2013) 2802–2807.
35. M.G. Pala, S. Brocard, Exploiting hetero-junctions to improve the performance of III–V nanowire tunnel-FETs, *IEEE Journal of the Electron Devices Society*, 3 (2015) 115–121.
36. D.S. Lemtur, J. Patel, P. Suman, C. Rajan, Two-stage op-amp and integrator realisation through GaAsP/AlGaSb nanowire CP-TFET, *Micro & Nano Letters*, 14 (2019) 980–985.
37. H. Carrillo-Nuñez, J. Lee, S. Berrada, C. Medina-Bailón, F. Adamu-Lema, M. Luisier, A. Asenov, V.P. Georgiev, Random dopant-induced variability in Si-InAs nanowire tunnel FETs: a quantum transport simulation study, *IEEE Electron Device Letters*, 39 (2018) 1473–1476.

38. E. Lind, E. Memišević, A.W. Dey, L.-E. Wernersson, III–V heterostructure nanowire tunnel FETs, *IEEE Journal of the Electron Devices Society*, 3 (2015) 96–102.
39. M. Visciarelli, E. Gnani, A. Gnudi, S. Reggiani, G. Baccarani, Impact of strain on tunneling current and threshold voltage in III–V nanowire TFETs, *IEEE Electron Device Letters*, 37 (2016) 560–563.
40. X. Zhao, A. Vardi, Jesús A. del Alamo, Sub-thermal subthreshold characteristics in top–down InGaAs/InAs hetero-junction vertical nanowire tunnel FETs, *IEEE Electron Device Letters*, 38 (2017) 855–858.
41. E. Gnani, A. Gnudi, S. Reggiani, G. Baccarani, Theory of the junctionless nanowire FET, *IEEE Transactions on Electron Devices*, 58 (2011) 2903–2910.
42. J.P. Duarte, S.-J. Choi, D.-I. Moon, Y.-K. Choi, A nonpiecewise model for long-channel junctionless cylindrical nanowire FETs, *IEEE Electron Device Letters*, 33 (2011) 155–157.
43. S.-J. Choi, D.-I. Moon, S. Kim, J.-H. Ahn, J.-S. Lee, J.-Y. Kim, Y.-K. Choi, Nonvolatile memory by all-around-gate junctionless transistor composed of silicon nanowire on bulk substrate, *IEEE Electron Device Letters*, 32 (2011) 602–604.
44. P. Singh, J. Miao, V. Pott, W.-T. Park, D.-L. Kwong, Piezoresistive sensing performance of junctionless nanowire FET, *IEEE Electron Device Letters*, 33 (2012) 1759–1761.
45. V. Rana, G. Ahmad, A.K. Ramesh, S. Das, P. Singh, Diameter-dependent piezoresistive sensing performance of junctionless gate-all-around nanowire FET, *IEEE Transactions on Electron Devices*, 67 (2020) 2884–2889.
46. P.-Y. Kuo, Y.-H. Lu, T.-S. Chao, High-performance GAA sidewall-damascened sub-10-nm in situ n+-doped poly-Si NWs channels junctionless FETs, *IEEE Transactions on Electron Devices*, 61 (2014) 3821–3826.
47. D. Shafizade, M. Shalchian, F. Jazaeri, Ultrathin junctionless nanowire FET model, including 2-D quantum confinements, *IEEE Transactions on Electron Devices*, 66 (2019) 4101–4106.
48. F. Yesayan J.-M. Jazaeri, Sallese, Analytical modeling of double-gate and nanowire junctionless ISFETs, *IEEE Transactions on Electron Devices*, 67 (2020) 1157–1164.
49. L. Ansari, B. Feldman, G. Fagas, C.M. Lacambra, M.G. Haverty, K.J. Kuhn, S. Shankar, J.C. Greer, First principle-based analysis of single-walled carbon nanotube and silicon nanowire junctionless transistors, *IEEE Transactions on Nanotechnology*, 12 (2013) 1075–1081.
50. Y.S. Yu, A unified analytical current model for N-and P-type accumulation-mode (junctionless) surrounding-gate nanowire FETs, *IEEE Transactions on Electron Devices*, 61 (2014) 3007–3010.
51. H. Carrillo-Nunez, M.M. Mirza, D.J. Paul, D.A. MacLaren, A. Asenov, V.P. Georgiev, Impact of randomly distributed dopants on $\Omega $-gate junctionless silicon nanowire transistors, *IEEE Transactions on Electron Devices*, 65 (2018) 1692–1698.
52. J.-M. Sallese, F. Jazaeri, L. Barbut, N. Chevillon, C. Lallement, A common core model for junctionless nanowires and symmetric double-gate FETs, *IEEE Transactions on Electron Devices*, 60 (2013) 4277–4280.
53. U.-S. Jeong, C.-K. Kim, H. Bae, D.-I. Moon, T. Bang, J.-M. Choi, J. Hur, Y.-K. Choi, Investigation of low-frequency noise in non-volatile memory composed of a gate-all-around junctionless nanowire FET, *IEEE Transactions on Electron Devices*, 63 (2016) 2210–2213.
54. A.K. Bansal, C. Gupta, A. Gupta, R. Singh, T.B. Hook, A. Dixit, 3-D LER and RDF matching performance of nanowire FETs in inversion, accumulation, and junctionless modes, *IEEE Transactions on Electron Devices*, 65 (2018) 1246–1252.
55. E. Simoen, A. Veloso, P. Matagne, N. Collaert, C. Claeys, Junctionless versus inversion-mode gate-all-around nanowire transistors from a low-frequency noise perspective, *IEEE Transactions on Electron Devices*, 65 (2018) 1487–1492.

56. R. Ragi, M.A. Romero, Fully analytical compact model for the I–V characteristics of large radius junctionless nanowire FETs, *IEEE Transactions on Nanotechnology*, 18 (2019) 762–769.
57. A.K. Bansal, M. Kumar, C. Gupta, T.B. Hook, A. Dixit, Series resistance reduction with linearity assessment for vertically stacked junctionless accumulation mode nanowire FET, *IEEE Transactions on Electron Devices*, 65 (2018) 3548–3554.
58. D.-I. Moon, S.-J. Choi, J.P. Duarte, Y.-K. Choi, Investigation of silicon nanowire gate-all-around junctionless transistors built on a bulk substrate, *IEEE Transactions on Electron Devices*, 60 (2013) 1355–1360.
59. K.-H. Goh, S. Yadav, K.L. Low, G. Liang, X. Gong, Y.-C. Yeo, Gate-all-around In 0.53 Ga 0.47 As junctionless nanowire FET with tapered source/drain structure, *IEEE Transactions on Electron Devices*, 63 (2016) 1027–1033.
60. L.-C. Chen, M.-S. Yeh, K.-W. Lin, M.-H. Wu, Y.-C. Wu, Junctionless poly-Si nanowire FET with gated raised S/D, *IEEE Journal of the Electron Devices Society*, 4 (2016) 50–54.
61. S. Sahay, M.J. Kumar, Controlling L-BTBT and volume depletion in nanowire JLFETs using core–shell architecture, *IEEE Transactions on Electron Devices*, 63 (2016) 3790–3794.
62. S. Sahay, M.J. Kumar, Insight into lateral band-to-band-tunneling in nanowire junctionless FETs, *IEEE Transactions on Electron Devices*, 63 (2016) 4138–4142.
63. S. Sahay, M.J. Kumar, Diameter dependence of leakage current in nanowire junctionless field effect transistors, *IEEE Transactions on Electron Devices*, 64 (2017) 1330–1335.
64. S. Singh, A. Raman, Gate-all-around charge plasma-based dual material gate-stack nanowire FET for enhanced analog performance, *IEEE Transactions on Electron Devices*, 65 (2018) 3026–3032.
65. J. Ajayan, T. Ravichandran, P. Mohankumar, P. Prajoon, J.C. Pravin, D. Nirmal, Investigation of DC and RF performance of novel MOSHEMT on silicon substrate for future submillimetre wave applications, *Semiconductors*, 52, No. 16 (2018) 1991–1997.
66. J. Ajayan, D. Nirmal, K. Dheena, P. Mohankumar, L. Arivazhagan, A.S. Augustine Fletcher, T.D. Subash, M. Saravanan, Investigation of the impact of gate underlap/overlap on the analog/RF performance of composite channel double gate MOSFETs, *Journal of Vacuum Science & Technology B*, 37, No. 6 (2019) 06221.
67. J. Ajayan, D. Nirmal, P. Mohankumar, L. Arivazhagan, M. Saravanan, S. Saravanan, LG= 20 nm high performance GaAs substrate based metamorphic metal oxide semiconductor high electron mobility transistor for next generation high speed low power applications, *Journal of Nanoelectronics and Optoelectronics*, 14, No. 8 (2019) 1133–1142.
68. J. Ajayan, D. Nirmal, 20-nm T-gate composite channel enhancement-mode metamorphic HEMT on GaAs substrates for future THz applications. *Journal of Computational Electronics*, 15 (2016) 1291–1296.
69. J. Ajayan, D. Nirmal, 20 nm high performance enhancement mode InP HEMT with heavily doped S/D regions for future THz application. *Superlattices and Microstructures*, 100 (2016) 526–534.
70. J. Ajayan, T. Ravichandran, P. Mohankumar, P. Prajoon, J.C. Pravin, D. Nirmal, Investigation of DC-RF and breakdown behaviour in Lg= 20 nm novel asymmetric GaAs MHEMTs for future submillimetre wave applications. *AEU- International Journal of Electronics and Communications*, 84 (2018) 387– 393.
71. J. Ajayan, D. Nirmal, A review of InP/InAlAs/InGaAs based transistors for high frequency applications. *Superlattices and Microstructures*, 86 (2015) 1–19.

72. J. Ajayan, T.D. Subash, D. Kurian, 20 nm high performance novel MOSHEMT on InP substrate for future high speed low power applications. *Superlattices and Microstructures*, 109 (2017) 183–193.
73. J. Ajayan, T. Ravichandran, P. Prajoon, J.C. Pravin, D. Nirmal, Investigation of breakdown performance in Lg = 20 nm novel asymmetric InP HEMTs for future high-speed high-power applications. *Journal of Computational Electronics*, 17, No. 1 (2018) 265–272.
74. J. Ajayan, D. Nirmal, P. Prajoon, J.C. Pravin, Analysis of nanometer-scale InGaAs/InAs/InGaAs composite channel MOSFETs using high-K dielectrics for high speed applications. *AEU-International Journal of Electronics and Communications*, 79 (2017) 151–157.
75. J. Ajayan, D. Nirmal, 22 nm In0:75Ga0: 25As channel-based HEMTs on InP/GaAs substrates for future THz applications. *Journal of Semiconductors*, 38 (2017) 27–32.
76. J. Ajayan, D. Nirmal, T. Ravichandran, P. Mohankumar, P. Prajoon, L. Arivazhagan, K.S. Chandan, InP high electron mobility transistors for submillimetre wave and terahertz frequency applications: A review. *International Journal of Electronics and Communications*, 94 (2018) 199–214.
77. J. Ajayan, D. Nirmal, 20-nm enhancement-mode metamorphic GaAs HEMT with highly doped InGaAs source/drain regions for high-frequency applications. *International Journal of Electronics*, 104 (2017) 504–512.
78. J. Ajayan, D. Nirmal, P. Mohankumar, K. Dheena, F. Augustine, L. Arivazhagan, B. Santhosh Kumar, GaAs metamorphic high electron mobility transistors for future deep space-biomedical-military and communication system applications: A review. *Microelectronics Journal*, 92 (2019) 104604.
79. J.C. Pravin, D Nirmal, P Prajoon, J Ajayan, Implementation of nanoscale circuits using dual metal gate engineered nanowire MOSFET with high-k dielectrics for low power applications, *Physical E: Low-Dimensional Systems and Nanostructures*, 83 (2016) 95–100.
80. J.C. Pravin, D Nirmal, P Prajoon, N.M. Kumar, J. Ajayan, Investigation of 6T SRAM memory circuit using high-k dielectrics based nano scale junctionless transistor, *Superlattices and Microstructures*, 104 (2017) 470–476.
81. J Ajayan., D Nirmal., P Mohankumar., L Arivazhagan.: Investigation of Impact of Passivation Materials on the DC/RF Performances of InP-HEMTs for Terahertz Sensing and Imaging, *Silicon*, 12 (2020) 1225–1230.
82. J. Ajayan, T. Ravichandran, P. Mohankumar, P. Prajoon, J.C. Pravin, & D. Nirmal, Investigation of RF and DC performance of E-mode $In_{0.80}Ga_{0.20}As/InAs/In_{0.80}Ga_{0.20}as$ channel based DG-HEMTs for future submillimetre wave and THz applications, *IETE Journal of Research*, (2018). DOI: 10.1080/03772063.2018.1553641.
83. S.-Y. Lee, H.-W. Chen, C.-H. Shen, P.-Y. Kuo, C.-C. Chung, Y.-E. Huang, H.-Y. Chen, T.-S. Chao, Experimental Demonstration of Stacked Gate-All-Around Poly-Si Nanowires Negative Capacitance FETs With Internal Gate Featuring Seed Layer and Free of Post-Metal Annealing Process, *IEEE Electron Device Letters*, 40 (2019) 1708–1711.
84. W. Huang, H. Zhu, Z. Wu, X. Yin, Q. Huo, K. Jia, Y. Li, Y. Zhang, Investigation of negative DIBL effect and miller effect for negative capacitance nanowire field-effect-transistors, *IEEE Journal of the Electron Devices Society*, 8 (2020) 879–884.
85. S.-Y. Lee, H.-W. Chen, C.-H. Shen, P.-Y. Kuo, C.-C. Chung, Y.-E. Huang, H.-Y. Chen, T.-S. Chao, Effect of seed layer on gate-all-around poly-Si nanowire negative-capacitance FETs with MFMIS and MFIS structures: Planar capacitors to 3-D FETs, *IEEE Transactions on Electron Devices*, 67 (2020) 711–716.
86. M. Simon, A. Heinzig, J. Trommer, T. Baldauf, T. Mikolajick, W.M. Weber, Top-down technology for reconfigurable nanowire FETs with symmetric on-currents, *IEEE Transactions on Nanotechnology*, 16 (2017) 812–819.

87. M. De Marchi, J. Zhang, S. Frache, D. Sacchetto, P.-E. Gaillardon, Y. Leblebici, G. De Micheli, Configurable logic gates using polarity-controlled silicon nanowire gate-all-around FETs, *IEEE Electron Device Letters*, 35 (2014) 880–882.
88. J. Trommer, A. Heinzig, T. Baldauf, S. Slesazeck, T. Mikolajick, W.M. Weber, Functionality-enhanced logic gate design enabled by symmetrical reconfigurable silicon nanowire transistors, *IEEE Transactions on Nanotechnology*, 14 (2015) 689–698.
89. T. Baldauf, A. Heinzig, J. Trommer, T. Mikolajick, W.M. Weber, Stress-dependent performance optimization of reconfigurable silicon nanowire transistors, *IEEE Electron Device Letters*, 36 (2015) 991–993.
90. W.M. Weber, A. Heinzig, J. Trommer, M. Grube, F. Kreupl, T. Mikolajick, Reconfigurable nanowire electronics-enabling a single CMOS circuit technology, *IEEE Transactions on Nanotechnology*, 13 (2014) 1020–1028.
91. O. Prakash, S. Maheshwaram, S. Beniwal, N. Gupta, N. Singh, S. Manhas, Impact of time zero variability and BTI reliability on SiNW FET-based circuits, *IEEE Transactions on Device and Materials Reliability*, 19 (2019) 741–750.
92. O. Prakash, S. Beniwal, S. Maheshwaram, A. Bulusu, N. Singh, S. Manhas, Compact NBTI reliability modeling in Si nanowire MOSFETs and effect in circuits, *IEEE Transactions on Device and Materials Reliability*, 17 (2017) 404–413.
93. J. Lai, Y. Su, J. Bu, B. Li, B. Li, G. Zhang, Study on degradation mechanisms of thermal conductivity for confined nanochannel in gate-all-around silicon nanowire field-effect transistors, *IEEE Transactions on Electron Devices*, 67 (2020) 4060–4066.
94. S. Sahay, M.J. Kumar, Physical insights into the nature of gate-induced drain leakage in ultrashort channel nanowire FETs, *IEEE Transactions on Electron Devices*, 64 (2017) 2604–2610.
95. C.-H. Shen, W.-Y. Chen, S.-Y. Lee, P.-Y. Kuo, T.-S. Chao, Nitride induced stress affecting crystallinity of sidewall damascene gate-all-around nanowire poly-Si FETs, *IEEE Transactions on Nanotechnology*, 19 (2020) 322–327.
96. C. Gupta, A. Gupta, S. Tuli, E. Bury, B. Parvais, A. Dixit, Characterization and modeling of hot carrier degradation in N-channel gate-all-around nanowire FETs, *IEEE Transactions on Electron Devices*, 67 (2019) 4–10.
97. D. Nagy, G. Espiñeira, G. Indalecio, A.J. García-Loureiro, K. Kalna, N. Seoane, Benchmarking of FinFET, Nanosheet, and Nanowire FET Architectures for Future Technology Nodes, *IEEE Access*, 8 (2020) 53196–53202.
98. S. Sahay, M.J. Kumar, Comprehensive analysis of gate-induced drain leakage in emergingFET architectures: Nanotube FETs versus nanowire FETs, *IEEE Access*, 5 (2017) 18918–18926.
99. D. Nagy, G. Indalecio, A.J. García-Loureiro, M.A. Elmessary, K. Kalna, N. Seoane, Metalgrain granularity study on a gate-all-around nanowire FET, *IEEE Transactions on Electron Devices*, 64 (2017) 5263–5269.
100. N.K. Rajan, D.A. Routenberg, J. Chen, M.A. Reed, $\hbox {1}/f $ noise of silicon nanowire BioFETs, *IEEE Electron Device Letters*, 31 (2010) 615–617.
101. J.-H. Ahn, J.-Y. Kim, K. Choi, D.-I. Moon, C.-H. Kim, M.-L. Seol, T.J. Park, S.Y. Lee, Y.-K. Choi, Nanowire FET biosensors on a bulk silicon substrate, *IEEE Transactions on Electron Devices*, 59 (2012) 2243–2249.
102. C.-H. Kuo, H.-C. Lin, I.-C. Lee, H.-C. Cheng, T.-Y. Huang, A novel scheme for fabricating CMOS inverters with poly-Si nanowire channels, *IEEE Electron Device Letters*, 33 (2012) 833–835.
103. X. Gong, R. Zhao, X. Yu, A 3-D-silicon nanowire FET biosensor based on a novel hybrid process, *Journal of Microelectromechanical Systems*, 27 (2018) 164–170.

104. X. Yang, S. Chen, H. Zhang, Z. Huang, X. Liu, Z. Cheng, T. Li, Trace level analysis of nerve agent simulant DMMP with silicon nanowire FET sensor, *IEEE Sensors Journal*, 20 (2020) 12096–12101.
105. D. Kim, C. Park, W. Choi, S.-H. Shin, B. Jin, R.-H. Baek, J.-S. Lee, Improved long-term responses of Au-decorated Si nanowire FET sensor for NH 3 detection, *IEEE Sensors Journal*, 20 (2019) 2270–2277.
106. N.N. Mojumder, K. Roy, Band-to-band tunneling ballistic nanowire FET: Circuit-compatible device modeling and design of ultra-low-power digital circuits and memories, *IEEE Transactions on Electron Devices*, 56 (2009) 2193–2201.
107. B.C. Paul, R. Tu, S. Fujita, M. Okajima, T.H. Lee, Y. Nishi, An analytical compact circuit model for nanowire FET, *IEEE Transactions on Electron Devices*, 54 (2007) 1637–1644.
108. W. Liu, J.J. Liou, Y. Jiang, N. Singh, G. Lo, J. Chung, Y. Jeong, Investigation of sub-10-nm diameter, gate-all-around nanowire field-effect transistors for electrostatic discharge applications, *IEEE Transactions on Nanotechnology*, 9 (2010) 352–354.
109. S.K. Yoo, S. Yang, J.-H. Lee, Hydrogen ion sensing using Schottky contacted silicon nanowire FETs, *IEEE Transactions on Nanotechnology*, 7 (2008) 745–748.
110. K.-S. Shin, K. Lee, J.-H. Park, J.Y. Kang, C.O. Chui, Schottky contacted nanowire field-effect sensing device with intrinsic amplification, *IEEE Electron Device Letters*, 31 (2010) 1317–1319.
111. S. Yoo, J. An, S. Yang, J. Lee, Subthreshold operation of Schottky barrier silicon nanowire FET for highly sensitive pH sensing, *Electronics Letters*, 46 (2010) 1450–1452.
112. S. Hamedi-Hagh, A. Bindal, Spice modeling of silicon nanowire field-effect transistors for high-speed analog integrated circuits, *IEEE Transactions on Nanotechnology*, 7 (2008) 766–775.
113. E. Stern, A. Vacic, M.A. Reed, Semiconducting nanowire field-effect transistor biomolecular sensors, *IEEE Transactions on Electron Devices*, 55 (2008) 3119–3130.
114. D.J. Baek, S.-J. Choi, J.-H. Ahn, J.-Y. Kim, Y.-K. Choi, Addressable nanowire field-effect-transistor biosensors with local back gates, *IEEE Transactions on Electron Devices*, 59 (2012) 2507–2511.
115. S.S. Sylvia, H.-H. Park, M.A. Khayer, K. Alam, G. Klimeck, R.K. Lake, Material selection for minimizing direct tunneling in nanowire transistors, *IEEE Transactions on Electron Devices*, 59 (2012) 2064–2069.
116. K.E. Moselund, M.T. Bjork, H. Schmid, H. Ghoneim, S. Karg, E. Lortscher, W. Riess, H. Riel, Silicon nanowire tunnel FETs: Low-temperature operation and influence of high-$ k $ gate dielectric, *IEEE Transactions on Electron Devices*, 58 (2011) 2911–2916.
117. M. Mescher, B. Marcelis, L.C. de Smet, E.J. Sudholter, J.H. Klootwijk, Pulsed method for characterizing aqueous media using nanowire field effect transistors, *IEEE Transactions on Electron Devices*, 58 (2011) 1886–1891.
118. S. Kim, M. Luisier, A. Paul, T.B. Boykin, G. Klimeck, Full three-dimensional quantum transport simulation of atomistic interface roughness in silicon nanowire FETs, *IEEE Transactions on Electron Devices*, 58 (2011) 1371–1380.
119. N.D. Akhavan, A. Afzalian, A. Kranti, I. Ferain, C.-W. Lee, R. Yan, P. Razavi, R. Yu, J.- P. Colinge, Influence of elastic and inelastic electron–phonon interaction on quantum transport in multigate silicon nanowire MOSFETs, *IEEE Transactions on Electron Devices*, 58 (2011) 1029–1037.
120. D.B. Razavieh, J. Janes, Appenzeller, Transconductance linearity analysis of 1-D, nanowire FETs in the quantum capacitance limit, *IEEE Transactions on Electron Devices*, 60 (2013) 2071–2076.

121. Á. Szabó, M. Luisier, Under-the-barrier model: An extension of the top-of-the-barrier model to efficiently and accurately simulate ultrascaled nanowire transistors, *IEEE Transactions on Electron Devices*, 60 (2013) 2353–2360.
122. Y.-R. Kim, S.-H. Lee, C.-W. Sohn, D.-Y. Choi, H.-C. Sagong, S. Kim, E.-Y. Jeong, D.-W. Kim, H. Hong, C.-K. Baek, Simple S/D series resistance extraction method optimized for nanowire FETs, *IEEE Electron Device Letters*, 34 (2013) 828–830.
123. T. Wang, L. Lou, C. Lee, A junctionless gate-all-around silicon nanowire FET of high linearity and its potential applications, *IEEE Electron Device Letters*, 34 (2013) 478–480.
124. T. Moh, M. Nie, G. Pandraud, L. de Smet, E. Sudhölter, Q. Huang, P. Sarro, Effect of silicon nanowire etching on signal-to-noise ratio of SiNW FETs for (bio) sensor applications, *Electronics Letters*, 49 (2013) 782–784.
125. J. Zhang, X. Tang, P.-E. Gaillardon, G. De Micheli, Configurable circuits featuring dual-threshold-voltage design with three-independent-gate silicon nanowire FETs, *IEEE Transactions on Circuits and Systems I: Regular Papers*, 61 (2014) 2851–2861.
126. M. Salmani-Jelodar, S.R. Mehrotra, H. Ilatikhameneh, G. Klimeck, Design guidelines for sub-12 nm nanowire MOSFETs, *IEEE Transactions on Nanotechnology*, 14 (2015) 210–213.
127. Martinez, J.R. Barker, M. Aldegunde, R. Valin, Study of local power dissipation in ultrascaled silicon nanowire FETs, *IEEE Electron Device Letters*, 36 (2014) 2–4.
128. K. Natori, Compact modeling of quasi-ballistic silicon nanowire MOSFETs, *IEEE Transactions on Electron Devices*, 59 (2011) 79–86.
129. J. Park, C. Shin, Impact of interface traps and surface roughness on the device performance of stacked-nanowire FETs, *IEEE Transactions on Electron Devices*, 64 (2017) 4025–4030.
130. M. Gaillardin, C. Marcandella, M. Martinez, O. Duhamel, T. Lagutere, P. Paillet, M. Raine, N. Richard, F. Andrieu, S. Barraud, Total ionizing dose response of multiple-gate nanowire field effect transistors, *IEEE Transactions on Nuclear Science*, 64 (2017) 2061–2068.
131. D. Nagy, G. Indalecio, A.J. García-Loureiro, M.A. Elmessary, K. Kalna, N. Seoane, FinFET versus gate-all-around nanowire FET: Performance, scaling, and variability, *IEEE Journal of the Electron Devices Society*, 6 (2018) 332–340.
132. D. Logoteta, N. Cavassilas, A. Cresti, M.G. Pala, M. Bescond, Impact of the gate and insulator geometrical model on the static performance and variability of ultrascaled silicon nanowire FETs, *IEEE Transactions on Electron Devices*, 65 (2018) 424–430.
133. X. Chen, S. Chen, S.-L. Zhang, P. Solomon, Z. Zhang, Low-noise Schottky junction trigate silicon nanowire field-effect transistor for charge sensing, *IEEE Transactions on Electron Devices*, 66 (2019) 3994–4000.
134. W. Fang, A. Veloso, E. Simoen, M.-J. Cho, N. Collaert, A. Thean, J. Luo, C. Zhao, T. Ye, C. Claeys, Impact of the effective work function gate metal on the low-frequency noise of gate-all-around silicon-on-insulator NWFETs, *IEEE Electron Device Letters*, 37 (2016) 363–365.
135. J. Appenzeller, J. Knoch, M.T. Bjork, H. Riel, H. Schmid, W. Riess, Toward nanowire electronics, *IEEE Transactions on Electron Devices*, 55 (2008) 2827–2845.
136. S.H. Lee, Y.S. Yu, S.W. Hwang, D. Ahn, A SPICE-compatible new silicon nanowire field-effect transistors (SNWFETs) model, *IEEE Transactions on Nanotechnology*, 8 (2009) 643–649.
137. P. Michetti, G. Mugnaini, G. Iannaccone, Analytical model of nanowire FETs in a partially ballistic or dissipative transport regime, *IEEE Transactions on Electron Devices*, 56 (2009) 1402–1410.

138. E.-S. Liu, N. Jain, K.M. Varahramyan, J. Nah, S.K. Banerjee, E. Tutuc, Role of metal–semiconductor contact in nanowire field-effect transistors, *IEEE Transactions on Nanotechnology*, 9 (2009) 237–242.
139. L. Choi, B.H. Hong, Y.C. Jung, K.H. Cho, K.H. Yeo, D.-W. Kim, G.Y. Jin, K.S. Oh, W.-S. Lee, S.-H. Song, Extracting mobility degradation and total series resistance of cylindrical gate-all-around silicon nanowire field-effect transistors, *IEEE Electron Device Letters*, 30 (2009) 665–667.
140. S. Poli, M.G. Pala, T. Poiroux, Full quantum treatment of remote Coulomb scattering in silicon nanowire FETs, *IEEE Transactions on Electron Devices*, 56 (2009) 1191–1198.
141. K. Rogdakis, E. Bano, L. Montes, M. Bechelany, D. Cornu, K. Zekentes, Rectifying source and drain contacts for effective carrier transport modulation of extremely doped SiC nanowire FETs, *IEEE Transactions on Nanotechnology*, 10 (2010) 980–984.
142. L. De Michielis, K.E. Moselund, L. Selmi, A.M. Ionescu, Corner effect and local volume inversion in SiNW FETs, *IEEE Transactions on Nanotechnology*, 10 (2010) 810–816.
143. R.-H. Baek, C.-K. Baek, H.-S. Choi, J.-S. Lee, Y.Y. Yeoh, K.H. Yeo, D.-W. Kim, K. Kim, D.M. Kim, Y.-H. Jeong, Characterization and modeling of 1/$ f $ noise in Si-nanowire FETs: Effects of cylindrical geometry and different processing of oxides, *IEEE Transactions on Nanotechnology*, 10 (2010) 417–423.
144. C.-W.L. Afzalian, N.D. Akhavan, R. Yan, I. Ferain, J.-P. Colinge, Quantum confinement effects in capacitance behavior of multigate silicon nanowire MOSFETs, *IEEE Transactions on Nanotechnology*, 10 (2010) 300–309.
145. W. Liu, J.J. Liou, Y. Jiang, N. Singh, G. Lo, J. Chung, Y. Jeong, Failure analysis of Si nanowire field-effect transistors subject to electrostatic discharge stresses, *IEEE Electron Device Letters*, 31 (2010) 915–917.
146. S. Bangsaruntip, G.M. Cohen, A. Majumdar, J.W. Sleight, Universality of short-channel effects in undoped-body silicon nanowire MOSFETs, *IEEE Electron Device Letters*, 31 (2010) 903–905.
147. K. Majumdar, N. Bhat, P. Majhi, R. Jammy, Effects of parasitics and interface traps on ballistic nanowire FET in the ultimate quantum capacitance limit, *IEEE Transactions on Electron Devices*, 57 (2010) 2264–2273.
148. N. Clément, K. Nishiguchi, A. Fujiwara, D. Vuillaume, Evaluation of a gate capacitance in the Sub-aF range for a chemical field-effect transistor with a Si nanowire channel, *IEEE Transactions on Nanotechnology*, 10 (2011) 1172–1179.
149. J. Chen, T. Saraya, T. Hiramoto, Experimental investigations of electron mobility in silicon nanowire nMOSFETs on (110) silicon-on-insulator, *IEEE Electron Device Letters*, 30 (2009) 1203–1205.
150. D. Sacchetto, Y. Leblebici, G. De Micheli, Ambipolar gate-controllable SiNW FETs for configurable logic circuits with improved expressive capability, *IEEE Electron Device Letters*, 33 (2011) 143–145.
151. J.-H. Ahn, S.-J. Choi, J.-W. Han, T.J. Park, S.Y. Lee, Y.-K. Choi, Investigation of size dependence on sensitivity for nanowire FET biosensors, *IEEE Transactions on Nanotechnology*, 10 (2011) 1405–1411.
152. L. Mu, Y. Chang, S.D. Sawtelle, M. Wipf, X. Duan, M.A. Reed, Silicon nanowire field-effect transistors: A versatile class of potentiometric nanobiosensors, *IEEE Access*, 3 (2015) 287–302.
153. K. Nayak, M. Bajaj, A. Konar, P.J. Oldiges, K. Natori, H. Iwai, K.V. Murali, V.R. Rao, CMOS logic device and circuit performance of Si gate all around nanowire MOSFET, *IEEE Transactions on Electron Devices*, 61 (2014) 3066–3074.

154. F. Puppo, M.-A. Doucey, J.-F. Delaloye, T.S. Moh, G. Pandraud, P.M. Sarro, G. DeMicheli, S. Carrara, SiNW-FET in-air biosensors for highly sensitive and specific detection in breast tumor extract, *IEEE Sensors Journal*, 16 (2015) 3374–3381.
155. D. Kwon, J.H. Lee, S. Kim, R. Lee, H. Mo, J. Park, D.H. Kim, B.-G. Park, Drift-free pH detection with silicon nanowire field-effect transistors, *IEEE Electron Device Letters*, 37 (2016) 652–655.
156. K. Shimanovich, T. Coen, Y. Vaknin, A. Henning, J. Hayon, Y. Roizin, Y. Rosenwaks, CMOS compatible electrostatically formed nanowire transistor for efficient sensing of temperature, *IEEE Transactions on Electron Devices*, 64 (2017) 3836–3840.
157. L.E. Froberg, C. Rehnstedt, C. Thelander, E. Lind, L.-E. Wernersson, L. Samuelson, Heterostructure barriers in wrap gated nanowire FETs, *IEEE Electron Device Letters*, 29(2008) 981–983.
158. J. Yang, J. He, F. Liu, L. Zhang, F. Liu, X. Zhang, M. Chan, A compact model of silicon-based nanowire MOSFETs for circuit simulation and design, *IEEE Transactions on Electron Devices*, 55 (2008) 2898–2906.
159. M. Curreli, R. Zhang, F.N. Ishikawa, H.-K. Chang, R.J. Cote, C. Zhou, M.E. Thompson, Real-time, label-free detection of biological entities using nanowire-based FETs, *IEEE Transactions on Nanotechnology*, 7 (2008) 651–667.
160. E. Gnani, A. Gnudi, S. Reggiani, G. Baccarani, Effective mobility in nanowire FETs under quasi-ballistic conditions, *IEEE Transactions on Electron Devices*, 57 (2009) 336–344.
161. J. Wang, A. Rahman, A. Ghosh, G. Klimeck, M. Lundstrom, On the validity of the parabolic effective-mass approximation for the IV calculation of silicon nanowire transistors, *IEEE Transactions on Electron Devices*, 52 (2005) 1589–1595.
162. Y. Li, H.-M. Chou, J.-W. Lee, Investigation of electrical characteristics on surrounding-gate and omega-shaped-gate nanowire FinFETs, *IEEE Transactions on Nanotechnology*, 4 (2005) 510–516.
163. H.-C. Lin, H.-H. Hsu, C.-J. Su, T.-Y. Huang, A novel multiple-gate polycrystalline silicon nanowire transistor featuring an inverse-T gate, *IEEE Electron Device Letters*, 29 (2008) 718–720.
164. N. Neophytou, A. Paul, M.S. Lundstrom, G. Klimeck, Bandstructure effects in silicon nanowire electron transport, *IEEE Transactions on Electron Devices*, 55 (2008) 1286–1297.
165. K.H. Cho, S.D. Suk, Y.Y. Yeoh, M. Li, K.H. Yeo, D.-W. Kim, D. Park, W.-S. Lee, Y.C. Jung, B.H. Hong, Temperature-dependent characteristics of cylindrical gate-all-around twin silicon nanowire MOSFETs (TSNWFETs), *IEEE Electron Device Letters*, 28 (2007) 1129–1131.
166. R. Wang, R. Huang, Y. He, Z. Wang, G. Jia, D.-W. Kim, D. Park, Y. Wang, Characteristics and fluctuation of negative bias temperature instability in Si nanowire field-effect transistors, *IEEE Electron Device Letters*, 29 (2008) 242–245.
167. S.C. Rustagi, N. Singh, Y. Lim, G. Zhang, S. Wang, G. Lo, N. Balasubramanian, D.-L. Kwong, Low-temperature transport characteristics and quantum-confinement effects in gate-all-around Si-nanowire N-MOSFET, *IEEE Electron Device Letters*, 28 (2007) 909–912.
168. L. Pecchia, L. Salamandra, B. Latessa, T. Aradi, A. Frauenheim, Di Carlo, Atomistic modeling of gate-all-around Si-nanowire field-effect transistors, *IEEE Transactions on Electron Devices*, 54 (2007) 3159–3167.
169. N. Neophytou, A. Paul, G. Klimeck, Bandstructure effects in silicon nanowire hole transport, *IEEE Transactions on Nanotechnology*, 7 (2008) 710–719.
170. E. Gnani, A. Gnudi, S. Reggiani, M. Luisier, G. Baccarani, Band effects on the transport characteristics of ultrascaled SNW-FETS, *IEEE Transactions on Nanotechnology*, 7 (2008) 700–709.

171. N. Singh, A. Agarwal, L. Bera, T. Liow, R. Yang, S. Rustagi, C. Tung, R. Kumar, G. Lo, N. Balasubramanian, High-performance fully depleted silicon nanowire (diameter/spl les/5nm) gate-all-around CMOS devices, *IEEE Electron Device Letters*, 27 (2006) 383–386.
172. W. Lu, P. Xie, C.M. Lieber, Nanowire transistor performance limits and applications, *IEEE Transactions on Electron Devices*, 55 (2008) 2859–2876.
173. D.K. Ferry, M.J. Gilbert, R. Akis, Some considerations on nanowires in nanoelectronics, *IEEE Transactions on Electron Devices*, 55 (2008) 2820–2826.
174. D. Kim, Y. Jung, M. Park, B. Kim, S. Hong, M. Choi, M. Kang, Y. Yu, D. Whang, S. Hwang, Electrical characteristics of the back gated bottom-up silicon nanowire FETs, *IEEE Transactions on Nanotechnology*, 7 (2008) 683–687.
175. M. Lenzi, P. Palestri, E. Gnani, S. Reggiani, A. Gnudi, D. Esseni, L. Selmi, G. Baccarani, Investigation of the transport properties of silicon nanowires using deterministic and Monte Carlo approaches to the solution of the Boltzmann transport equation, *IEEE Transactions on Electron Devices*, 55 (2008) 2086–2096.
176. B.H. Hong, Y.C. Jung, J.S. Rieh, S.W. Hwang, K.H. Cho, K. Yeo, S. Suk, Y. Yeoh, M. Li, D.-W. Kim, Possibility of transport through a single acceptor in a gate-all-around silicon nanowire PMOSFET, *IEEE Transactions on Nanotechnology*, 8 (2009) 713–717.
177. T.-Y. Liow, K.-M. Tan, R.T. Lee, M. Zhu, B.L.-H. Tan, N. Balasubramanian, Y.-C. Yeo, Germanium source and drain stressors for ultrathin-body and nanowire field-effect transistors, *IEEE Electron Device Letters*, 29 (2008) 808–810.
178. H.-H. Hsu, H.-C. Lin, L. Chan, T.-Y. Huang, Threshold-voltage fluctuation of double-gated poly-Si nanowire field-effect transistor, *IEEE Electron Device Letters*, 30 (2009) 243–245.
179. G. Liang, D. Kienle, S.K. Patil, J. Wang, A.W. Ghosh, S.V. Khare, Impact of structure relaxation on the ultimate performance of a small diameter, n-Type $\langle 110\rangle $ Si-Nanowire MOSFET, *IEEE Transactions on Nanotechnology*, 6 (2007) 225–229.
180. Y. Jiang, T. Liow, N. Singh, L. Tan, G.Q. Lo, D.S. Chan, D.L. Kwong, Nickel salicided source/drain extensions for performance improvement in ultrascaled (sub 10 nm) Si–nanowire transistors, *IEEE Electron Device Letters*, 30 (2009) 195–197.
181. E. Gnani, A. Gnudi, S. Reggiani, G. Baccarani, Quasi-ballistic transport in nanowire field-effect transistors, *IEEE Transactions on Electron Devices*, 55 (2008) 2918–2930.
182. W.-W. Fang, N. Singh, L.K. Bera, H.S. Nguyen, S.C. Rustagi, G. Lo, N. Balasubramanian, D.-L. Kwong, Vertically stacked SiGe nanowire array channel CMOS transistors, *IEEE Electron Device Letters*, 28 (2007) 211–213.
183. M. Shin, Quantum simulation of device characteristics of silicon nanowire FETs, *IEEE Transactions on Nanotechnology*, 6 (2007) 230–237.
184. E.B. Ramayya, D. Vasileska, S.M. Goodnick, I. Knezevic, Electron mobility in silicon nanowires, *IEEE Transactions on Nanotechnology*, 6 (2007) 113–117.
185. E. Gnani, S. Reggiani, M. Rudan, G. Baccarani, Effects of high-$\kappa $(HfO $ _2 $) gate dielectrics in double-gate and cylindrical-nanowire FETs scaled to the ultimate technology nodes, *IEEE Transactions on Nanotechnology*, 6 (2007) 90–96.
186. H.-C. Lin, M.-H. Lee, C.-J. Su, S.-W. Shen, Fabrication and characterization of nanowire transistors with solid-phase crystallized poly-Si channels, *IEEE Transactions on Electron Devices*, 53 (2006) 2471–2477.
187. D. Son Myeong, H. Kim, H. Shin, Analysis of self-heating effect in DC/AC mode in multi-channel GAA-field effect transistor, *IEEE Transactions on Electron Devices*, 66 (2019) 4631–4637.
188. B. Liu, H.-S. Wong, M. Yang, Y.-C. Yeo, Strained silicon nanowire p-channel FETs with diamond-like carbon liner stressor, *IEEE Electron Device Letters*, 31 (2010) 1371–1373.

189. A.B. Sachid, H.-Y. Lin, C. Hu, Nanowire FET with corner spacer for high-performance, energy-efficient applications, *IEEE Transactions on Electron Devices*, 64 (2017) 5181–5187.
190. J.-H. Bae, D. Kwon, N. Jeon, S. Cheema, A.J. Tan, C. Hu, S. Salahuddin, Highly scaled, high endurance, Ω-gate, nanowire ferroelectric FET memory transistors, *IEEE Electron Device Letters*, 41 (2020) 1637–1640.
191. P.R. Kumar, S. Mahapatra, Quantum threshold voltage modeling of short channel quad gate silicon nanowire transistor, *IEEE Transactions on Nanotechnology*, 10 (2009) 121–128.
192. Y. Su, J. Lai, L. Sun, Investigation of self-heating effects in vacuum gate dielectric gate-all-around vertically stacked silicon nanowire field effect transistors, *IEEE Transactions on Electron Devices*, 67 (2020) 4085–4091.
193. G. Indalecio, A.J. García-Loureiro, M.A. Elmessary, K. Kalna, N. Seoane, Spatial sensitivity of Silicon GAA nanowire FETs under line edge roughness variations, *IEEE Journal of the Electron Devices Society*, 6 (2018) 601–610.
194. D. Nagy, G. Indalecio, A.J. García-Loureiro, G. Espiñeira, M.A. Elmessary, K. Kalna, N. Seoane, Drift-diffusion versus Monte Carlo simulated ON-current variability in nanowire FETs, *IEEE Access*, 7 (2019) 12790–12797.
195. W. Liu, J.J. Liou, A. Chung, Y.-H. Jeong, W.-C. Chen, H.-C. Lin, Electrostatic discharge robustness of Si nanowire field-effect transistors, *IEEE Electron Device Letters*, 30 (2009) 969–971.
196. S. Poli, M.G. Pala, T. Poiroux, S. Deleonibus, G. Baccarani, Size dependence of surface-roughness-limited mobility in silicon-nanowire FETs, *IEEE Transactions on Electron Devices*, 55 (2008) 2968–2976.
197. H.-H. Hsu, T.-W. Liu, L. Chan, C.-D. Lin, T.-Y. Huang, H.-C. Lin, Fabrication and characterization of multiple-gated poly-Si nanowire thin-film transistors and impacts of multiple-gate structures on device fluctuations, *IEEE Transactions on Electron Devices*, 55 (2008) 3063–3069.
198. C. Buran, M.G. Pala, M. Bescond, M. Dubois, M. Mouis, Three-dimensional real-space simulation of surface roughness in silicon nanowire FETs, *IEEE Transactions on Electron Devices*, 56 (2009) 2186–2192.
199. M.G. Cresti S. Pala, M. Poli, G. Mouis, Ghibaudo, A comparative study of surface-roughness-induced variability in silicon nanowire and double-gate FETs, *IEEE Transactions on Electron Devices*, 58 (2011) 2274–2281.
200. W.-T. Lai, C.-W. Wu, C.-C. Lin, P.-W. Li, Analysis of carrier transport in trigate Si nanowire MOSFETs, *IEEE Transactions on Electron Devices*, 58 (2011) 1336–1343.
201. M.G. Bekaddour, N.-E. Pala, G. Chabane-Sari, Ghibaudo, Deterministic method to evaluate the threshold voltage variability induced by discrete trap charges in Si-nanowire FETs, *IEEE Transactions on Electron Devices*, 59 (2012) 1462–1467.
202. R. Coquand, M. Casse, S. Barraud, D. Cooper, V. Maffini-Alvaro, M.-P. Samson, S. Monfray, F. Boeuf, G. Ghibaudo, O. Faynot, Strain-induced performance enhancement of trigate and omega-gate nanowire FETs scaled down to 10-nm width, *IEEE Transactions on Electron Devices*, 60 (2012) 727–732.
203. K. Nayak, S. Agarwal, M. Bajaj, K.V. Murali, V.R. Rao, Random dopant fluctuation induced variability in undoped channel Si gate all around nanowire n-MOSFET, *IEEE Transactions on Electron Devices*, 62 (2015) 685–688.
204. K. Nayak, S. Agarwal, M. Bajaj, P.J. Oldiges, K.V. Murali, V.R. Rao, Metal-gate granularity-induced threshold voltage variability and mismatch in Si gate-all-around nanowire n-MOSFETs, *IEEE Transactions on Electron Devices*, 61 (2014) 3892–3895.
205. M. Bajaj, K. Nayak, S. Gundapaneni, V.R. Rao, Effect of metal gate granularity induced random fluctuations on Si Gate-All-Around nanowire MOSFET 6-T SRAM cell stability, *IEEE Transactions on Nanotechnology*, 15 (2016) 243–247.

206. J.-S. Yoon, K. Kim, T. Rim, C.-K. Baek, Performance and variations induced by single interface trap of nanowire FETs at 7-nm node, *IEEE Transactions on Electron Devices*, 64 (2016) 339–345.
207. G. Espiñeira, D. Nagy, G. Indalecio, A. García-Loureiro, K. Kalna, N. Seoane, Impact of gate edge roughness variability on FinFET and gate-all-around nanowire FET, *IEEE Electron Device Letters*, 40 (2019) 510–513.
208. N. Martinez, A.R. Seoane, J.R. Brown, A. Barker, Asenov, Variability in Si nanowire MOSFETs due to the combined effect of interface roughness and random dopants: A fully three-dimensional NEGF simulation study, *IEEE Transactions on Electron Devices*, 57 (2010) 1626–1635.
209. L.C. Hong, Y. Jung, S. Hwang, K. Cho, K. Yeo, D.-W. Kim, G. Jin, D. Park, S. Song, Temperature dependent study of random telegraph noise in gate-all-around PMOS silicon nanowire field-effect transistors, *IEEE Transactions on Nanotechnology*, 9 (2010)754–758.
210. S. Poli, M.G. Pala, Channel-length dependence of low-field mobility in silicon-nanowire FETs, *IEEE Electron Device Letters*, 30 (2009) 1212–1214.
211. E. Anju, I. Muneta, K. Kakushima, K. Tsutsui, H. Wakabayashi, Relaxation of self-heating-effect for stacked-nanowire FET and p/n-stacked 6T-SRAM layout, *IEEE Journal of the Electron Devices Society*, 6 (2018) 1239–1245.
212. R. Singh, K. Aditya, A. Veloso, B. Parvais, A. Dixit, Experimental evaluation of self-heating and analog/RF FOM in GAA-nanowire FETs, *IEEE Transactions on Electron Devices*, 66 (2019) 3279–3285.
213. A. Dasgupta Y.S. Agarwal, Chauhan, Unified compact model for nanowire transistors including quantum effects and quasi-ballistic transport, *IEEE Transactions on Electron Devices*, 64 (2017) 1837–1845.
214. S. Sahay, M.J. Kumar, Spacer design guidelines for nanowire FETs from gate-induced drain leakage perspective, *IEEE Transactions on Electron Devices*, 64 (2017) 3007–3015.
215. D.K. Sharma, A. Datta, Oxide edge trap density extraction in silicon nanowire MOSFET from tunnel current noise measurement in gated diode like arrangement, *IEEE Transactions on Device and Materials Reliability*, 20 (2020) 512–516.
216. J. Riffaud, M. Gaillardin, C. Marcandella, N. Richard, O. Duhamel, M. Martinez, M. Raine, P. Paillet, T. Lagutere, F. Andrieu, TID response of nanowire field-effect transistors: Impact of the back-gate bias, *IEEE Transactions on Nuclear Science*, 67(2020) 2172–2178.
217. M. Assif, G. Segev, Y. Rosenwaks, Dynamic and power performance of multiple state electrostatically formed nanowire transistors, *IEEE Transactions on Electron Devices*, 64 (2016) 571–578.
218. J. Park, H. Lee, S. Oh, C. Shin, Design for variation-immunity in sub-10-nm stacked-nanowire FETs to suppress LER-induced random variations, *IEEE Transactions on Electron Devices*, 63 (2016) 5048–5054.
219. A.K. Bansal, I. Jain, T.B. Hook, A. Dixit, Series resistance reduction in stacked nanowire FETs for 7-nm CMOS technology, *IEEE Journal of the Electron Devices Society*, 4 (2016) 266–272.
220. S. Sahay, M.J. Kumar, A novel gate-stack-engineered nanowire FET for scaling to the sub-10-nm regime, *IEEE Transactions on Electron Devices*, 63 (2016) 5055–5059.
221. S. Pregl, A. Heinzig, L. Baraban, G. Cuniberti, T. Mikolajick, W.M. Weber, Printable parallel arrays of Si nanowire Schottky-barrier-FETs with tunable polarity for complementary logic, *IEEE Transactions on Nanotechnology*, 15 (2016) 549–556.
222. P. Zheng, D. Connelly, F. Ding, T.-J.K. Liu, FinFET evolution toward stacked-nanowire FET for CMOS technology scaling, *IEEE Transactions on Electron Devices*, 62 (2015) 3945–3950.

223. C.-C. Yang, W.-H. Huang, T.-Y. Hsieh, T.-T. Wu, H.-H. Wang, C.-H. Shen, W.-K. Yeh, J.-H. Shiu, Y.-H. Chen, M.-C. Wu, High gamma value 3D-stackable HK/MG-stacked tri-gate nanowire poly-Si FETs with embedded source/drain and back gate using low thermal budget green nanosecond laser crystallization technology, *IEEE Electron Device Letters*, 37 (2016) 533–536.
224. K. Natori, Compact modeling of ballistic nanowire MOSFETs, *IEEE Transactions on Electron Devices*, 55 (2008) 2877–2885.
225. K. Rogdakis, S.-Y. Lee, M. Bescond, S.-K. Lee, E. Bano, K. Zekentes, 3C-silicon carbide nanowire FET: an experimental and theoretical approach, *IEEE Transactions on Electron Devices*, 55 (2008) 1970–1976.
226. W. Yang, S. Lee, G. Liang, R. Eswar, Z. Sun, D. Kwong, Temperature dependence of carrier transport of a silicon nanowire Schottky-barrier field-effect transistor, *IEEE Transactions on Nanotechnology*, 7 (2008) 728–732.

6 Investigation of Tunnel Field-Effect Transistors (TFETs) for Label-Free Biosensing

Suneet Kumar Agnihotri, Chithraja Rajan, Anil Lodhi, and Dip Prakash Samajdar

CONTENTS

6.1 INTRODUCTION

Human tissue is composed of a wide variety of biological materials or biomolecules, such as water, fat, and proteins, which exhibit distinguishable dielectric properties under the influence of an electromagnetic wave (Mahalaxmi et al., 2020). Therefore, accurate detection of biomolecules is extremely helpful in disease recognition and treatment. The COVID-19 pandemic taught us that early infection recognition and isolation is the primary step towards preventing such communal diseases (Zaki et al., 2012). Accuracy and speed are the two prime factors for determining biosensor sensitivity. In this regard, Bio-FETs (Im et al., 2007) are considered label-free biosensors in the healthcare industry; hence, transistor technology can aid with smaller, low-cost, and low-power biosensor production.

In 1970, the first FET-based biosensor using an ion-sensitive FET (ISFET) biosensor, proposed by Bergveld, detected an electrical property variation in the presence of charged biomolecules (Bergveld, 1985). However, it was unable to detect neutral biomolecules and so dielectrically modulated (DM) FETs were developed, inspired by the working principle of MOSFETs, where a potential change over the gate electrode drives the device into accumulation, depletion, and inversion states.

DOI: 10.1201/9781003155751-6

Similarly, biomolecules carrying a positive or negative charge inside the cavity region, situated above the gate region, shift the device potential and generate a drain current in a positive or negative direction. The variation of the dielectric constant value in the biomolecule changes the effective dielectric constant value near the channel/oxide interface, and hence causes fluctuation in drain current and threshold voltages. These drain current or threshold voltage changes are considered to be biosensor sensitivity parameters. The sensitivity of a Bio-FET is mainly affected by three factors: (i) biomolecule concentration, (ii) area of contact, and (iii) dielectric or/and charge values (Syu et al., 2018). Therefore, in a DMFET, biomolecules are introduced into a cavity located below the gate electrode inside the dielectric oxide (Kalra et al., 2016), and electrical device characteristics change with the variation in dielectric constant (K) and/or charge (ρ) value with respect to the gate voltage. However, the effective benefits of Bio-FETs are difficult to achieve because of the short channel effects (SCEs) that rise with the continuous downscaling of MOSFETs (Narang et al., 2015). Drain-induced barrier lowering (DIBL), hot carrier effect (HCE), velocity saturation, threshold roll-off, and punch through are some of the SCEs that drastically reduce gate control, increase OFF-current, limit subthreshold swing (SS) > 60 mV/decade, and hence higher power consumption restricting the use of biosensors in low-power biomedical applications. Also, small devices need complicated fabrication procedures that induce process variations and hence reduce sensor accuracy (Narang et al., 2012). Alternatively, emerging low-power devices like TFETs provide a prominent solution to these problems, since they work on band-to-band tunneling (BTBT) rather than the drift–diffusion phenomenon (Singh et al., 2016). Therefore, a biosensor-based Bio-TFET exhibits better SS, I_{ON}/I_{OFF} ratio, sensitivity, and response than a Bio-MOSFET (Dwivedi and Kranti, 2018). However, TFETs suffer from low ON-current and higher ambipolarity, which reduces biosensor efficiency and hence advanced amplifying stages are required at the sensor output (Das et al., 2018). Instead, optimized TFET structures such as the use of hetero-dielectric oxides, exploration of III-V hetero-materials in the source and drain/channel regions, gate all around (GAA)/ nanowire structures and doppingless/junctionless devices, will be a breakthrough for biosensing applications (Lodhi et al., 2020; Reddy et al., 2020). Therefore, in this chapter, dual-gate source electrode dielectrically modulated TFETs (DG-SE-DM-TFETs) and hetero-material source electrode dielectrically modulated TFETs (HM-SE-DM-TFETs) for biosensing applications are investigated in details.

6.2 SENSITIVITY ANALYSIS

Bio-TFET sensitivity toward biomolecules is obtained by analyzing electrical device characteristics for various dielectric constants and charge values for changing gate voltage. Drain current and threshold voltage are some of the measurable quantities used to detect biomolecules (Wadhwa et al., 2019). Whenever a biomolecule is brought into contact with the device, a surface current flow is initiated by virtue of the gate voltage as compared to the idle condition. The idle condition means there is an absence of biomolecule or there is the presence of air. Ideally, in the absence of a biomolecule, biosensors should produce no drain current and dissipate no power (Soni and Sharma, 2019). However, a Bio-FET produces some ON-current in air or $K = 1$ condition and is always to be kept small. Once a biomolecule is introduced, depending on its K or ρ value, the drain current initiates, which increases with the rising gate voltage. It is observed that as the K or ρ value increases, the drain current improves, and device sensitivity is obtained using Eq. (6.1):

$$S_d = \left(\frac{I_{ds}^{bio} - I_{ds}^{air}}{I_{ds}^{air}} \right) \tag{6.1}$$

Similarly, for threshold voltage or any other electrical characteristic, the drain current term is replaced with the corresponding electrical term, and sensitivity analysis is performed in a similar way as in Kumar et al (2020).

6.3 CASE STUDY 1: DUAL-GATE SOURCE ELECTRODE DIELECTRICALLY MODULATED TFETs FOR BIOSENSING APPLICATION

Figures 6.1(a) and 6.1(b) demonstrate cross-sectional views of a DG-SE-DM-TFET biosensor and a short-gate dual-metal source electrode dielectrically modulated (SDM SE DM) TFET biosensor, respectively. Table 6.1 shows the device dimensions and design parameters of the proposed and conventional models (Soni and Sharma, 2019). Two parameters, namely the dielectric constant and charge density, the governing parameters for gate modulation, are used to detect biomolecules in the nanogap cavity. The nanogap cavity is initially devoid of biomolecules, implying that it is filled with air ($K=1$), while $K>1$ indicates the presence of biomolecules in the cavity. The enhancement of gate capacitance with an increase in dielectric constant from $K=1$ (air) to $K=3$ and 5 (biomolecules) causes an increase in the drain current. For DNA analysis, the action of the biomolecules in the nanogap cavity is studied in terms of negative charge density (ρ). The gate area is divided into two parts, G1 and G2, with separate work functions; the work function of G1 is higher than the

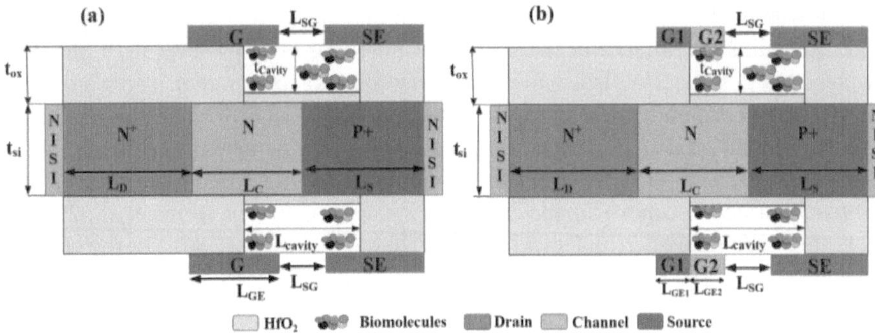

FIGURE 6.1 *Schematic* of the device architecture of (a) conventional device and (b) proposed device.

TABLE 6.1
The Parameters and Device Dimensions Used in Simulation

Parameters	Unit	Conventional	Proposed
Source Doping (N_A)	cm^{-3}	1×10^{20}	1×10^{20}
Channel Doping (N_D)	cm^{-3}	1×10^{17}	1×10^{17}
Drain Doping (N_D)	cm^{-3}	1×10^{20}	1×10^{20}
Source length (L_S)	nm	100	100
Channel length (L_C)	nm	50	50
Drain length (L_D)	nm	100	100
Gate electrode length (L_{GE})	nm	50	-
Gate1-electrode length (L_{GE1})	nm	-	15
Gate2-electrode length (L_{GE2})	nm	-	15
Silicon Thickness (t_{si})	nm	10	10
Oxide Thickness (t_{ox})	nm	6	6
Gate-to-source electrode space (L_{GS})	nm	4	4
Gate electrode work function ($Ø_G$)	eV	4.53	-
Gate1 electrode work function ($Ø_{G1}$)	eV	-	4.53
Gate2 electrode work function ($Ø_{G2}$)	eV	-	3.8
Source electrode work function ($Ø_{SE}$)	eV	4.53	4.53
Voltage at source electrode (V_{SE})	V	−1.2	−1.2
Thickness of cavity (t_{cavity})	nm	5.5	5.5
Length of cavity (L_{cavity})	nm	30	30

work function of G2 to make the junction abrupt and attain a high ON-current and low OFF-current for the dual-metal concept used in this chapter. Also, the negative voltage applied to the source area produces holes at the Si/HfO_2 interface, forming a plasma layer of holes, and thereby overcoming the solubility limit. The layer of holes in the source area produces an abrupt source/channel junction, which raises the drain current. A TCAD 2D Device ATLAS SILVACO simulator is used for simulation (ATLAS, 2014). The tool includes a nonlocal BTBT and bandgap narrowing (BGN) model, which calculates the carrier generation rate at the tunneling junction in general. Tunneling takes place at the source/channel interface. As a result, a quantum tunneling region is established to allow tunneling. Recombination and a minority charge carrier are both taken care of by the SRH and AUGER models, respectively (Agnihotri et al., 2020). Furthermore, the quantum confinement model for the quantum effect, as well as the Fermi Dirac and Newton trap models, ensure that the simulated effects are more numerically efficient and accurate (Verma et al., 2017).

Figures 6.2(a) and 6.2(b) depict the energy band diagrams (EBDs) of the conventional and proposed devices for dielectric constants $K=1$ (air) and $K>1$ (biomolecules), along the X-axis. Because of the rise in dielectric constant as the biomolecules reach the cavity, band bending rises. This band bending happens when biomolecules in the nanogap cavity become immobilized, causing the barrier width to decrease. However, band bending increases in the proposed system because the work function

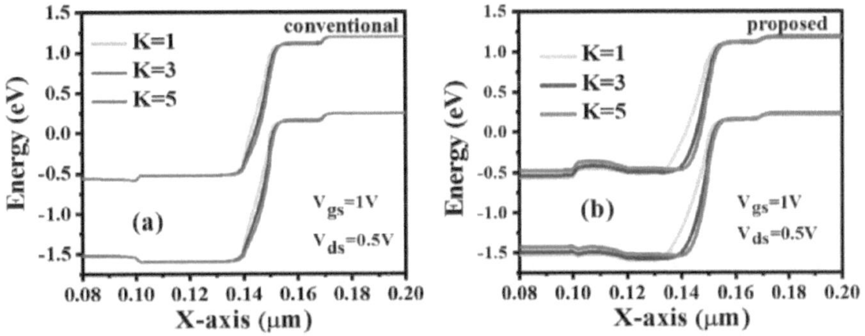

FIGURES 6.2 EBDs along the X-axis of (a) conventional and (b) proposed biosensors for different dielectric constants.

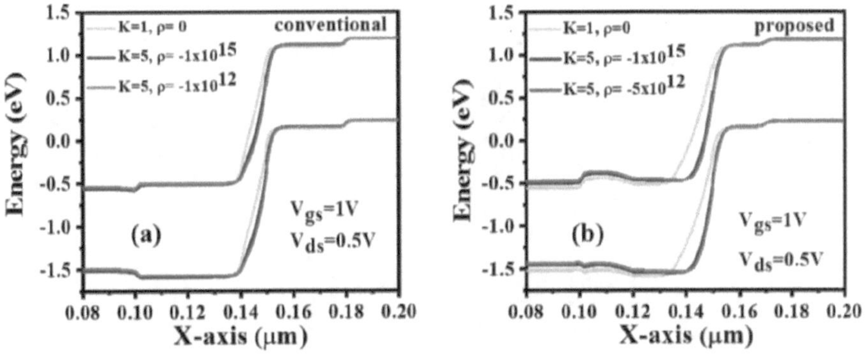

FIGURES 6.3 EBDs along the X-axis of (a) conventional and (b) proposed devices for different charge densities.

of G2 at the source/channel junction is lower than that of G1, lowering the tunneling barrier and improving the gate-control ability at the source/channel junction.

The superiority of TFET-based biosensors over FET-based biosensors lies in their ability to detect charged molecules. The energy band diagrams of conventional and proposed models for charge density variance are seen in Figures 6.3(a) and 6.3(b). However, as seen in Figure 6.3, the barrier width increases as negative charge density increases. The steep increment in band bending is one of the highlights of the proposed device, which is significant compared to the conventional one.

Figures 6.4(a) and 6.4(b) display the I_d–V_{gs} characteristics for the conventional and proposed devices with variable dielectric constants and charge densities, respectively. I_d of the conventional device is less than that of the proposed biosensor to indicate the presence of biomolecules ($K > 1$) since the work function of the gate electrode G2 at the source end is less than that of the gate electrode G1 at the drain end. This reduces the tunnel barrier width and improves the gate-control capability at the source/channel interface. Figure 6.4(b) displays the I_d–V_{gs} characteristics for

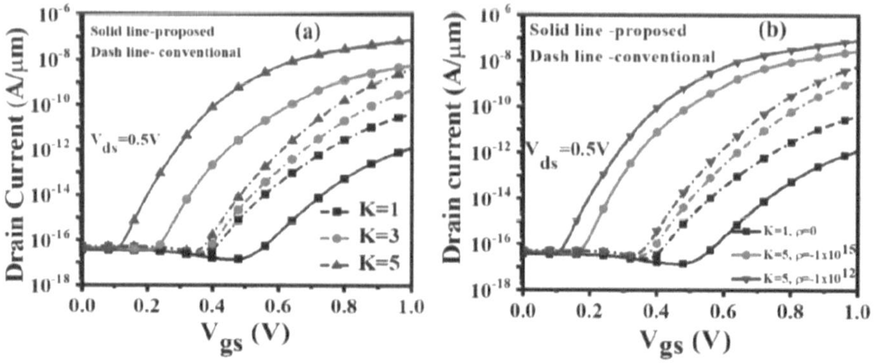

FIGURES 6.4 I_d–V_{gs} characteristics of conventional and proposed devices with (a) different dielectric constants and (b) different charge densities for $K = 5$.

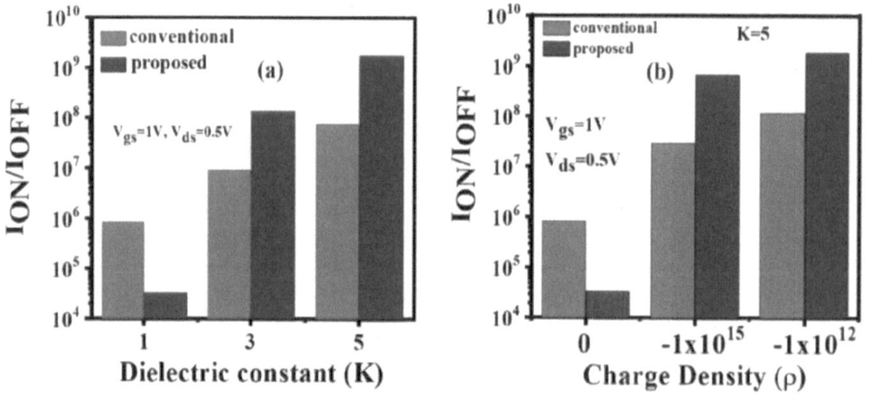

FIGURES 6.5 I_{ON}/I_{OFF} ratio with different (a) dielectric constant and (b) charge density for the conventional and proposed device.

various charge densities in the cavity, demonstrating that the proposed system has a higher current driving capacity.

For TFET devices, the I_{ON}/I_{OFF} ratio is an important parameter to note. Figures 6.5(a) and 6.5(b) depicts I_{ON}/I_{OFF} ratio plots as a function of the dielectric constant and charge density, respectively. We have calculated the I_{ON}/I_{OFF} current with the help of I_{ds}–V_{gs} characteristics. Note that the I_{ON}/I_{OFF} ratio shows a sharp increase as the K value (dielectric constant) increases and a higher I_{ON}/I_{OFF} ratio can be observed for the proposed device compared to the conventional device. Also, the I_{ON}/I_{OFF} ratio increases as negative charge density increases and better results are obtained for the proposed device.

Sensitivity is calculated in Eq. (6.1), as discussed in Section 6.2. Figures 6.6(a) and 6.6(b) illustrate the comparative study of I_{ds} sensitivity over V_{gs} for various dielectric constants (K) and different charge densities, respectively. The sensitivity

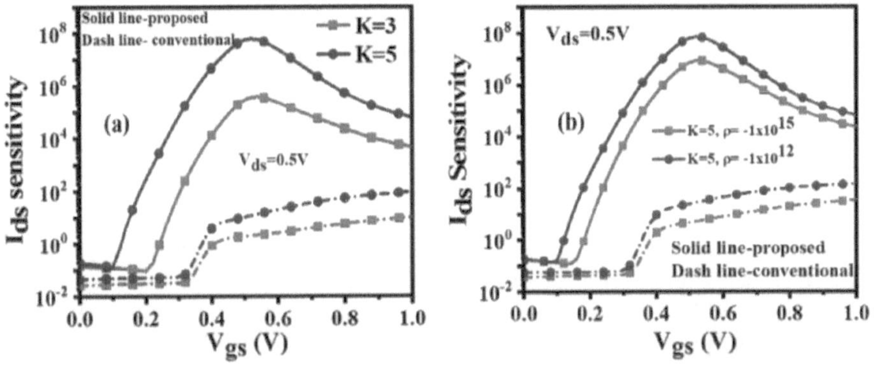

FIGURES 6.6 *Sensitivity* characteristics of I_d over V_{gs} of conventional and proposed devices with (a) different dielectric constants and (b) different charge densities for $K = 5$.

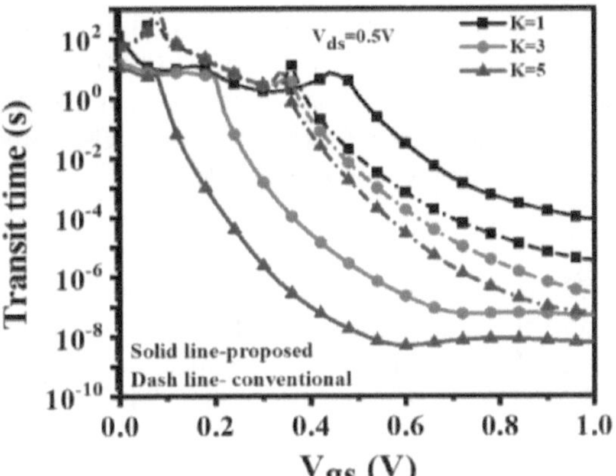

FIGURE 6.7 Transit time as a function of gate voltage for different dielectric constants for the conventional and proposed device.

is analyzed with respect to the drain current when the nanogap cavity is devoid of biomolecules ($K = 1$). When the cavity is completely filled with biomolecules with different dielectric constants, we can see substantial changes in the sensitivity parameter. With an increase in the dielectric constant of the biomolecules in the nanogap cavity, sensitivity increases accordingly. The same results can be seen for the different negatively charged biomolecules for fixed $K = 5$, and it can be seen that sensitivity is higher in the case of the proposed device because the work function of G2 at the source/channel junction is lower than that of G1. This lowers the tunnel barrier and improves the gate-control ability at the source/channel junction, which increases the drain current.

Transit time is another important parameter to study. It is defined as the time required to detect bio-molecular substances in the cavity region. Figure 6.7 depicts a plot of transit time over V_{gs} with different dielectric constants, and it can be observed that our proposed device detects the biomolecules in the nanogap cavity much faster than the conventional device for dielectric constant $K > 1$.

6.4 CASE STUDY 2: SOURCE ELECTRODE HETERO-MATERIAL DIELECTRICALLY MODULATED TFETs FOR BIOSENSING APPLICATION

This section is dedicated to HM TFET-based biosensors, a potential alternative to the double gate biosensor discussed in the previous section. It is obvious that constructing multiple gates in a device needs complex fabrication techniques. Also, uncontrollable process variations cause work function fluctuations, which severely affect biosensor performance. Hence, using an HM with a low band gap material in the source region and a high band gap material in drain/channel regions allows band gap steepening at the source/channel junction and hence allow more tunnel current for higher *K* and charge values as compared to the idle condition. Therefore, a HM-SE-DM-TFET provides better sensitivity with lesser fabrication complexities. Figure 6.8 shows the device structure of SiGe/Si-SE-DM-TFET, where SiGe is used in the source region and Si is used in drain/channel regions (Rajan et al., 2021). The device dimensions and models used are the same as in Case Study 1 in the previous section (Section 6.3). Here, a source electrode (SE) is also established over the source region, in which negative potential accumulates more holes near SiGe/oxide interface.

This further improves source/channel junction abruptness. Figure 6.9 shows that the gate/channel coupling strength improves as the *K* value increases and band steepness increases in the HM-SE-DM-TFET as compared to the conventional device. Therefore, Figure 6.10(a) shows that the change in drain current with respect to gate voltage is higher in a HM-SE-DM-TFET as compared to the conventional TFET and hence has a better sensing capability, as shown in Figure 6.10 (b).

A similar effect has been observed for positive and negative charge variations. Figure 6.11(a) shows that as the charge value increases, the drain current also

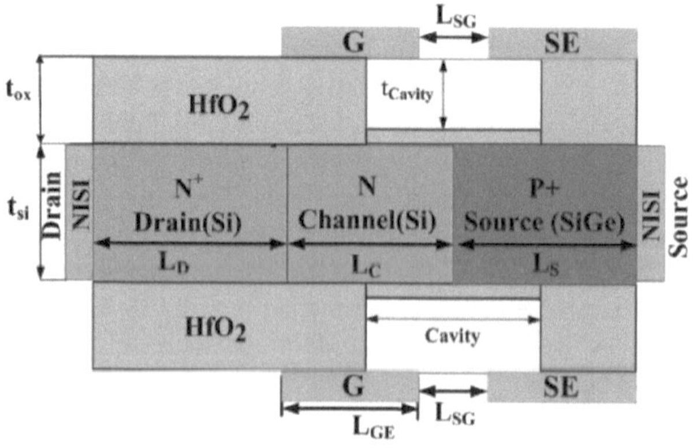

FIGURE 6.8 HM-SE-DM-TFET biosensor device structure.

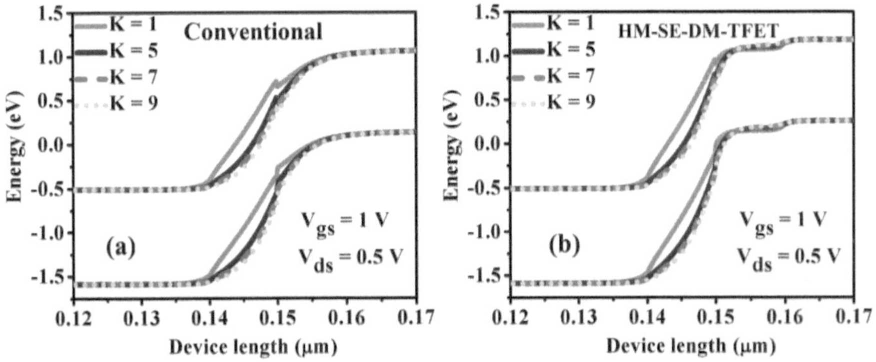

FIGURE 6.9 Energy band diagrams of (a) conventional and (b) HM-SE-DM-TFET biosensors for varying K values.

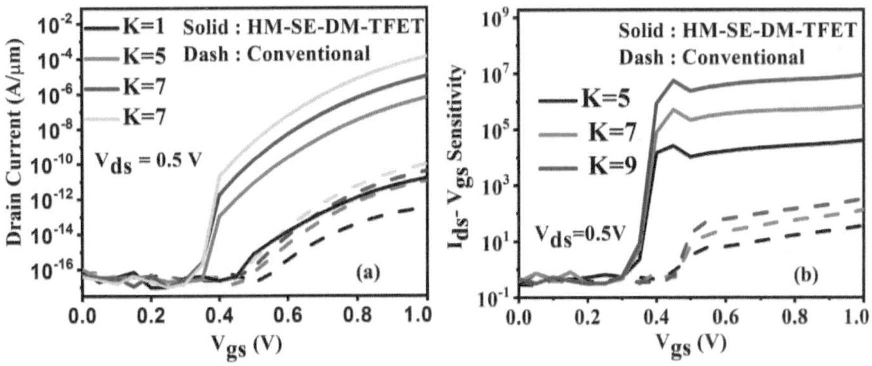

FIGURE 6.10 Comparison between conventional and HM-SE-DM-TFET (a) drain current and (b) I_{ds}–V_{gs} sensitivity for varying K value.

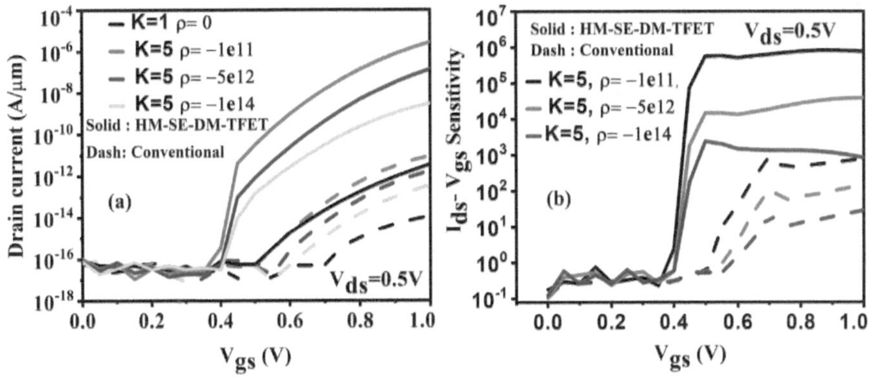

FIGURE 6.11 Comparison between conventional and HM-SE-DM-TFET (a) drain current and (b) I_{ds}–V_{gs} sensitivity for varying charge value.

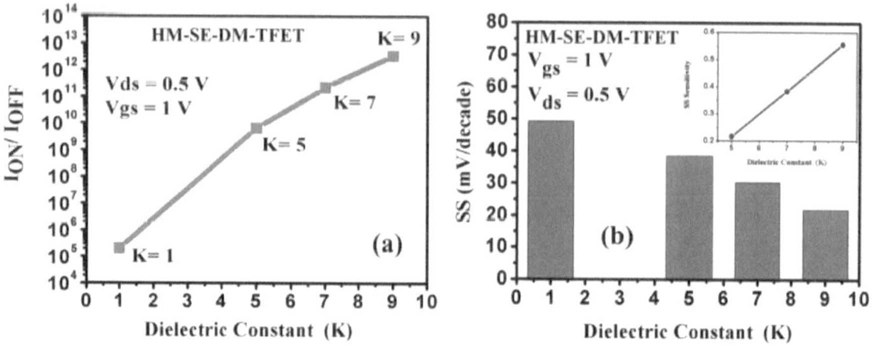

FIGURE 6.12 Variation of (a) I_{ON}/I_{OFF} ratio and (b) SS in HM-SE-DM-TFET for increasing *K* values.

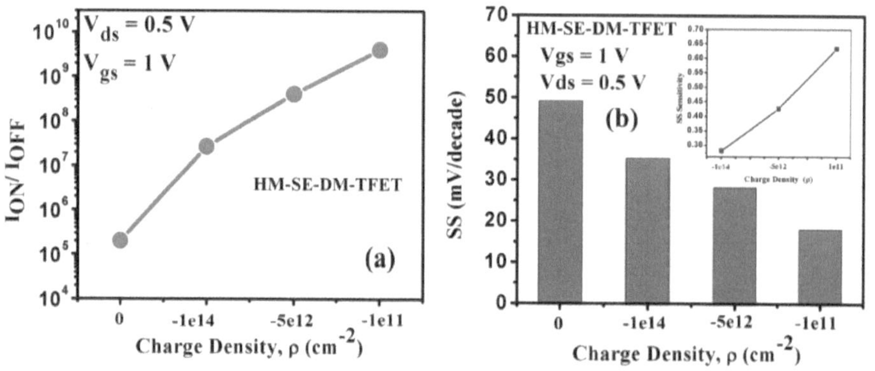

FIGURE 6.13 Variation of (a) I_{ON}/I_{OFF} ratio and (b) SS in HM-SE-DM-TFET for increasing ρ values.

increases; hence, the drain current sensitivity improves in the proposed device compared to the conventional device, as shown in Figure 6.11(b).

Finally, Figures 6.12 and 6.13 show that the I_{ON}/I_{OFF} ratio and SS improve for the proposed device as the *K* and ρ values increase, confirming that a HM-SE-DM-TFET is a power-efficient device with better sensing capabilities.

6.5 CONCLUSIONS

The importance of a Bio-FET-based sensor, which can be used to detect a wide variety of positively or negatively charged biomolecules with varied dielectric properties, is discussed. When body fluid is introduced into the biosensor cavity, the Bio-FET can detect and distinguish biomolecules through drain current or threshold voltage variation as the effective charge value within the device changes and shifts device characteristics. Also, a Bio-FET is a label-free, low-cost, low-power, and low-area device that can be easily used in bio-implants. However, CMOS Bio-FETs face SCEs and are therefore not recommended for sub-90 nm technology. Also, the sensitivity offered by CMOS Bio-FETs is not up to the mark. Alternatively, TFET-based biosensors provide better sensitivity along with low-power and high-density features. Therefore, this chapter first discussed DG-SE-DM-TFETs, which promise to provide better sensitivity, I_{ON}/I_{OFF} ratio, and SS. Instead of fabricating dual gates in a Si-TFET, a hetero-material SE TFET-based biosensor was also discussed that promises to provide better sensitivity at low-power consumption with reduced fabrication complexity.

REFERENCES

Agnihotri, Suneet Kumar, Dip Prakash Samajdar, Chithraja Rajan, Ankam Srujan Yadav, and Gowri Gnanesh. "Performance analysis of gate engineered dielectrically modulated TFET biosensors." *International Journal of Electronics* 108 (2020): 607–622.

ATLAS Device Simulation Software, Silvaco Int., Santa Clara, CA, 2014.

Bergveld, Piet. "The impact of MOSFET-based sensors." *Sensors and Actuators* 8, no. 2 (1985): 109–127.

Das, Gyan Darshan, Guru Prasad Mishra, and Sidhartha Dash. "Impact of source-pocket engineering on device performance of dielectric modulated tunnel FET." *Superlattices and Microstructures* 124 (2018): 131–138.

Dwivedi, Praveen, and Abhinav Kranti. "Dielectric modulated biosensor architecture: Tunneling or accumulation based transistor?." *IEEE Sensors Journal* 18, no. 8 (2018): 3228–3235.

Im, H., X. J. Huang, B. Gu, and Y. K. Choi, "A dielectric-modulated field-effect transistor for biosensing," *Nature Nanotechnology* 2, no. 7 (2007): 430–434. DOI: 10.1038/nnano.2007.180

Kalra, Sumeet, M.J. Kumar and A. Dhawan, "Dielectric-modulated field effect transistors for DNA detection: Impact of DNA orientation", *IEEE Electron Device Letters* 37, no. 11 (2016): 1485–1488.

Kumar, Sandeep, Yashvir Singh, Balraj Singh, and Pramod Kumar Tiwari. "Simulation study of dielectric modulated dual channel trench gate TFET-based biosensor." *IEEE Sensors Journal* 20, no. 21 (2020): 12565–12573.

Lodhi, Anil, Chithraja Rajan, Amit Kumar Behera, Dip Prakash Samajdar, Deepak Soni, and Dharmendra Singh Yadav. "Sensitivity and sensing speed analysis of extended nano-cavity and source over electrode in Si/SiGe based TFET biosensor." *Applied Physics A* 126, no. 11 (2020): 1–8.

Mahalaxmi, Acharya, Bibhudendra, and Guru Prasad Mishra. "Design and analysis of dual-metal-gate double-cavity charge-plasma-TFET as a label free biosensor." *IEEE Sensors Journal* 20, no. 23 (2020): 13969–13975.

Narang, Rakhi, KV Sasidhar Reddy, Manoj Saxena, R. S. Gupta, and Mridula Gupta. "A dielectric-modulated tunnel-FET-based biosensor for label-free detection: Analytical modeling study and sensitivity analysis." *IEEE Transactions on Electron Devices* 59, no. 10 (2012): 2809–2817.

Narang, Rakhi, Manoj Saxena, and Mridula Gupta. "Comparative analysis of dielectric-modulated FET and TFET-based biosensor." *IEEE Transactions on Nanotechnology* 14, no. 3 (2015): 427–435.

Rajan, Chithraja, Dip Prakash Samajdar, and Anil Lodhi. "Investigation of DC, RF and linearity performances of III–V semiconductor-based electrically doped TFET for mixed signal applications." *Journal of Electronic Materials* 50, no. 4 (2021): 2348–2355.

Reddy, N. Nagendra, and Deepak Kumar Panda. "Simulation study of dielectric modulated dual material gate TFET based biosensor by considering ambipolar conduction." *Silicon* 13 (2021): 4545–45517.

Singh, Deepika, Sunil Pandey, Kaushal Nigam, Dheeraj Sharma, Dharmendra Singh Yadav, and Pravin Kondekar. "A charge-plasma-based dielectric-modulated junctionless TFET for biosensor label-free detection." *IEEE Transactions on Electron Devices* 64, no. 1 (2016): 271–278.

Soni, Deepak, and Dheeraj Sharma. "Design of NW TFET biosensor for enhanced sensitivity and sensing speed by using cavity extension and additional source electrode." *Micro & Nano Letters* 14, no. 8 (2019): 901–905.

Syu, Yu-Cheng, Wei-En Hsu, and Chih-Ting Lin. "Field-effect transistor biosensing: Devices and clinical applications." *ECS Journal of Solid State Science and Technology* 7, no. 7 (2018): Q3196.

Verma, Madhulika, Sukeshni Tirkey, Shivendra Yadav, Dheeraj Sharma, and Dharmendra Singh Yadav. "Performance assessment of a novel vertical dielectrically modulated TFET-based biosensor." *IEEE Transactions on Electron Devices* 64, no. 9 (2017): 3841–3848.

Wadhwa, Girish, Priyanka Kamboj, and Balwinder Raj. "Design optimisation of junctionless TFET biosensor for high sensitivity." *Advances in Natural Sciences: Nanoscience and Nanotechnology* 10, no. 4 (2019): 045001.

Zaki, Ali M., S. V. Boheemen, Theo M. Bestebroer, A. DME Osterhau, and Ron AM Fouchier, "Isolation of a novel coronavirus from a man with pneumonia in Saudi Arabia," *New England Journal of Medicine* 367, no. 19 (2012): 1814–1820. DOI: 10.1056/NEJMoa1211721

7 Analog and Linearity Analysis of Vertical Nanowire TFET

Sahil Sankhyan and Tarun Chaudhary

CONTENTS

7.1 INTRODUCTION

The high density of low-dimension devices is a positive trend toward supporting many usable devices with minimal power consumption. The semiconductor industry's principal goal is to reduce the size of transistors. [1] According to Moore's law, gadgets scale at an exponential rate. As a result, power dissipation and leakage currents increase as the system size is reduced to the submicron zone or below 10 nm. The increase in spontaneous dopant fluctuations (RDFs) and short-channel effects is another drawback to system growth. As a result, a low-power unit with lower short-channel effect strength is in high demand (SCEs) [2]. Though, to date, academics have tried a variety of solutions to address the aforementioned problems, there are a variety of further tactics or procedures that might be used to address these problems. One example of a technology based on MOS design engineering is tunnel field-effect transistors (TFETs), which have extremely small subthreshold characteristics and superior device behavior [3].

Traditional MOS-based devices have a subthreshold slope (SS) of 60 mV/decade, representing theoretical limits or Boltzmann limits, as well as the physics underpinning electron harmonic emission. TFETs are a promising technology with a low SS that can be used in energy-efficient circuits. The band-to-band tunneling principle underpins this TFET-based system's mechanics (BTBT). These devices have minimal SCEs and RDFs but a low ON-current value [4]. Eminent experts are working on a number of effective methods to address TFETs' major flaws. Low-bandgap materials and numerous unique topologies, such as dual-gate architecture, hetero-gate

DOI: 10.1201/9781003155751-7

material, III V compound material, cylindrical structure, gate-all-around (GAA) architecture, and nanotubes, can boost TFETs' low ON-current (I_{ON}) to a desirable range (NT). Among these ground breaking architectures, the GAA nanowire design is especially relevant and appropriate. Nanowire-based TFETs have a smaller area, a smaller diameter, and a greater rate of band-to-band tunneling than conventional TFETs (BTBT). Nanowire TFETs' steeper SS and higher I_{ON} are likewise comparable to planer units. [5]

According to a recent study, GAA nanowire TFETs give better gate control than nanotube (NT) designs with core gates. TFETs carry out charge transfer via interband tunneling, allowing the system to be conveniently turned on and off by managing the band-bending phenomenon. Quantum mechanical tunneling, in contrast to classical systems with leakage, frequently allows for a sharper turn-on [6].

The proposed design in this study is a nanowire GAA TFET, which integrates the work functions of two different drain source materials. I_{ON} is improved as a result of the proposed approach. Because of this, the tunneling area of nanowire TFETs is increased, resulting in a greater I_{ON}. [7] The proposed (NW-GAA-TFET) design features a higher I_{ON}/I_{OFF} current ratio (10^{14}), a higher I_{ON} of 3×10^{-5} Amp, and a lower I_{OFF} current of 10^{-19} Amp. With reference to the device radius, channel length, and V_{GS}, a simulation study of system aspects such as energy band diagram, hole concentration, electron concentration, tunneling rate of the band-to-band, electric field, and analog parameters output analysis is also explored.

7.2 TYPES OF DEVICES

I. **Junction FET:** A reverse-biased diode junction provides the gate connection for a junction FET or JFET. The structure is made up of an N-type or P-type semiconductor channel. The channel is then constructed with a semiconductor diode, with the diode's voltage influencing the FET channel. It is reverse biased when in use, thus isolating it from the channel and enabling only the diode to reverse the current passing between them. The JFET is the most basic FET and was the first to be manufactured [8]. However, it continues to provide excellent service in a variety of technological disciplines.

II. **Insulated Gate FET:** Any form of FET with an insulated gate is known as an insulated gate FET (IGFET). The silicon MOSFET is the most prevalent type of IGFET (metal oxide silicon FET). The gate is formed by depositing a metal layer on top of the silicon oxide, which is then placed on top of the silicon channel. MOSFETs are widely employed in electrical applications, especially integrated circuits. The IGFET's most important feature is the extraordinarily high gate impedance that these FETs can provide. [9] However, a corresponding capacitance rises when the frequency rises, lowering the input impedance.

III. **Dual-Gate MOSFET:** Two gates are connected in series along the channel in this form of MOSFET. This offers significant performance advantages over single-gate devices, especially at RF frequencies. A MOSFET's second gate provides additional isolation between the input and output, as well as mixing and multiplication capabilities.

IV. **MESFET:** Because it is often composed of gallium arsenide, the metal silicon FET is also known as a GaAs FET. Because of its high-gain and low-noise characteristics, GaAsFETs are extensively used in RF applications. The highly narrow gate structure of GaAsFET technology is one of its drawbacks, since it makes it extremely susceptible to damage from static electricity, or ESD. Working with these devices necessitates extreme care.

V. **HEMT:** Although the high electron mobility transistor and pseudomorphic high electron mobility transistor are variations on the basic FET idea, they were designed to operate at extremely high frequencies. They are expensive but allow for extremely high frequencies and excellent performance. HEMT is an abbreviation for high electron mobility transistor. This device is a field-effect transistor (FET) with the rare property of a very narrow channel, allowing it to function at extremely high frequencies. The HEMT not only offers a high frequency response, but also a low-noise response. [10,11] The device is essentially a field-effect transistor with a hetero-junction channel (a junction between two materials with distinct band gaps) rather than the doped region of a traditional MOSFET. Due to its construction, the HEMT is also known as a hetero-junction FET, HFET, or a modulation doped FET, MODFET.

VI. **FinFET:** Integrated circuits are increasingly using technology to enable higher degrees of integration by allowing for smaller feature sizes. FinFET technology is becoming more widely used as higher density levels are required, and ever-smaller feature sizes become more difficult to achieve. FinFET technology addressed several difficulties that prevented planar bulk CMOS from scaling further. [12,13] Other lithography challenges, including limitations related to double-patterning mask alignment and reliability and performance concerns associated with aging, become more significant with each scaling step.

VII. **Carbon Nanotube FET:** Carbon nanotubes (CNTs) are hollow nanotubes made up of one or more layers of graphene sheets coiled around a central axis. They're light and have a hexagonal linking structure that's great. Because of their unique electrical transport characteristics, CNTs could be useful in nanodevices. For example, CNTs are atomically thin, allowing for good electrostatic control of the channel, which is crucial when scaling down the device. Because of their atomically thin structure, the electrical performance advantage of CNT-based field-effect transistors (CNTFETs) has been extended to various chemical and biological sensors. CNTFET-based sensors provide high sensitivity, high selectivity, easy operation, low operating temperature, fast reaction speed, short recovery time, label-free detection, and superior stability compared to other detecting technologies.

VIII. **Gate-All-Around FET:** As the fin width in a FinFET approaches 5nm, channel width changes could cause undesired variability and mobility loss. A potential and futuristic transistor rival, the gate-all-around FET, may be able to fix the problem [14]. A gate-all-around CMOS device has a gate on each of the four sides of the channel. In terms of electrostatics, it is regarded as the ultimate CMOS device. It's essentially a silicon nanowire with a gate surrounding it. In some cases, InGaAs or other III-V materials could be employed in the gate-all-around FET's channels [15].

IX. **Nanowire FET:** Nanosheets (NS) gate-all-around (GAA) FET devices, compared to FinFETs, promise to be viable solutions for reaching a superior power-performance metric for logic applications in advanced sub-5 nm technology nodes. Vertical NW/NS GAA FETs appear to be particularly promising for dense memory cells such as SRAMs (due to their improved read and write stability) and as selection devices for ultra-scaled MRAMs with lower energy consumption values. These cells can be made using a low-cost co-integration technology with a triple-gate FinFET or a lateral NW/NS GAA FET high-performance logic platform for larger on-chip memory content.

X. **Device structure and simulation parameters:**

This section covers the basic device structure of a tunnel field-effect transistor with a nanowire gate (NW-GAA-TFET). Figure 7.1 shows the device's two-dimensional construction. In this figure, a colored construction is seen, with different colors representing various materials utilized in the design. The following are the physical dimensions of certain parameters: The device's channel length (L_C) is 20 nm; intrinsic doping across the source, drain, and channel is maintained, and the length of the region from source to channel or drain to channel is 9 nm, gate oxide thickness (T_{ox}) is 3.5 nm, body thickness or radius (T_{Si}) is 35 nm, source voltage (V_S) is –1.5 V, and gate electrode working function is 4 eV. Inverted charge carriers assist early tunneling of minority charge carriers from the source region in this device. Inverted charge carriers are generated when metal (M1) with a

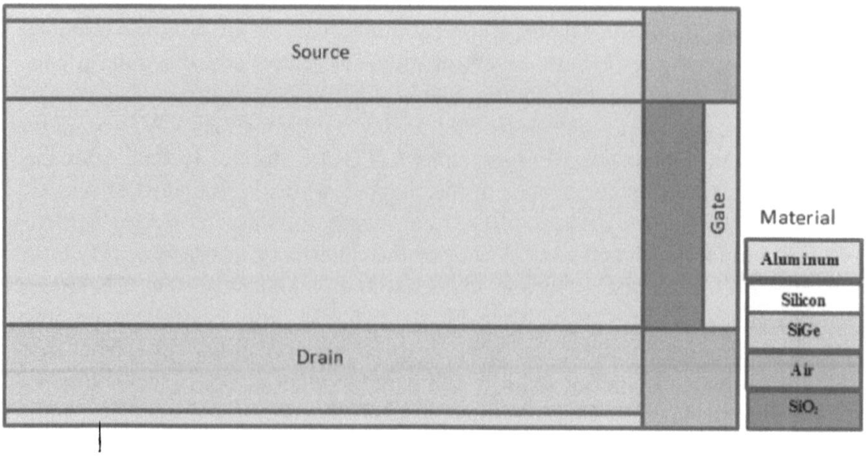

FIGURE 7.1 2D structure of a NW-GAA-TFET.

TABLE 7.1
Parameters Used in the Proposed Simulation of the Device

s.no.	Parameter	Symbol	Dimension
1	Channel length	L_C	20 nm
2	Source length	L_S	80 nm
3	Drain length	L_D	80 nm
4	Gate Oxide	$\emptyset_{MG}$	3.5 nm
5	Gate work Function	T_{ox}	4eV
6	Device radius	T_{si}	35 nm

reduced working feature is present at the source-channel interface. The M1/M2 interface's built-in greater electric field aids in the rapid entry of electrons into the channel. The early rise in drain current aided in improving the structure's subthreshold slope and threshold voltage. The performance of the proposed NW-GAA-TFET system is investigated using the Silvaco Atlas TCAD tool [16].

Silicon dioxide (SiO_2) is used to introduce an insulation layer between the gate and conducting channel. SiO_2 is a strong dielectric substance with superior insulating properties. By oxidizing the already created 'Si' layer, isolation with SiO_2 is straightforward to achieve. The SRH and AUGER models are utilized for minority carrier recombination and high current density recombination, respectively. To assess mobility, the FLDMOB and CONMOB models are being used. The energy band shift is explained using the BGN model. For appropriate tunneling, an electron's effective mass is 0.22 m_e while a hole's effective mass is 0.12 m_h. The physical parameters used in the simulation of NW-GAA-TFET-dependent electrostatic plasma are listed in Table 7.1.

7.3 RESULTS AND DISCUSSION

The discussion of the device linear parameter and analog parameter of the nanowire gate-all-around tunnel field-effect transistor (NW-GAA-TFET) is carried out in this section with a difference in the source material and the material of the drain. Silicon (Si) is used at the source and germanium silicon (GeSi) is used on the drain side. Adopting the different source-drain materials attains much-improved device characteristics. In the NW-GAA-TFET, the electron concentration increased near the transition point of the working function, generating a rise in driving current [17,18]. The linear parameter was discussed relating to the energy band diagram, hole/electron concentration, non-local BTBT tunneling rate, electric field, and SRH recombination rate at equilibrium and ON-state. The analog parameter was discussed relating to the variation of the drain current (I_D) in relation to the gate voltage (V_{GS}) and drain voltage (V_{DS}) at three distinct voltages of 0.1 V, 0.75 V, and 0.5 V.

7.3.1 Linear Parameters Analysis

Figure 7.2 shows the different linear parameters driven by the NW-GAA-TFET structure. The biasing scheme describes the operation of the nanowire gate-all-around the transistor of the tunnel field effect (NW-GAA-TFET), such as the state of equilibrium [$V_{DS} = 0$ and $V_{GS} = 0$], the state of ON [$V_{DS} > 0$ and $V_{GS} = 0$]. The BTBT-based NW-GAA-TFET process includes carrier tunneling from the valence band to the conduction band through the band gap. In Figure 7.2(a), the potential variation over the system length of the NA-GAA-TFET is defined. On the source side, the potential is lowest, and on the unit drain side, it is highest. The function of the drain and source metal connectors determines the device's potential. Due to the external supply voltage being negative ($V_S= -1.5$ V), the negative potential on the source side is negative.

The energy band in equilibrium and ON-state is defined by Figure 7.2(b) along the length of the system. In the ON-state, the valance band lies above the channel conduction band near the source-channel interface, resulting in significantly increased band-bending. The tunneling width is less in the ON-state than in the equilibrium state, allowing more electron carriers to cross from the source valence band (E_V) to the channel conduction band (E_C).

Figure 7.2(c) depicts the fluctuation in electron concentration along the length of the system. Because the source is P-type doped and the drain is N-type doped, the concentration of electrons at the source side in the ON-state is lower and greatest at the drain side. The concentration of electrons in the channel remains constant throughout its length.

Figure 7.2(d) depicts the variation of the electric field along the system length of the NW-GAA-TFET. A fast peak develops at the source/channel interface. In comparison to the equilibrium state, the electric field in the ON-state is higher. In the ON-state, the depletion and tunneling widths are lower than equilibrium, resulting in a larger electric field at the source-channel interface.

Along the length of the system, the non-local electron band-to-band tunneling (BTBT) electron/hole ratio varies as defined in Figure 7.2(e) and 7.2(f). The ON-state is represented by the black color and the equilibrium state by the red color. In the non-local BTBT model, the carriers are tunneled through the depletion layer between the source and channel interface. The material characteristics, such as electron tunneling mass (*me.tunnel*) and the hole tunneling mass (*mh.tunnel*), are considered for proper calibration. The SRH recombination rate is shown in Figure 7.2(g). It represents the minority carrier lifetime and is used to include a mobility calculation. SRH recombination should also be used in most simulations.

7.3.2 Analog Parameters Analysis

Figure 7.3 shows the variation of the drain current (I_{DS}) with respect to the gate voltage (V_{GS}) at 0.01 V, 0.75 V, and 0.5 V. The conduction band of the channel moves upward for the higher gate metal work function. In order to force it down into the source valance band, more gate voltage is needed to start the tunneling, which

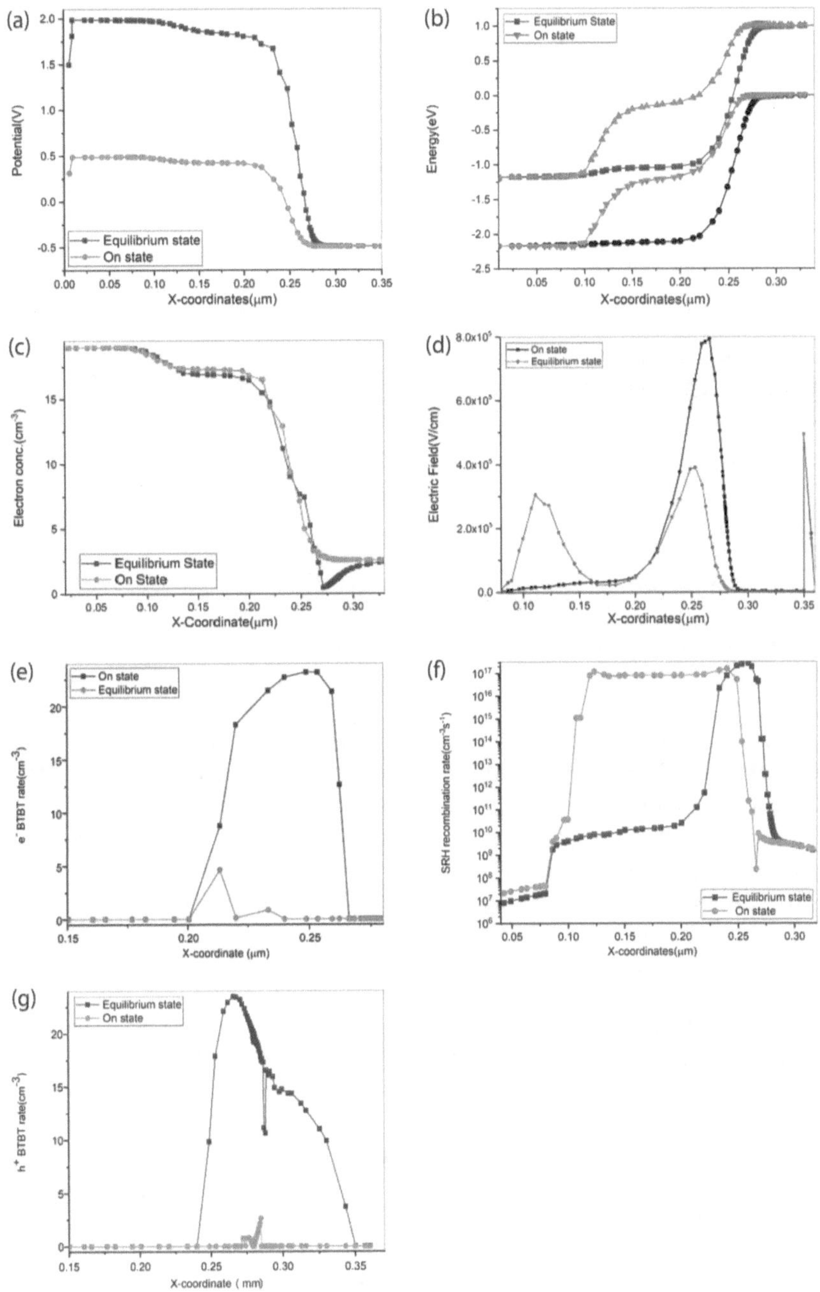

FIGURE 7.2 (a) Electric potential variation. (b) Energy band diagram. (c) Electron concentration variation. (d) Electric field variation. (e) and (f) Electron/hole non-local BTBT rate. (g) SRH recombination rate.

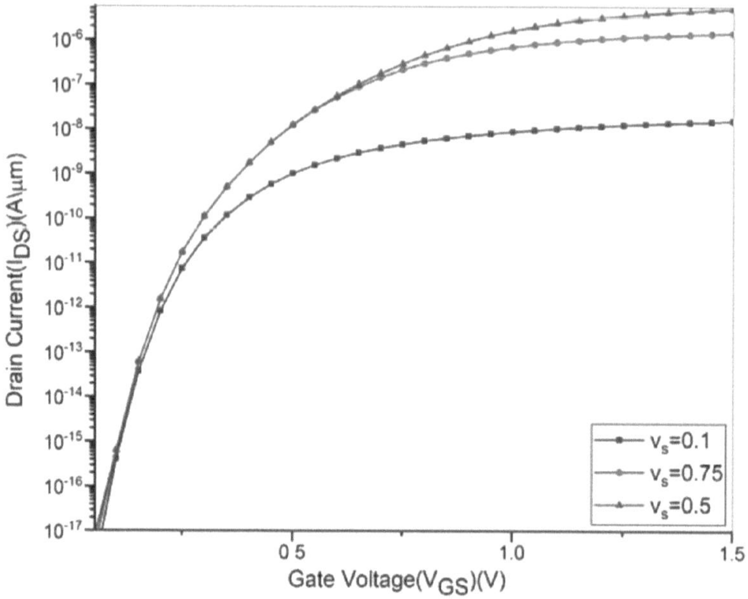

FIGURE 7.3 Drain current (I_{DS}) variation with respect to gate-to-source (V_{GS}) voltage at different voltages.

increases the device's VT. It is clearly shown from the figure that TFET demonstrates its steep threshold behavior for lower gate voltages. In standby mode, this decreased the subthreshold leakage for electronic devices.

Figure 7.4 shows the variation of the drain current (I_{DS}) with respect to the drain voltage (V_{DS}) at 0.01 V, 0.75 V, and 0.5 V. With regard to drain voltage, the ON-current variance is approximately linear.

7.4 CONCLUSIONS

The analysis of various analog parameters and linearity parameters of the nanowire gate-all-around tunnel field-effect transistor (NW-GAA-TFET) architecture is presented in this chapter. The outcome of this work against TFET contains a lower swing subthreshold, confirming the device's suitability for low-power and low-noise applications. The ON-current in the NW-GAA-TFET is 310^{-5} A/um, while the OFF current is 1.5 10^{-17} A/um. Analog parameters are obtained on different voltages. The narrow nanotube structure reduces substrate material while improving electrostatic control over the route, resulting in enhanced interface properties. Because of the different source and drain content, the system can produce a higher ON-state current, a lower cutoff frequency, and a subthreshold slope. Furthermore, the linearity and analog/RF parameters were prominent in the case of the NW-GAA-TFET, confirming the motivation behind the suggested system's employment in low-power circuit applications.

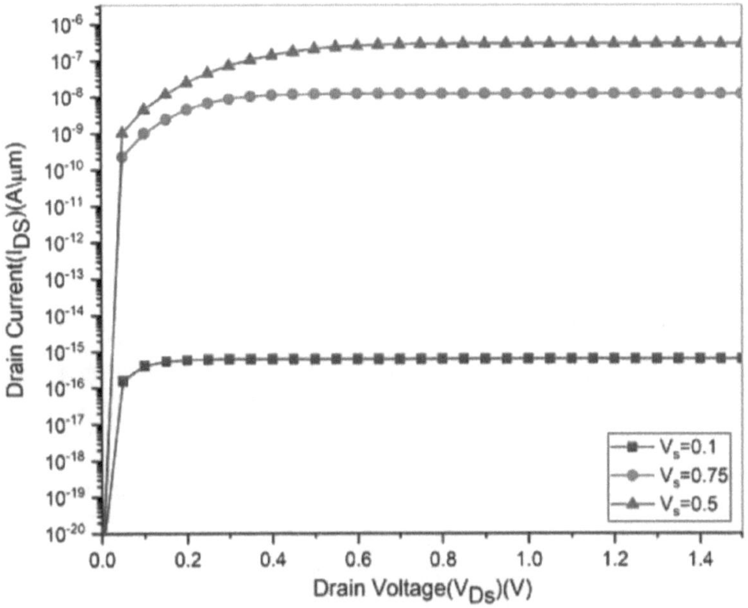

FIGURE 7.4 Drain current (I_{DS}) variation with respect to drain-to-source (V_{DS}) voltage at different voltages.

REFERENCES

1. Pop, E, 2010. Energy dissipation and transport in nanoscale devices. *Nano Research*, *3*(3), pp.147–169.
2. Shakouri, A., 2004, March. Nanoscale devices for solid state refrigeration and power generation. In Twentieth Annual IEEE Semiconductor Thermal Measurement and Management Symposium (IEEE Cat. No. 04CH37545) (pp.1–9). IEEE, San Jose, CA, USA.
3. Gupta, A. K. and Raman, A., 2020. Performance analysis of electrostatic plasma-based dopingless nanotube TFET. *Applied Physics A*, *126*(7), pp.1–10.
4. Kumar, N., Mushtaq, U., Amin, S. I. and Anand, S., 2019. Design and performance analysis of dual-gate all around core-shell nanotube TFET. *Superlattices and Microstructures*, *125*, pp.356–364.
5. Gupta, A. K. and Raman, A., 2020. Electrostatic-doped nanotube TFET: Proposal, design, and investigation with linearity analysis. *Silicon*, 13, pp.2401–2413.
6. Gupta, A. K., Raman, A. and Kumar, N., 2019. Design and investigation of a novel charge plasma-based core-shell ring-TFET: analog and linearity analysis. *IEEE Transactions on Electron Devices*, *66*(8), pp.3506–3512.
7. Amato, M. and Rurali, R., 2016. Surface physics of semiconducting nanowires. *Progress in Surface Science*, *91*(1), pp.1–28.
8. Huang, J., Momenzadeh, M. and Lombardi, F., 2007. An overview of nanoscale devices and circuits. *IEEE Design & Test of Computers*, *24*(4), pp.304–311.
9. Gupta, A. K., Raman, A. and Kumar, N., 2019. Design and investigation of a novel charge plasma-based core-shell ring-TFET: Analog and linearity analysis. *IEEE Transactions on Electron Devices*, *66*(8), pp.3506–3512.

10. Qian, F., Li, Y., Gradecak, S., Wang, D., Barrelet, C. J. and Lieber, C. M., 2004. Gallium nitride-based nanowire radial heterostructures for nanophotonics. *Nano Letters*, *4*(10), pp.1975–1979.
11. Gupta, A. K., Raman, A. and Kumar, N., 2019. Cylindrical nanowire-TFET with core-shell channel architecture: Design and investigation. *Silicon*, 12, pp.2329–2336.
12. Singh R. 2006 May 1. Reliability and performance limitations in SiC power devices. *Microelectronics Reliability*, *46*(5–6), 713–30.
13. Ozpineci, B. and Tolbert, L. M., 2003. *Comparison of Wide-Bandgap Semiconductors for Power Electronics Applications*, Oak Ridge National Laboratory Report.
14. Singh, S. and Raman, A., 2018. Gate-all-around charge plasma-based dual material gate-stack nanowire FET for enhanced analog performance. *IEEE Transactions on Electron Devices*, *65*(7), pp.3026–3032.
15. Chow, T. P., 2014. Progress in high voltage SiC and GaN power switching devices. In Proceedings of the Materials Science Forum, Las Vegas, USA, pp.1077–1082.
16. Chow, T. P., 2015. Wide bandgap semiconductor power devices for energy efficient systems. In Proceedings of the IEEE Workshop on Wide Bandgap Power Devices and Applications (WiPDA), Las Vegas, USA, pp.402–405
17. Chaudhary, T. and Khanna, G., 2016. Analytical modeling of drain current and short channel effects of junctionless double gate vertical slit field effect transistor. *Journal of Nanoelectronics and Optoelectronics*, *11*(6), pp.738–744.
18. Chaudhary, T. and Khanna, G., 2017. A 2D potential based threshold voltage model analysis and comparison of junctionless symmetric double gate vertical slit field effect transistor. *IETE Journal of Research*, *63*(4), pp.451–460.

8 Effects of Variation in Gate Material on Enhancement Mode P-GaN AlGaN/GaN HEMTs

Shamshad Alam, Mamta Khosla, Ashish Raman, and Ravi Ranjan

CONTENTS

8.1 INTRODUCTION

Electronic devices have truly blended into our lives and expanded our capability and potential. Researchers are continuously looking for novel devices with optimized performance in terms of power and size. For low-power applications, there are a number of novel devices like MOSFET [1], TFET [2, 3], Omega FET [4], GAA nanowire FET [5], and many more devices, which are introduced here along with their improved performances. Many semiconductor materials have been introduced for high-power and high-frequency application devices, like GaN, Ga_2O_3, GaAs, SiC, etc. Solid-state devices are used in electronics to monitor or transform electrical energy into various forms. As silicon power devices approach their theoretical limits,

DOI: 10.1201/9781003155751-8

converter designers face numerous challenges in terms of increasing converter ratings, viz., operating voltage, temperature, and performance. These silicon limitations lead researchers to go for wide band gap materials like silicon carbide (SiC) [6] and gallium nitride (GaN) [7]. Compared with silicon, these materials' main benefits are high critical electric field, high-temperature ranges, and high saturation velocity. HEMT is a device which can be used for high-power applications [8]. HEMT devices work in depletion mode, but we want our device to operate in enhancement mode or normally-off condition for switching applications. Several methods, such as recessed gate structure and P-GaN gate structure [9, 10], and fluorine ion treatment and gate injection transistor, have been proposed to overcome the challenges of achieving normally-off HEMTs. But unfortunately, while introducing a cap layer or recessing the gate, coming close to the interface means decreasing the barrier thickness, which strongly affects the density of the two-dimensional electron gas (2DEG) [11]. If fluorine is implanted, getting closer will increase the likelihood of fluorine ions entering the tube, reducing the mobility of the 2DEG. A P-GaN-based HEMT is the most effective device for normally-off operation [12].

8.2 POWER DEVICES

For ultimate performance in power conversion, an ideal switch is required. An ideal switch has almost zero voltage drop, and no limit on current level carried during the ON-state, while during the OFF-state, it should have infinite resistance and be able to sustain unlimited voltage. The time between the ON-state and OFF-state should be zero. In practical switches, there is a trade-off between voltage, current and frequency rating. Tech engineers are faced with the task of increasing the ratings of converters in terms of operating voltage, operating temperature, and performance as silicon control devices approach their theoretical limits in terms of temperature and power operation. Researchers are turning to big bandgap materials like silicon carbide and gallium nitride for solutions to silicon limitations [13]. Wide-bandgap materials have many advantages over silicon, including a larger bandgap, a higher critical breakdown field resistance, and, in some cases, higher thermal conductivity.

8.3 GALLIUM NITRIDE

Gallium nitride is a semiconductor device belonging to the III-V compound semiconductor. Its band energy varies with their different crystal structures, namely Wurtzite and blend structure. GaN is one of the materials that are on the rise to replace silicon. It is anticipated that GaN is a next-generation power semiconductor with much faster switching than silicon [14]. It's high thermal stability, high conductivity, and lesser on resistance make it promising for high-power-based applications. AlGaN/GaN HEMTs are grown on a Wurtzite structure. Aside from the fact that the Wurtzite structure is stable at room temperature, it also has a built-in polarization field known as spontaneous polarization.

8.4 HIGH ELECTRON MOBILITY TRANSISTOR (HEMT)

8.4.1 Conventional HEMT Structure

The HEMT is a field-effect transistor with two layers of materials with separate bandgaps, AlGaN/GaN [15–17]. Two-dimensional electron gas (2DEG) is formed at the interface of AlGaN/GaN because of polarization. When larger lattice material (AlGaN) is grown upon lower lattice material (GaN), piezoelectric polarization is observed on the AlGaN side. Due to polarization, 2DEG is formed at the interface. The 2DEG represents the HEMT's channel, and the device gate controls the current flow between the drain and the source. The cross-section structure of conventional HEMT is shown in Figure 8.1.

Currently, HEMTs are grown on silicon, sapphire, silicon carbide, and GaN substrate. A comparison of these four substrates can be made considering their size, cost, thermal conductivity, heat expansion variation, and lattice constant. The first substrates to be investigated were silicon carbide and sapphire, taking advantage of the LED industry's established expertise and tool collection [18, 19]. However, the industry is drawn to GaN growth on silicon substrates due to the low cost of large silicon wafers. While using a GaN substrate, there is an advantage of attaining no mismatch problem [20]; still, this substrate is less desirable because of its small size and extremely high cost.

8.4.2 Source of Two-Dimensional Electron Gas

Traditionally, AlGaN layers are grown on a GaN buffer. This AlGaN layer causes polarization, resulting in a positive charge on the AlGaN/GaN interface and an opposite charge on the top of the AlGaN layer. This creates an electric field within the AlGaN layer by differently charging various areas. The negative charge on the top of the AlGaN layer is termed 2DEG. So basically, 2DEG is formed due to polarization.

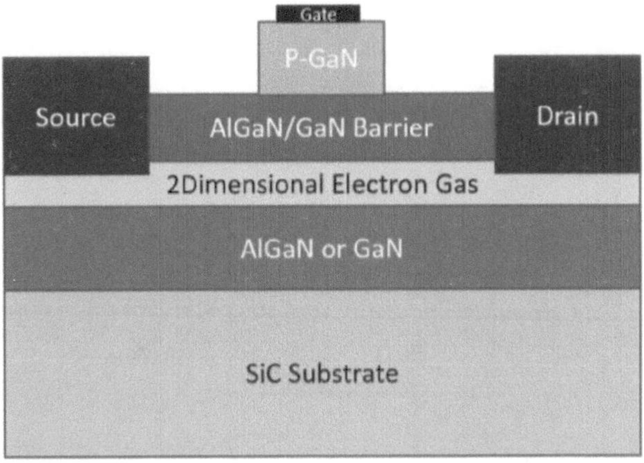

FIGURE 8.1 Schematic structure of a high electron mobility transistor (HEMT).

HEMTs are called normally on devices because 2DEG is treated as a channel without an applied gate bias, as shown in Figure 8.2.

8.4.3 Band Diagram of a Conventional HEMT

The behavior of HEMTs can be understood through their energy band diagram. HEMT is a hetero-junction device with two different semiconductor materials with distinct bandgaps. Due to a distinct band gap, there is a spike at the interface below the Fermi level. The alignment of the energy band diagram of conventional HEMT is shown in Figure 8.3.

The spike below the Fermi level at the interface of AlGaN/GaN is the reason for polarization, resulting in the formation of 2DEG, as shown in Figure 8.3, which acts as a channel through which current flows from source to drain.

8.4.4 Normally-Off HEMT

Conventionally a HEMT is operated in depletion mode (normally when $V_{TH} < 0$ V). To switch off the unit, we must apply a negative voltage. However, normally-off

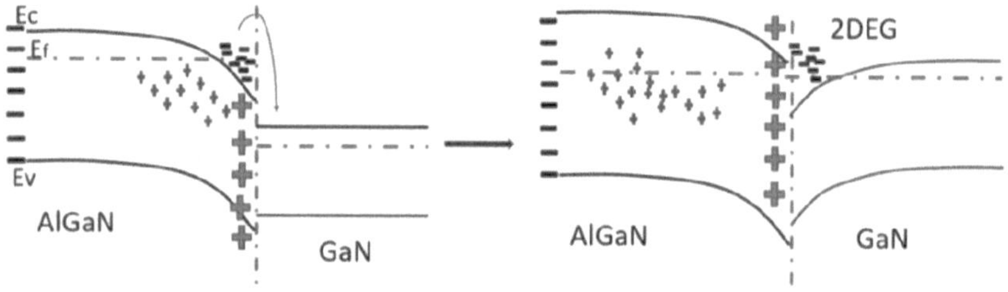

FIGURE 8.2 Formation of 2D electron gas (2DEG) [17].

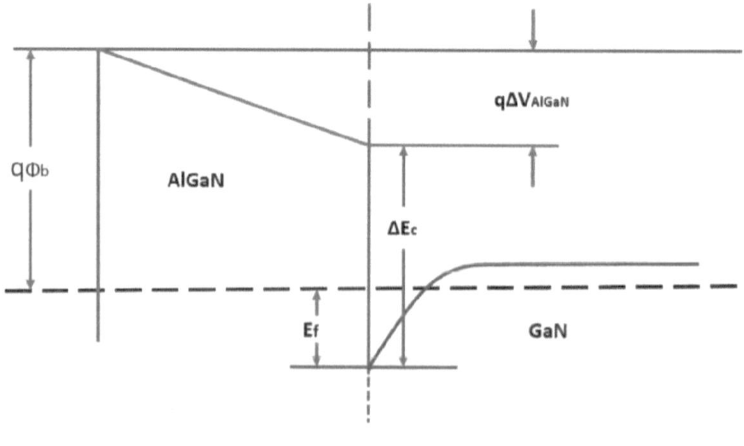

FIGURE 8.3 Energy band diagram of a conventional AlGaN/GaN HEMT [17].

HEMTs (V_{TH} > 0 V) are necessary to minimise circuit complexity and eliminate standby power consumption. So there are several techniques that have been proposed, like gate recess structure, thin barrier layer, gate injection transistor, P-GaN gate HEMT, and fluorine implantation [9, 10].

8.5 SIMULATION AND RESULTS

Developing theoretical research and simulating the I-V characteristic of HEMTs is the best way to understand their electrical behavior. HEMT works primarily in depletion mode due to 2DEG forming at the heterostructure interface. So, to operate our device in enhancement mode (normally-on), P-GaN material is introduced near the gate. Introducing the P-GaN layer below the gate forms a depletion region and depletes the channel in OFF-state conditions. In enhancement mode we have to apply some positive gate voltage to turn on the device, as shown in Figure 8.4. This figure shows a transfer characteristic where the gate voltage is applied from 0 V to 5 V.

The device is turned on when we apply some positive voltage at the gate. There is no current below 0 V, and the threshold voltage of the proposed device is positive. On increasing the work function of gate metal, the OFF-state current reduces while small changes in ON-state current are observed. Below $\varphi = 5.1$ eV, there is a significant current at 0 V, while on increasing φ above 5.7 eV the device operates completely in enhancement mode without affecting the ON-state current. From Figure 8.5 it can be observed that the small change in the I_{ON} is observed with an increase in the work function of the gate. Gate work function variation mostly affects the OFF-state and threshold voltage of the device. On increasing gate work function, our I_{OFF} is reduced, as shown in Figure 8.4. So overall the I_{ON}/I_{OFF} ratio increases with gate work function 4.8 eV to 5.8 eV, as shown in Figure 8.6.

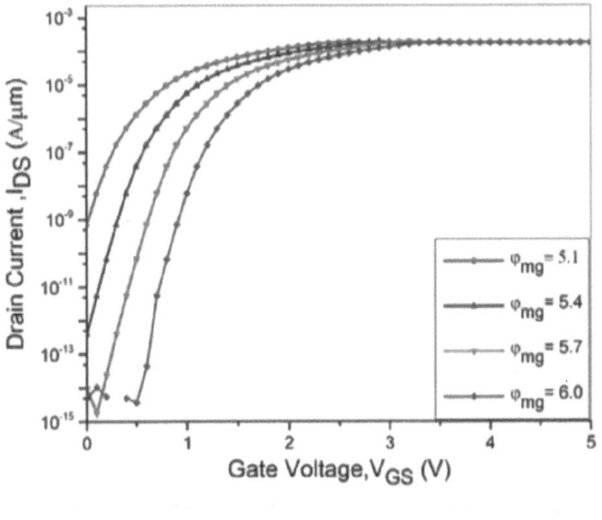

FIGURE 8.4 I_{DS}/V_{GS} characteristic of an enhancement mode P-GaN HEMT.

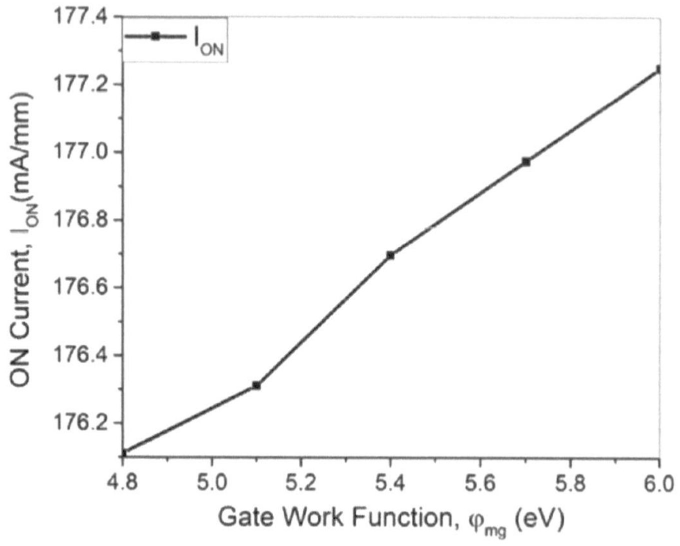

FIGURE 8.5 Variation in I_{ON} with work function.

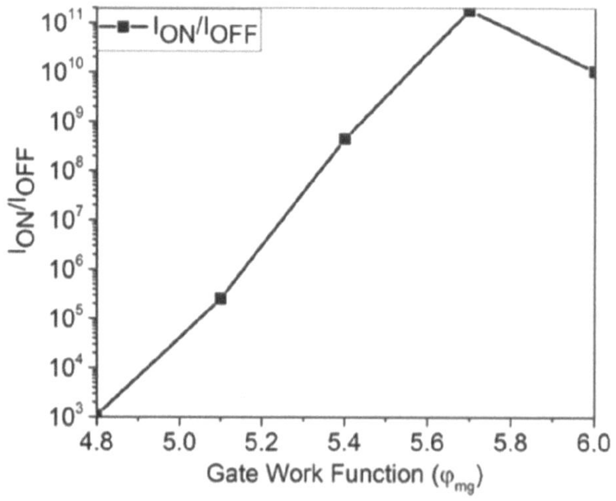

FIGURE 8.6 I_{ON}/I_{OFF} ratio with variation in gate work function.

The threshold voltage of any device is the minimum voltage required to turn on the device, but in power devices there is a requirement of a larger threshold voltage. As our device is working in enhancement mode, we have to apply voltage to turn on the device. Figure 8.7 shows the variation in threshold voltage with gate work function. In our device, 0.6 V is the minimum threshold voltage at a work function of 5.1 eV; on increasing gate work function, threshold voltage increases and reaches to 1.9 V. So, platinum metal is used as a gate metal in power devices.

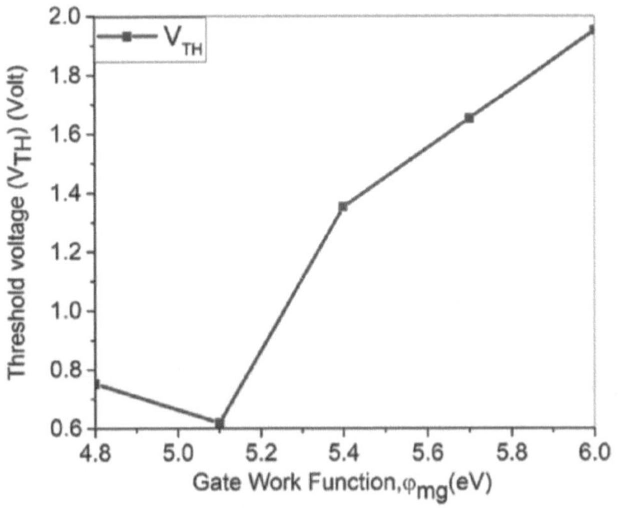

FIGURE 8.7 Variation in threshold voltage with gate work function.

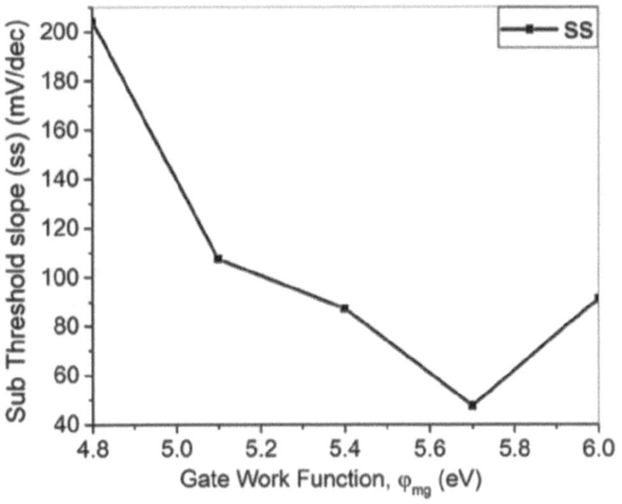

FIGURE 8.8 Variation of subthreshold (SS) slop with gate work function.

The subthreshold slope is defined as the amount of voltage required to change the current within one decade; its unit is mV/dec. The subthreshold slope is reciprocal to the subthreshold swing. The lesser the subthreshold of the device, the better its performance will be. So subthreshold slope should be as low as possible. In Figure 8.8, the variation of the subthreshold slope with the work function of the gate is shown. On increasing the work function, the subthreshold slope is reduced. It gives a minimum subthreshold slope at $\varphi = 5.7$ eV.

8.6 CONCLUSION

A P-GaN-based HEMT was proposed, and analog parameters of the device were investigated with different gate metal work functions. The P-GaN technique is used to make the HEMT device operate in enhancement mode. A P-GaN HEMT creates a depletion region at the channel or at the interface of AlGaN and GaN. Due to the formation of the depletion region, the proposed device will work in normally-off mode. With increasing work function of the gate, it was analyzed that the threshold voltage of the device increases and OFF current reduces. The subthreshold slope of the device also reduces with increasing gate work function. The higher gate work function of the proposed device is suitable for the high-power application.

8.7 FUTURE SCOPE

In this work, we analyzed the effect of variations in gate material on the performance enhancement mode P-GaN-based AlGaN/GaN HEMT. On increasing the work function of gate material, overall the I_{ON}/I_{OFF} also increased. This device can use the delta (δ) doping technique on the source side. Insertion of the delta doping technique on the source side enhances electron transport efficiency; hence its performance is enhanced. It works in enhancement mode so it can be used for designing high-power analog circuits.

REFERENCES

1. Gowar, J., 1989. *Power MOSFETs: Theory and Applications.* Wiley.
2. Alam, S., Raman, A., Raj, B. and Kumar, N., 2019, May. Design and analysis of gate underlapped/overlapped surround gate nanowire TFET for analog performance. In 2019 4th International Conference on Recent Trends on Electronics, Information, Communication & Technology (RTEICT) (pp.454–458). IEEE.
3. Ferhati, H., Djeffal, F. and Drissi, L.B., 2020. Performance assessment of a new infrared phototransistor based on JL-TFET structure: Numerical study and circuit level investigation. *Optik*, 223, p.165471.
4. Lázaro, A., Pradell, L. and O'Callaghan, J.M., 1999. FET noise-parameter determination using a novel technique based on 50-/spl Omega/noise-figure measurements. *IEEE Transactions on Microwave Theory and Techniques*, 47(3), pp.315–324.
5. Alam, S., Raman, A., Raj, B., Kumar, N. and Singh, S., 2021. Design and analysis of gate overlapped/underlapped NWFET-based lable-free biosensor. *Silicon*, 14(3), pp.989–996
6. Casady, J.B. and Johnson, R.W., 1996. Status of silicon carbide (SiC) as a wide-bandgap semiconductor for high-temperature applications: A review. *Solid-State Electronics*, 39(10), pp.1409–1422.
7. del Alamo, J.A. and Joh, J., 2009. GaN HEMT reliability. *Microelectronics Reliability*, 49(9–11), pp.1200–1206.
8. Gangwani, P., Pandey, S., Haldar, S., Gupta, M. and Gupta, R.S., 2007. Polarization dependent analysis of AlGaN/GaN HEMT for high power applications. *Solid-State Electronics*, 51(1), pp.130–135.
9. Hwang, I., Kim, J., Choi, H.S., Choi, H., Lee, J., Kim, K.Y., Park, J.B., Lee, J.C., Ha, J., Oh, J. and Shin, J., 2013. p-GaN gate HEMTs with tungsten gate metal for high threshold voltage and low gate current. *IEEE Electron Device Letters*, 34(2), pp.202–204.

10. Roccaforte, F., Greco, G., Fiorenza, P. and Iucolano, F., 2019. An overview of normally-off GaN-based high electron mobility transistors. *Materials*, 12(10), p.1599.
11. Yuan, L., Chen, H. and Chen, K.J., 2011. Normally off AlGaN/GaN metal–2DEG tunnel-junction field-effect transistors. *IEEE Electron Device Letters*, 32(3), pp.303–305.
12. Meneghini, M., Hilt, O., Wuerfl, J. and Meneghesso, G., 2017. Technology and reliability of normally-off GaN HEMTs with p-type gate. *Energies*, 10(2), p.153.
13. Kaminski, N. and Hilt, O., 2014. SiC and GaN devices–wide bandgap is not all the same. *IET Circuits, Devices & Systems*, 8(3), pp.227–236.
14. Hamady, S., 2014. *New concepts for normally-off power gallium nitride (GaN) high electron mobility transistor (HEMT)* (Doctoral dissertation, Universite Toulouse III Paul Sabatier).
15. Bahat-Treidel, E., 2012. *GaN-Based HEMTs for High Voltage Operation: Design, Technology and Characterization* (Vol. 22). Cuvillier Verlag.
16. Shen, L., Heikman, S., Moran, B., Coffie, R., Zhang, N.Q., Buttari, D., Smorchkova, I.P., Keller, S., DenBaars, S.P. and Mishra, U.K., 2001. AlGaN/AlN/GaN high-power microwave HEMT. *IEEE Electron Device Letters*, 22(10), pp.457–459.
17. Xiao, L., Zhao, J., Zhang, B., Chen, D. and Wu, L., 2020. A unified electrothermal behavior modeling method for both SiC MOSFET and GaN HEMT. *IEEE Transactions on Industrial Electronics*, 68(10), pp. 9366–9375.
18. Xiao-Guang, H., De-Gang, Z. and De-Sheng, J., 2015. Formation of two-dimensional electron gas at AlGaN/GaN heterostructure and the derivation of its sheet density expression*. *Chinese Physics B*, 24(6), p.067301
19. Li, G., Wang, W., Yang, W., Lin, Y., Wang, H., Lin, Z. and Zhou, S., 2016. GaN-based light-emitting diodes on various substrates: A critical review. *Reports on Progress in Physics*, 79(5), p.056501.
20. Pal, S. and Jacob, C., 2004. Silicon: A new substrate for GaN growth. *Bulletin of Materials Science*, 27(6), pp.501–504.

9 Electrical Modeling of One Selector-One Resistor (1S-1R) for Mitigating the Sneak-Path Current in a Nano-Crossbar Array*

J. Arya Lekshmi, T. Nandha Kumar,
A. F Haider, and K.B. Jinesh

CONTENTS

* This chapter includes the contents adopted from [39], [40], and [47] published by same authors. Some figures, equations, analysis … etc. in this chapter are extracted from the Ph.D. thesis (under review) authored by the same authors of this chapter.

DOI: 10.1201/9781003155751-9

9.1 INTRODUCTION TO THE RRAM MODELS AND THEIR POTENTIAL CHALLENGES

In recent years, many memory devices have evolved to handle the ever-increasing data explosion due to the latest trending technologies, such as big data analytics, cloud computing, artificial intelligence, and neuromorphic computing. With a focus on utilizing the memory and storage systems in such applications, the prevailing memory technologies like DRAM, SRAM, and NAND are replaced with SDRAM, 3D NAND, and emerging non-volatile memories [1–3]. The broad classifications of semiconductor memories, including the existing memory technologies and developing memory technologies, are illustrated in Figure 9.1. The entire class of memory devices is classified into volatile memories and non-volatile memories at the first stage. Subsequent branching of non-volatile memory includes emerging memories, ROM, and flash memories.

The state-of-the-art non-volatile memory technology should combine the best features of existing memory technologies with CMOS compatibility, scalability beyond the present limits of SRAM and flash memories, and also expected to have features like non-volatility, high speed, high density, low cost, and low power. From the aforementioned memory technologies, emerging non-volatile memory technologies are more successful in attaining these features.

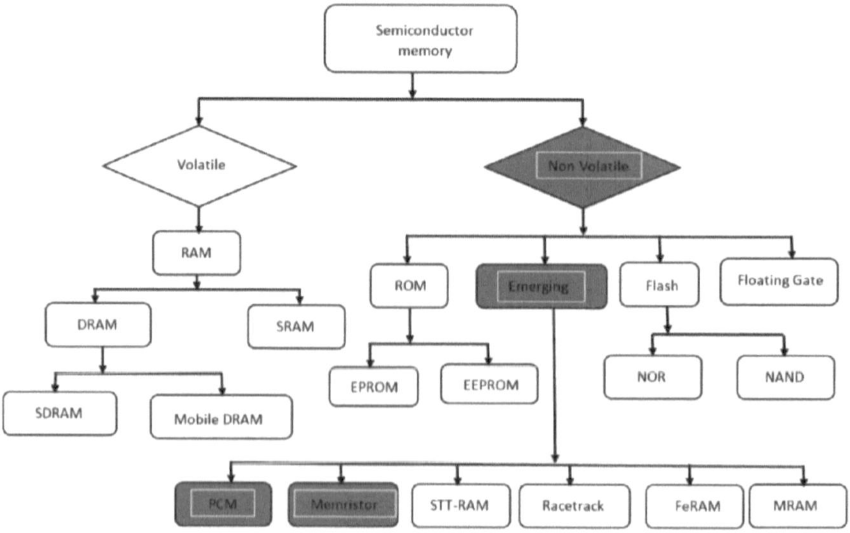

FIGURE 9.1 Classification of semiconductor memory technologies.

According to the *International Technology Roadmap for Semiconductors (ITRS) Report* [4], the emerging non-volatile memory includes prototypical memory technology such as phase-change memory (PCM), ferroelectric random-access memory (FeRAM), magneto-resistive random-access memory (MRAM), spin transfer torque magnetic random-access memory (STT RAM), and resistive random-access memory (RRAM) [5–7]. RRAM has emerged as a promising candidate among these memory technologies because it possesses all the essential features, like scalability, low power density, fast read-write, and non-volatility [8].

As shown in Table 9.1, memristor (RRAM) outperforms other emerging non-volatile devices due to its high memory density, low-energy density per bit, fast read/write time, low operating voltages, nanoscale dimension, high endurance, and high R_{off}/R_{on} ratio. In addition, RRAM provides good reliability measures such as endurance and retention compared to its counterparts. Compatibility with the CMOS is another feature that makes it distinct among other emerging memory technologies, such as PCM, MTT-RAM, etc.

9.1.1 RRAM Technology – Background Studies and Current Research Development

RRAM devices are futuristic devices, alternatively known as memristors, which were first postulated by Leon Chua in 1971 [9], and the prototype was invented by researchers at HP labs [10]. Memristors are two-terminal emerging non-volatile memory devices that work based on the resistance switching (RS) mechanism, where the device exhibits two different resistance levels controlled with current or voltage. The hysteresis behavior exhibited by the HP memristor device ($Pt/TiO_2/Pt$) is shown in Figure 9.2 [11, 12].

A variable resistor model of the above-mentioned memristor device proposed by HP labs is illustrated in Figure 9.2. The device structure consists of a thin oxide film of thickness D sandwiched between two metal contacts to form a metal/insulator/metal (MIM) structure, as shown in Figure 9.3(a). A region with a high concentration of dopants (marked as doped in the figure) and a low concentration of dopants

TABLE 9.1
Comparison of Various Parameters Across Different Emerging Non-Volatile Memory Devices [1]

	Memristor	PCM	STTRAM	DRAM	Flash	HDD
Density(F^2)	4	8–16	14–64	6–10	4–6	2–3
Energy per bit (pJ)	0.1–3	2–27	0.1	2	10000	$1–10 \times 10^9$
Read time (ns)	< 10	20–70	10–30	10-50	25000	$5–8 \times 10^6$
Write time (ns)	~ 20	50–500	13–95	10-50	200000	$5–8 \times 10^6$
Retention	years	years	weeks	<< second	years	years
Endurance (cycles)	10^{12}	10^7	10^{15}	10^{16}	$10^3–10^6$	10^{15}

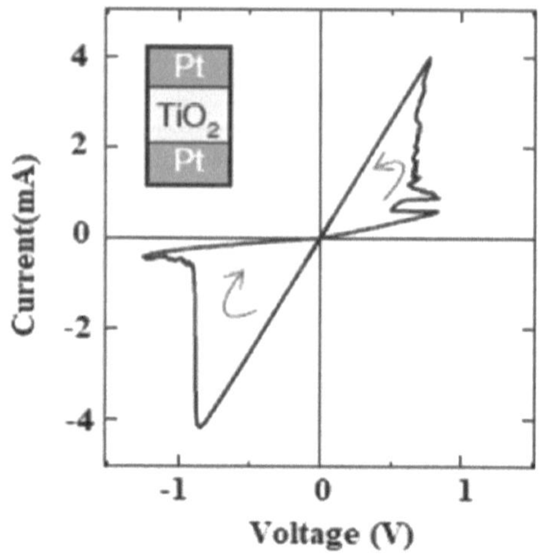

FIGURE 9.2 I-V characteristics obtained from a Pt/TiO_2/Pt HP memristor device [10].

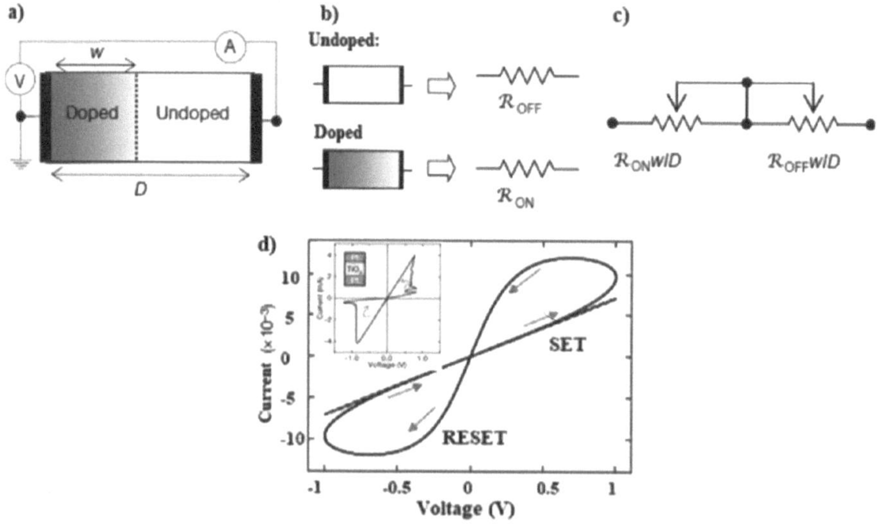

FIGURE 9.3 Basic workings of the HP memristor model: a) model with doped and undoped regions, b) equivalent resistance of doping regions, c) moving boundary between doped and undoped regions, d) I-V characteristics obtained from the model (inset showing corresponding actual characteristics obtained from the device) [10].

(marked as undoped in the figure). The concentration of dopants in the film affects the resistance of that region where the highly doped regions show low resistivity. The less doped region shows high resistivity; hence the total resistance of the device is contributed to by two variable resistors connected in series. The resistance values of the variable resistors are R_{on} and R_{off}, as shown in Figure 9.3(b). The application of an external bias v(t) across the device will move the boundary between the two resistance regions due to the charged dopants drifting, as depicted in Figure 9.3(c), whereby the model showed a hysteresis curve for current-voltage characteristics as given in Figure 9.3(d), (the inset shows the I-V characteristics obtained from the actual device). However, the device requires some atomic rearrangement that modulates the electronic current to attain the hysteresis behavior. Here the atomic arrangement refers to the movement of oxygen vacancies across the semiconducting film. In the given device, the device switches to the ON-state with the application of a positive voltage. The switched state of the device is referred to as the low resistance state (LRS), and the application of a negative voltage switches it back to OFF-state. The OFF-state of the device is referred to as a high resistance state (HRS). The device remains in the ON-state as long as a positive voltage is applied, and with a negative bias it will switch back to the OFF-state [10].

Later, the evolution of the RRAM device continues with different materials, such as transitional metal oxides and organic metal oxides, such as Ta_2O_5, Al_2O_3, HfO_2, NbO_2, and graphene. In addition to the development of a simple MIM structure, as shown in Figure 9.4(a) [14, 15], memristor devices are developed with a bilayer (Figure 9.4(b)) as well as multilayer stacking morphology (Figure 9.4(c)). The bilayer device structure is developed with metal–insulator–semiconductor–metal (MISM) stacking, where two layers can be made of the same metal oxide through its stoichiometric and sub-stoichiometric phases [13], or bilayers can be made from hetero-oxides [14, 15]. If a RRAM device is designed with more than two resistive switching layers, it is called a multilayer device [16].

Yet another classification of RRAM memory is based on the type of switching element; these can be valance change mechanism (VCM) based oxide-RRAM (ox-RRAM) and metallic cations-based conduction bridge RRAM (CBRAM) memories.

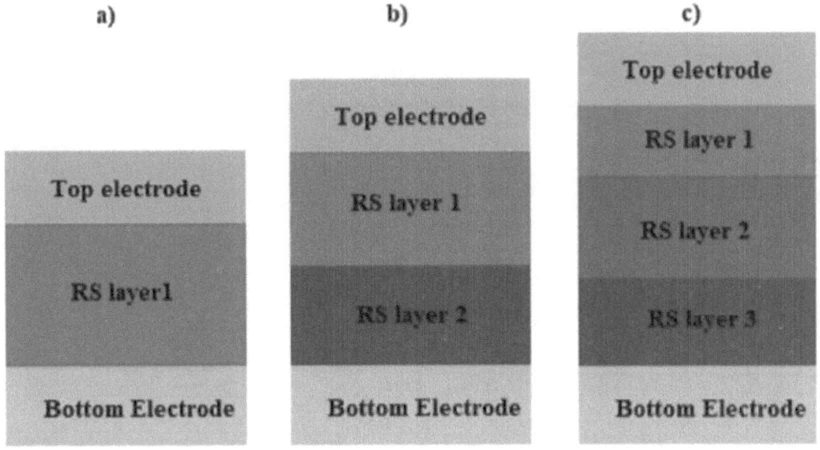

FIGURE 9.4 Structure of RRAM devices: a) single-layer MIM structure, b) bilayer structure with two resistance switching layers, c) Multilayer structure with multiple resistive switching layers.

In addition, based on the nature of the current profile variation with the input voltage, RRAM devices can be of two types, analog and digital. The analog devices show analog I-V characteristics where the switching between states is gradual. On the other hand, devices exhibiting digital characteristics show an abrupt switching process. According to the nature of device characteristics, memristive devices can be used in various applications. For instance, the analog memristor is suitable for neuromorphic applications. It emulates the step-by-step spiking of neurons, whereas abruptly switching digital memristive devices is beneficial for digital applications.

Thus, RRAM is gaining momentum in the field of emerging non-volatile memory industry, owing to its versatility in applications such as its ability to be used in in-memory computing technologies, storage systems, the realization of Boolean logic functions, such as NAND, NOR, look-up table (LUT) realization for FPGA, and sequential circuit implementations, such as D-latch, implementation of reconfigurable logic operations, crossbar adders, etc. The key application of RRAM memory is in neuromorphic computing, where oxide-based resistive switching devices (RRAM) have emerged as the leading candidate in the realization of synapse functions. Even though the technology is in its research and development stage, progressions achieved in the last few years are remarkable. The SanDisk Corporation demonstrated a 32Gb RRAM test chip with a 24 nm CMOS process in 2013. In 2014, researchers at IMEC and Panasonic created a 2-Mbit chip made of a 40 nm node with a $Ir/Ta_2O_5/TaO_x/TaN$ structure.

Further, a 16-Mbit array of a HfO_x/TaO_x bilayer RRAM stack was reported in 2015. In addition, 'Crossbar' reported a 4-Mbit 1S-1R chip and 1T-1R IP core are targeted for high-density memories [4]. Another milestone laid by IBM is the proposal of a resistive processing unit (RPU) for the storage and processing of data to improve deep neural network training (DNN) [4]. Further, in 2018, The Intel corporation introduced the non-volatile Optane DIMM with 3DXpoint technology (shown in Figure 9.5), which is considered a milestone for the non-volatile technology.

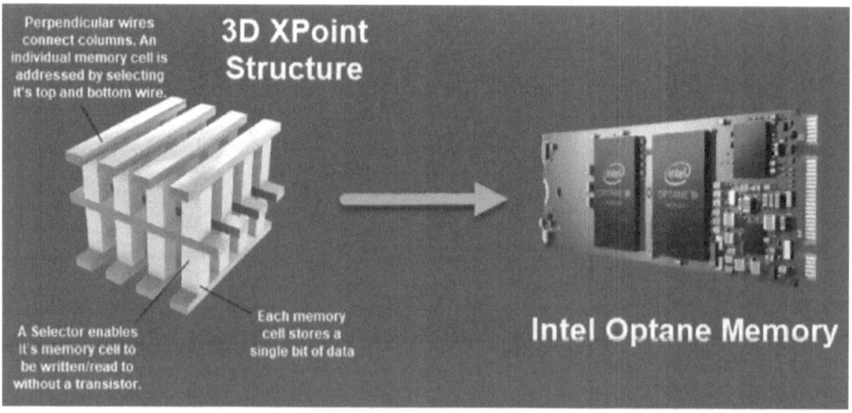

FIGURE 9.5 An Intel optane memory developed with a 3DX structure [4].

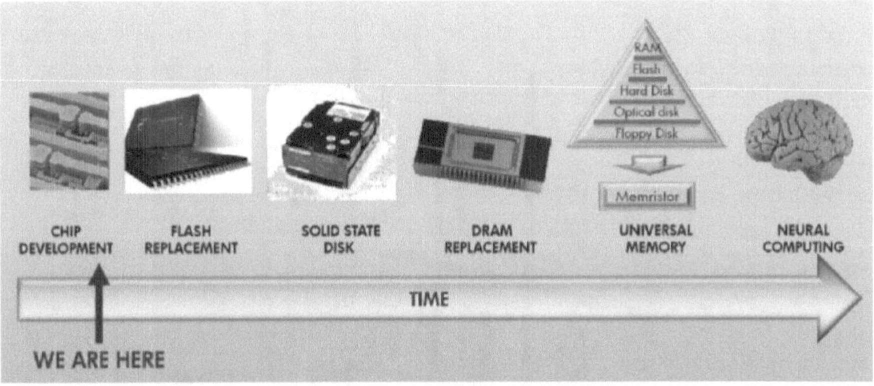

FIGURE 9.6 The present state of memristor research and the milestones to reach in the foreseeable future [8].

Moreover, it has been recently reported that RRAM devices can be used in IoT, AI, and data center storage. However, all these devices are in the research phase, and none have yet reached the consumer market.

Figure 9.6 gives an insight into the current state of research and the milestones that will be reached in the foreseeable future. The technology is still in its infant stage. As time progresses, the potential memristive device will replace flash technology and DRAM as a universal memory forage and high-end applications like neuromorphic computing [8].

9.1.2 RRAM Crossbar Array Formation and Sneak-Path Current

For developing dense RRAM memory, the devices should be configured in 2D or 3D nano-crossbar arrays (CBA). A nano-crossbar array consists of horizontal word lines and vertical bit lines where the bidirectional memory element sits between their cross points [17–19], though the significant challenge encountered with RRAM devices while forming an array is the sneaking of leakage current, and this unforeseen current component is known as the sneak-path current, which is caused by the low resistance path offered by the unselected cells remaining in the LRS in the crossbar array. An example of a crossbar array is shown in Figure 9.7, where the vertical bit lines are C1, C2, C3, and C4; horizontal word lines are denoted with R1 and R2. Note that in Figure 9.7, V_{dd} is applied to the selected word line, and the selected bit line is connected to the ground potential. In the figure, V1 and V2 are arbitrary voltage biases, and their magnitude depends on the adopted writing scheme.

9.2 METHODS TO MITIGATE SNEAK-PATH CURRENT IN RRAM ARRAYS

Various techniques are employed to circumvent the problem of sneak-path current, such as complementary resistive switching (CRS) devices [8, 9] and selector

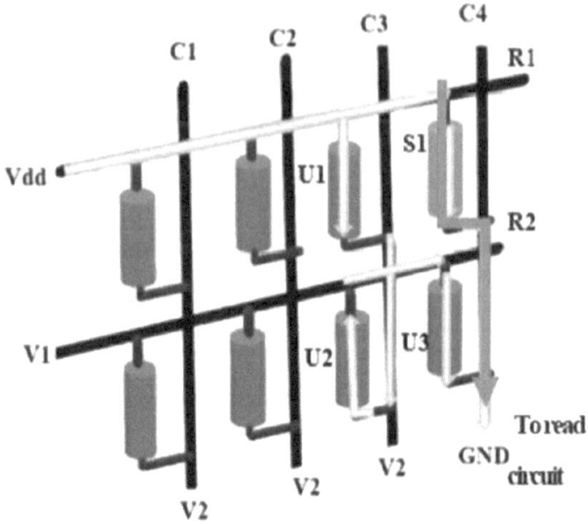

FIGURE 9.7 A simple crossbar array of RRAMs showing the traces of sneak-path current and actual expected current.

devices [10, 11]. To date, many selector devices, such as diodes [20], transistors [21, 22], and volatile bilayer devices are reported as being used in association with RRAM. Nevertheless, the unipolar behavior of the diodes and scaling concerns of transistor switches impede their use in CBA. Several nonlinear selectors have been realized [23, 24] to overcome these shortcomings, which are fabricated either as a single-layer or as a multilayer material stack with advanced engineered designs [25, 26]. Moreover, recent studies show that a selector device with a trilayer structure provides more stable selector characteristics [20] with significant nonlinearity. Similarly, various CRS devices are also available, where the most common configuration is the anti-serially connected RRAMs. This chapter focuses on mitigating sneak-path current by using a selector switch. The following section discusses a one selector-one resistor (1S-1R) device configuration in detail.

9.2.1 Associating a Selector Device With Memory Unit (1S-1R): A Literature Review

The recent progress in emerging non-volatile technology has been based around implementing selector devices in cooperation with RRAM or CBRAM devices to alleviate issues caused by the sneak-path current. Selectors are volatile devices that work with the electron population and do not show switching characteristics.

The desired characteristics expected from a good selector are high nonlinearity in I-V characteristics, low off-current, high selectivity, high current density, and voltage compatibility with the associated memory device. In a 1S-1R crossbar array, a highly nonlinear selector connects in series with RRAM memory at each cross-point [27] (as shown in Figure 9.8), where the selector acts as a switch to control the current that flows through the memristors while reading and writing. At low-voltage levels, the selector provides high impedance and thus acts as an open switch. At high input voltages, due to inherent high conductivity, the selector acts as a closed switch or fully conducting device, thus controlling the selection of the memory cell while blocking the sneak-path current [28–30].

The writing scheme also has significance in a 1S-1R array performance. Various writing schemes are present, such as the floating scheme, *V*/2 scheme, and *V*/3 scheme. In all three schemes, the selected word line and bit line are connected to the V_{dd} and ground, respectively. However, the unselected bit/word lines remain open/floating in the floating scheme. In the *V*/2 and *V*/3 schemes, the unselected word line will connect to *V*/2 and *V*/3 of the V_{dd}, respectively, whereas in *V*/2 and *V*/3 schemes, the bit lines will connect to *V*/2 and *V*/3 voltages, respectively.

Based on the type of application, two variants of selectors are present: selectors with analog I-V characteristics and digital I-V characteristics. The first category finds application in neuromorphic computing platforms, whereas the second can be used in digital applications with discrete resistance levels in which the abrupt transition of resistance occurs at a specific threshold voltage. $Ta_2O_5/TaO_x/TiO_2$ [23] trilayer stacks and $TiN1_{+x}/Ta_2O_5/TiN_{1+x}$ [31] are examples of analog selectors, while a field-assisted superliner threshold selector is an example of a selector with digital characteristics [32]. In a 1S-1R approach, various factors are considered before connecting the selector with the RRAM, such as the compatibility of electrodes, i.e., the top electrode of RRAM and the bottom electrode of the selector should be compatible. Then, the operating voltage and current of both devices should be in a comparable range. There are several 1S-1R combinations available in ox-RRAM memories and

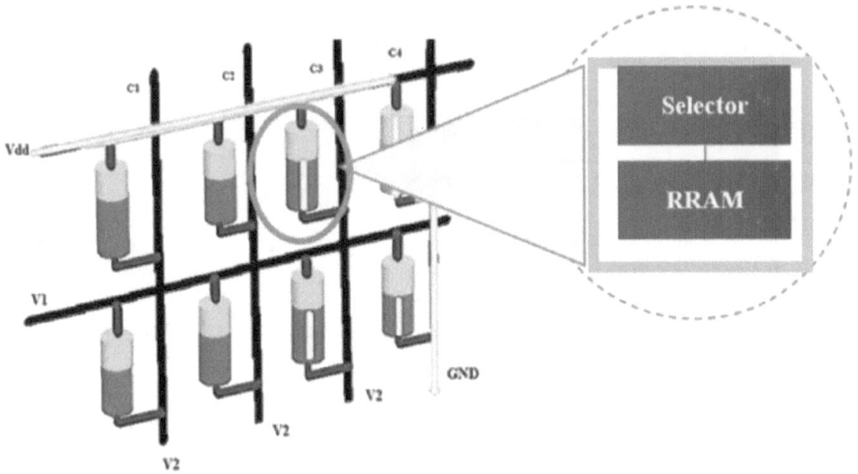

FIGURE 9.8 Crossbar structure with a selector-RRAM device.

conduction bridge resistive random-access memory (CBRAM) devices. The following sections review some of the selector devices and their combined characteristics with memory.

Various fabrication strategies are adopted to attain the desired selector characteristics, such as crusted barrier formation or tunneling barrier formation with oxides of nanometer thickness. The crusted oxide barrier has been utilized to suppress the sneak-path current in the $TaO_x/TiO_2/TaO_x$-based selector device, which offers a high current density (1.6*10^7 A/cm) and nonlinearity of 10^4. Also, it exhibits a better voltage compatibility with RRAM devices [33]. The sneak-current reduction is demonstrated in [33] where the selector device has been fabricated in a Pt/Cu doped HfO_2/Pt RRAM stack and tested in a crossbar array. Up until a bias potential of –0.9 V to 0.7 V, the selector device shows high impedance to the current, and beyond that the selector shows increased conduction. At the same time, the RRAM starts to SET or RESET according to the polarity. In that work, it is reported that the sneak-path current is suppressed by more than four orders of magnitude compared to RRAM. Besides, the device combination showed a 37 nA current during the $V_{read}/2$ bias, confirming the 1S-1R circuit's efficiency in limiting the sneak current. The selector characteristics and integrated selector-RRAM characteristics are given in Figures 9.9(a) and 9.9(b), respectively [34], where Figure 9.9(a) shows asymmetric selector characteristics with a nonlinearity factor of 10^4. Figure 9.9(b) shows the integrated characteristics showing the memory window with low leakage current at low voltages (37 nA).

Another implementation of a 1S-1R was based on a vertical stacking of a Ta_2O_5/TaO_x/TiO_2 selector device and Cu/TiO_2/Pt CBRAM, as reported in [35], in which the integrated device also reproduced robust bipolar switching behavior with an abrupt transition between states, and the effect of the selector on the 1S-1R circuit was the suppression of leakage current by a factor of 10^3. It has been reported

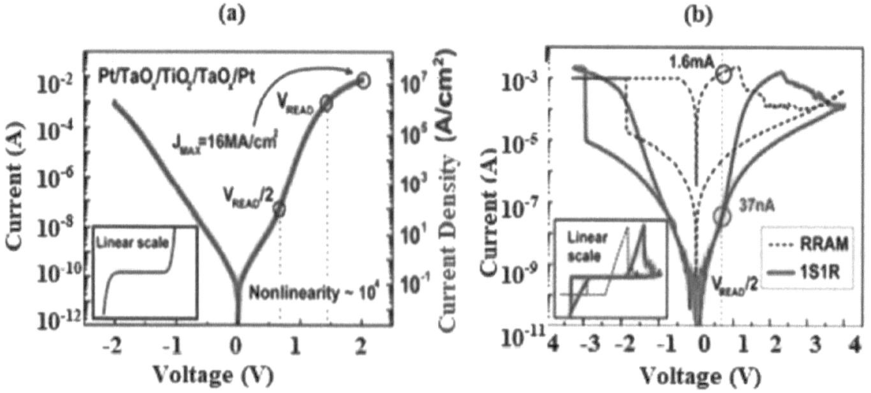

FIGURE 9.9 a) Typical I-V characteristics of a $TaO_x/TiO_2/TaO_x$ selector, b) characteristics of the selector with a Pt/Cu-HfO_2/Pt device [33].

that the presence of parasitic resistance affects device characteristics. In the combined circuit, the sensing margin was improved compared to the CBRAM device. Moreover, stable endurance is observed for the integrated devices while maintaining the on/off ratio (> 10). The selector I-V curve and the combined characteristics of the device and selector are shown in Figure 9.10 (a) and Figure 9.10(b), respectively.

A high-performance Ni/TiO_2/Ni bipolar selector with Ni/HfO_2/Pt memory element was integrated, and its characteristics were reported in [36], where the device structure of Ni/TiO_2/Ni/HfO_2/Pt with a crossbar array of dimension 8 × 8 was fabricated to demonstrate the reduction in sneak current. Also, it was reported that its V_{SET}/V_{RESET} was below ± 4 V and I_{RESET} was less than 150 μA. Another example of the 1S-1R configuration with an analog selector is the integration of a $TiN_{1+x}/Ta_2O_5/TiN_{1+x}$ material stack with Ta/TaO_x/Pt RRAM to achieve high nonlinearity, high endurance, low variability, and low dependence on temperature [24].

There are several 1S-1R devices formed by the connecting selector and ox-RRAM/CBRAM externally (which means not etched in the same substrate). The Ru/TaOx/W selector presented in [37] is externally connected with Cu/HfO_2/Pt RRAM to form the 1S-1R circuit. It is created by tuning defects and having a trapezoidal band structure that exhibits certain features, such as selectivity (5 × 10^4), low OFF-state current (~ 10 pA), robust endurance (> 10^{10}), and high current density (1 MA/cm^2). Besides this, the combined circuit shows self-compliance and excellent uniformity. When connecting 1S and 1R serially, the circuit shows a set voltage under 4 V and a reset voltage under –5 V as suggested for practical applications.

Yet another selector-RRAM combination connected externally to obtain the combined characteristics can be seen in [24], where a $TaN_{1+x}/Ta_2O_5/TaN_{1+x}$ stack is used as the selector device, and its I-V curve is given in Figure 9.11(a). In the experiment,

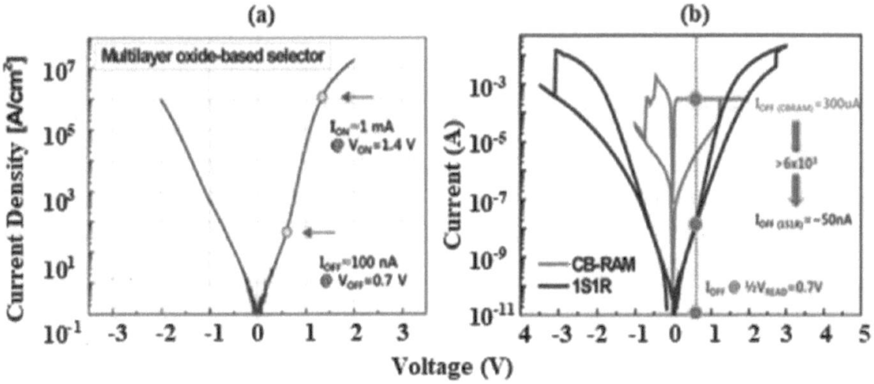

FIGURE 9.10 Selector characteristics for a) $Ta_2O_5/TaO_x/TiO_2$ device [35] and b) integrated device characteristics.

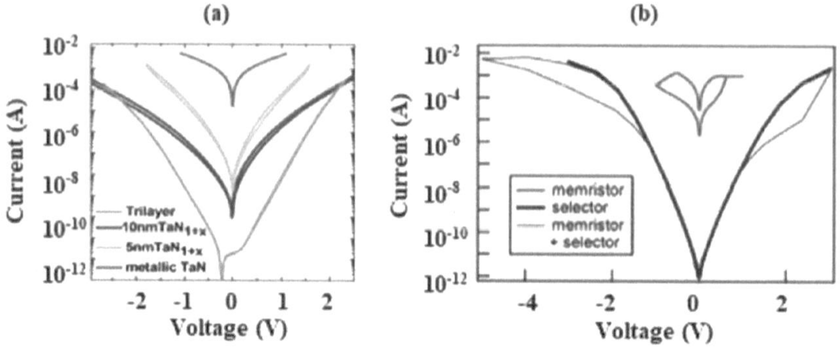

FIGURE 9.11 (a) Various selector characteristics, (b) combined characteristics with RRAM [24].

Ta/TaN$_{1+x}$/Pt is used as a RRAM device. And the resulting characteristics, as plotted in Figure 9.11(b), shows the memory window, though it is not very significant.

A field-assisted superliner selector exhibiting digital characteristics was introduced in [38] and claims a highest selectivity of $> 10^{10}$, where its threshold can be modulated by controlling the insulator thickness; moreover, this device provides a faster switch transition between ON and OFF.

From the analysis of selectors and their integration with various RRAMs (1S-1R), it can be concluded that the trilayer or multilayer structures of a tunnel or crusted barrier offer better selector functionality with significant nonlinearity. In addition, the device's resistance increases with an increase in the thickness of the layers in the multilayer structure, and the bidirectional functionality of these selector devices provides better performance to attain the memory feature. Experimental evidence shows that both externally connected selectors with RRAM and those fabricated in a single substrate offer high nonlinearity. Furthermore, it is observed that the minimum SET and RESET voltage required to program the combined circuits is $|> 3$ V. Also, it is noted that the readout margin is significantly decreased as the array size increases [25].

9.3 IMPLEMENTATION OF AN ELECTRICAL MODEL OF SELECTOR DEVICE AND RRAM DEVICE

Incorporating a selector device with RRAM memory (1S-1R) either by configuring it in the same substrate or by external connection is an experimentally proven solution to mitigate sneak-path current. Apart from the experimental investigations [25, 34], the modeling of 1S-1R is a timely requirement that enables verification of a device's compatibility in a CBA for reducing the sneak-path current.

However, to understand 1S-1R model functionality, a selector model is necessary. Among the available selectors, the $Ta_2O_5/TaO_x/TiO_2$ selector, which has excellent selectivity ($\sim 10^4$) and high current density [23], has been chosen for electrical modeling. The model considers electric field-driven electron tunneling mechanisms co-existing in thin multilayer devices. Hereafter the $Ta_2O_5/TaO_x/TiO_2$ selector will be referred to as the 'selector' throughout the chapter. The following sections explain the electrical model of the Ta_2O_5/TaO_x RRAM selector device.

9.3.1 Electrical Modeling of a Selector Device

To develop 1S-1R, a selector device and a RRAM device need to be connected in series. The compatibility of electrodes, voltages, and currents of both devices should be considered while choosing devices. The electron tunneling-based electrical model is proposed to study the electrical characteristics of the nonlinear multilayer $Ta_2O_5/TaO_x/TiO_2$ selector stack. The structure of the selector device is depicted in Figure 9.12, where the top and bottom electrode is Platinum, and the dielectric tunneling layers are Ta_2O_5 /TaO_x and TiO_2.

The current conduction in the device is explained after the tunneling mechanism, where the electric field-driven tunneling of electrons through the device depends on the formation of barriers across the thickness of the insulating layers. In the analytical model [39], the tunneling occurs across the triangular barrier at high electric fields. It can be modeled with Fowler Nordheim tunneling. In contrast, the electronic tunneling occurs across a trapezoidal barrier at low electric fields and follows direct tunneling or hopping across the bulk, showing an exponential relation with the tunneling distance [39]. However, the hopping mechanism is adopted in the electrical model to fit well with the proposed selector. Subsequently, these current conduction mechanisms are mapped to their equivalent current and voltage sources. The current components in the proposed electrical model are represented with parallelly connected dependent sources, which depend on their corresponding voltage sources;

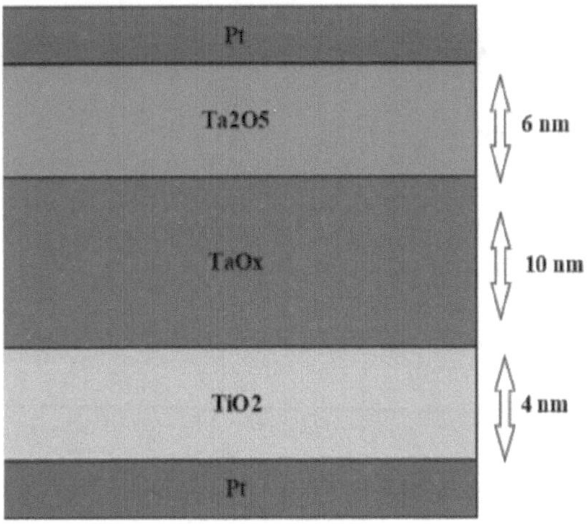

FIGURE 9.12 Multilayer stack selector device [39].

therefore, the current sources are considered voltage-dependent. Since the current through the device is controlled by the input voltage (electric field) developed across the layers, voltage-controlled voltage sources (VCVS) and voltage-controlled current sources (VCCS) available in the SPICE tool are utilized in the proposed electrical model.

For simulation, E-type voltage sources and G-type current sources available in the SPICE library are utilized to implement dependent voltage sources and current sources, respectively. The current source's transconductance is defined by the current magnitude obtained from tunneling/hopping models. Figure 9.13(a) shows the equivalent circuit model of the selector with two dependent current sources representing different current conduction mechanisms, and Figure 9.13(b) shows the corresponding VCVS. For simplicity, the current sources representing the forward and reverse bias are not shown separately. The current sources representing the tunneling and hopping conduction mechanisms are connected in parallel across the top and bottom electrodes marked with *G_tun* and *G_hop*, respectively. As a result, these current sources provide the sum of current from two sources to describe the total current, and the magnitude of the current provided by the current sources varies with the corresponding dependent voltage source. The voltage-controlled voltage sources shown in Figure 9.13(b) are the voltage sources used to derive the current components. *EE1*, *EI1_on/off*, and *EI2_on/off* are the voltage sources where *EE1* is the voltage source that depends on the electric field across the device, which is a function of applied input voltage *V(V)*, *EI1_on/off* represents the voltage source that depends on the electric field across the device, and *EI2_on/off* is the voltage source governing the hopping conduction.

Pseudocode is developed for simulating the circuit model, which is shown in Figure 9.14, and the code is based on the current port equations given in Eqs. (9.1)

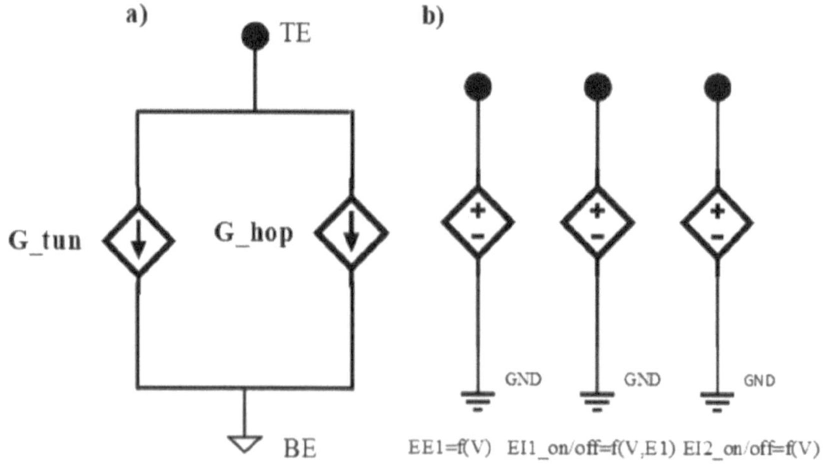

FIGURE 9.13 Electrical model of the selector: (a) Voltage-controlled current sources and (b) Voltage-controlled voltage sources.

```
******* Parameters

.SUBCKT sel plus minus PARAMS:
+k1=1.54e-7 k2=4.31e8  Io=3.9n Io'=2.5n
+c1=2.1 c2=9e-2 v0=0.2 A=1e-9 d0=0.3e-9
+D2=10e-9 ε_Ta2O5=80 ε_TiO2=26 D1=6e-9 D3=4e-9
*******Current Path SPICE Subcircuit
EV V 0 value={V(plus)-V(minus)}
EE1 E1 0 value={V(V)/(D1+D3*(ε_Ta2O5/ε_TiO2))}
EE2 E2 0 value={-V(V)/(D1+D3*(ε_Ta2O5/ε_TiO2))}
EI1_on I1_on 0 value={c1* ((  k1*V(E1)**2))
                    *exp(((-k2)/abs(V(E1))))*A}
EI1_off  I1_off 0 value={-c2*((  k1*V(E2)**2))
                    *exp(((-k2)/abs(V(E2))))*A}
EI2_on  I2_on 0 value={Io*exp(-(10n)/d0)
                    *sinh(V(V)/v0)}
EI2_off  I2_off 0 value={Io'*exp(-(10n)/d0)
                    *sinh(V(V)/v0)}
********Device current
G_tun plus minus value={(V(V) >= 0)  ? {V( I1_on)} : {V(I1_off)} }
G_hop plus minus value={(V(V) >= 0)  ? {V(I2_on)} : {V(I2_off)} }
```

FIGURE 9.14 Pseudocode For The Electrical Model Of A Selector.

and (9.2), in which dependent source functions are defined for the derivation of current components [40].

$$FN = A * \frac{q^3 m_{eff} E^2}{8\pi m_{diel} hq\varphi} e^{\frac{-4\sqrt{2m}\varphi^{3/2}}{3hqE}} \tag{9.1}$$

$$I(d,V) = I0 * e^{\left(\frac{-d}{d0}\right)} \sinh^{\left(\frac{V}{V0}\right)} \tag{9.2}$$

In (9.2), d_0 and v_0 are constants, d and V are the thickness of the dielectric layer and voltage. In the pseudocode, *c1* and *c2* are the fitting parameters, and *D1*, *D2*, and *D3* are the thicknesses of dielectric layers Ta_2O_5, TaO_x, and TiO_2, respectively, as shown in Figure 9.12

Where $k1 = \frac{q^3 m_{eff}}{8\pi m_{diel} hq\varphi}$ and $k2 = \frac{-4\sqrt{2m}\varphi^{3/2}}{3hq}$

A sinusoidal input of amplitude 3 V and frequency of 100 Hz is asserted to simulate the subcircuit of the proposed electrical model and analyzed for a transient of

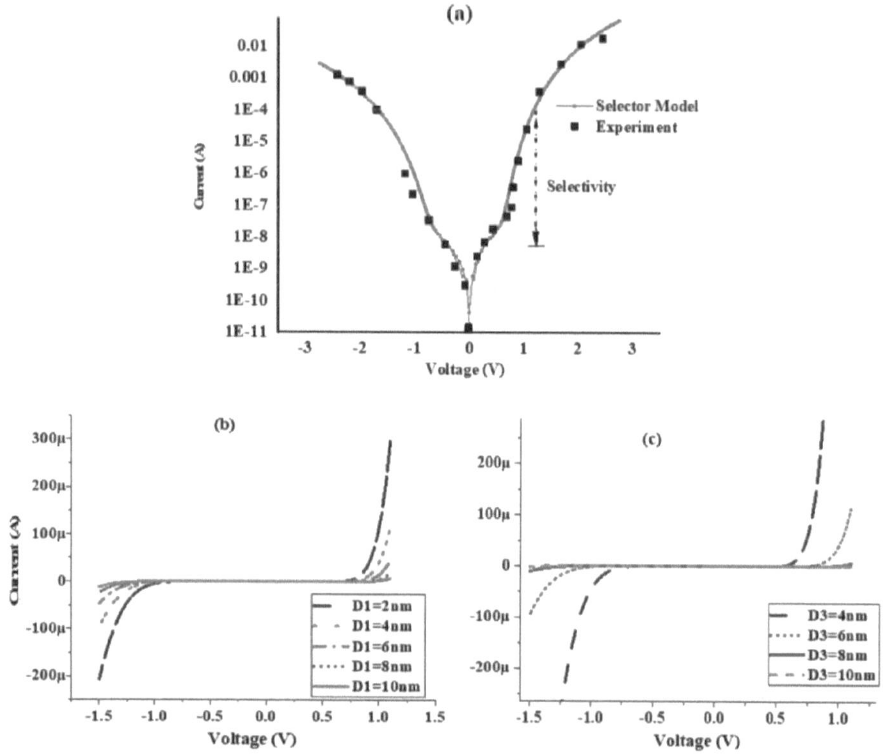

FIGURE 9.15 (a) Semi logarithmic I-V characteristics of the selector device and proposed model where the model is represented with a solid line and squares denote the measured data [34], (b) Variations in current value with variations in the thickness of Ta_2O_5 plotted in linear scale, (c) Drastic change in the current level with variations in TiO_2 thickness.

0.01 seconds; the resulting characteristics are depicted in Figure 9.15(a), showing congruence with the measured data as reported in [23]. From the simulated curve, it was noticed that below the threshold voltage (~ 0.7 V), the model shows a low leakage current attributed to the open switch, while at an input voltage > 1.2 V, the characteristics exponentially increase is attributed to closed switch behavior. Moreover, the device current is asymmetric at both polarities, with ~ 0.01 A appearing at 3 V, and 2 mA at –3 V.

Further observations from the proposed selector model are a high selectivity of ~ 10^4 with a nonlinear variation of current at low ($V_{read}/2$) and high voltages (V_{read}), with a high current density of ~ 10^7 A/cm^2.

Moreover, variations in current value for various thicknesses of tunneling oxides were analyzed (given in Figure 9.15(b)) and it was found that when the thickness of Ta_2O_5 increases, the current across the device decreases but in a linear manner. In contrast, an increase in the thickness of TiO_2 showed a drastic reduction in the current (as shown in Figure 9.15 (c)), and it was observed that beyond 8 nm, the current variations seen in simulations were negligible.

9.3.2 Electrical Model Implementation of a Pt/Ta_2O_5/TaO_x/Pt RRAM Device Incorporating the Current Through the Outside of the Conduction Filament (OCF)

Most of the materials used for developing RRAM are susceptible to a sneak-path current, though one of the prominent materials is Ta_2O_5 [41–43]. When a crossbar array of Ta_2O_5/TaO_x devices is subjected to a data write and subsequent read process, it has been observed that the sneak-path current originates in the device in the micro-Amperes range. Therefore, a Pt/Ta_2O_5/TaO_x/Pt RRAM device was selected to study the sneak-path current.

The experimental characteristics of the Ta_2O_5/TaOx device are adopted from [44], where the device showed a peculiar behavior of substantial reduction in resistance beyond the RESET voltage. Nevertheless, many analytical and electrical models of the device are available, and most of the simulated characteristics correlate well with the experimental result. But it is conspicuous that those models could not replicate the device characteristics beyond the RESET voltage. Moreover, it was observed that the current conduction in those models was realized by the current component that only conducts through the conduction filament. Therefore, the proposed model incorporates an additional current that occurs beyond the RESET voltage by considering current conduction through the outside of the filament (OCF). This concept has been introduced to the model by implementing the trap-assisted tunneling (TAT) mechanism supported by the oxygen vacancy traps present outside the filament, where these traps act as an intermediate jumping site to facilitate the conduction of electrons [39]

A two-terminal SPICE model with TAT is realized at the circuit level by implementing additional voltage-dependent current sources in the existing two-terminal circuit model of the device [13, 45]. In the existing device model, the current conduction is modeled with Schottky barrier tunneling, and the state dynamics of resistive switching are modeled with the oxygen vacancy drift present at the conduction filament, as depicted in Figure 9.16, where the capacitor *Cw* and the resistor *R* take values as given in [46]. G_{on} and G_{off} are the current sources representing the ionic current (*Iw*) during reverse and forward bias, respectively. Finally, the integration of *Iw* provides the state evolution of the RRAM model. The details of the realization of the current path subcircuit are explained in the following sub-sections.

In the proposed model, the increase in current conduction at the RESET phase of the Pt/Ta_2O_5/TaO_x/Pt device is attained by incorporating TAT, which is realized in the circuit model by integrating an additional voltage-dependent current source in the port equation given in [46]. In Figure 9.17, a single current source associated with TAT (*GTri*) has been connected parallel to the current sources representing Schottky conduction during forward (*GItunOFF*) and reverse bias (*GItunON*), respectively [47].

In addition to the current source, two voltage sources are also implemented in the design to indicate the contribution of the triangular barrier at high electric fields. These voltage-dependent voltage sources are *Etri* and *Eul*, which are added to the existing SPICE model port equation to obtain the RESET phase characteristics

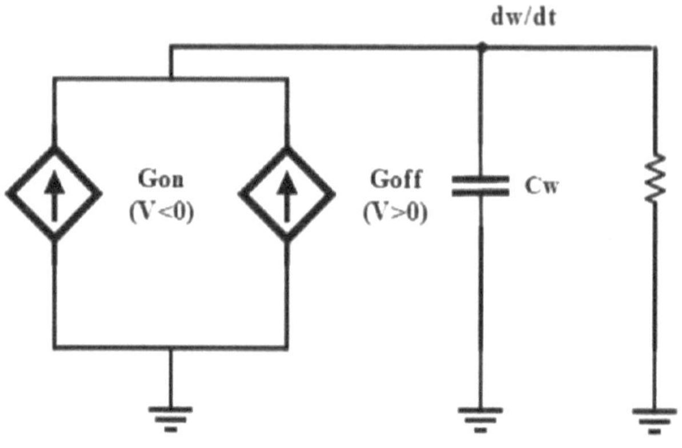

FIGURE 9.16 Subcircuit showing the evolution of state dynamics of $Pt/Ta_2O_5/TaO_x/Pt$ [13].

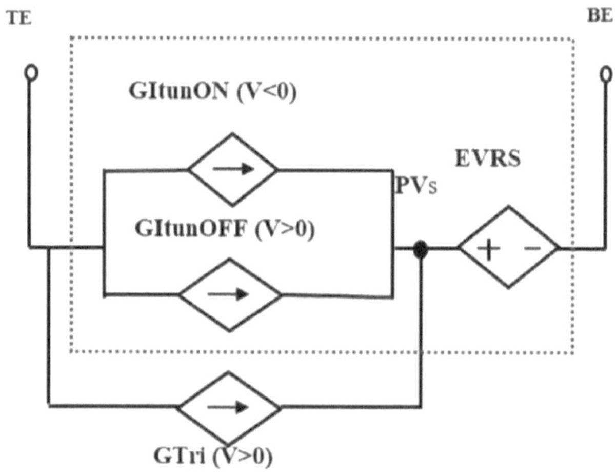

FIGURE 9.17 LTSPICE implementation of the proposed SPICE model for a TAT [47]. The two-terminal (current path) SPICE implementation of a single $Pt/Ta_2O_5/TaO_x/Pt$ device with three current components. Dotted lines show the port representation from the existing model [13], where PVs denote the internal node.

controlled by the voltage drop across the switching layer (VD). Also, the additional current source *GTri* is a state-independent TAT current.

The proposed device model is simulated with the sinusoidal input of an amplitude of 3 V and frequency of 100 Hz; then, the characteristics are analyzed for a transient of 0.01 seconds. The simulation result, as shown in Figure 9.18(a), demonstrates an increase in the current profile after 2 V. It demonstrates the reduction in the entire resistance of the device owing to the additional current path at OCF. Here, the assumption is that the number of oxygen vacancies in the OCF region is constant. The maximum current attained for TAT at 3 V is ~ 80 μA, matching the experimental results.

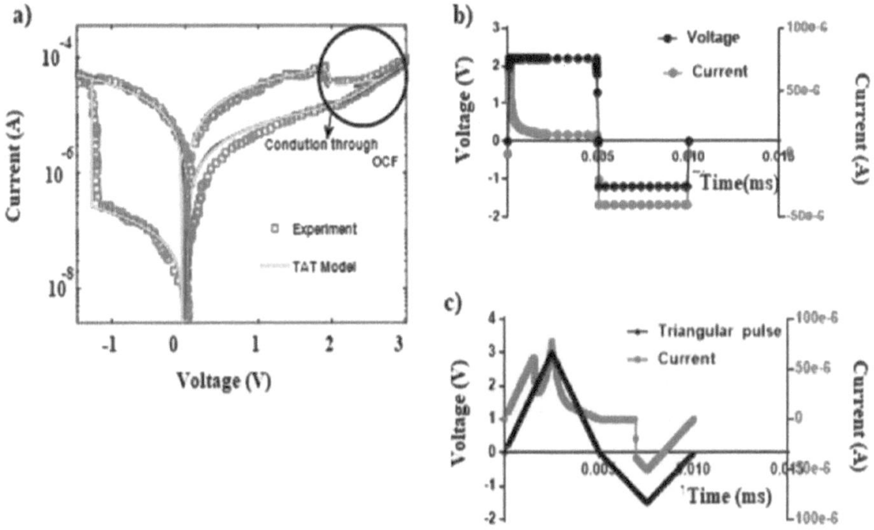

FIGURE 9.18 (a) The I-V characteristics obtained from the proposed model with the TAT approach; the region highlighted in a black circle denotes the increased current conduction beyond RESET due to the current conduction path through OCF [48]. The plotted experiment value is taken from [45]. b) The model response to step input with 2V amplitude and 100Hz frequency. c) The model response to triangular input with 2V and –1.5V voltage swings and frequency.

Next, the model's response is tested with triangular and rectangular pulses. An input pulse signal with a 2 V amplitude and 100 Hz is applied, and its output shows a ~ 80 μA current beyond the RESET voltage at 3 V as in Figure 9.18(b). Similarly, the model's response to triangular input is tested with 3 V and –1.5 V input swings with 100 Hz frequency. The result, as plotted in Figure 9.18(c), shows that the current at the RESET phase of the device increases up to a level of ~ 80 μA.

9.4 INTEGRATED MODELING OF THE PROPOSED SELECTOR MODEL WITH THE RRAM MODEL

Both RRAM and selector models are integrated to form a 1S-1R, comprising a Pt/Ta_2O_5/TaO_x/TiO_2/Pt selector, and an oxygen vacancy-based (V_0) Ta_2O_5/TaO_x RRAM is established to demonstrate its compatibility in reducing the sneak-path current. The following section discusses the electrical model of the 1S-1R circuit and its contribution to reducing the sneak-path current.

The complete circuit diagram representing the 1S-1R electrical model configuration is shown in Figure 9.19. The first part of the circuit model illustrates the selector, and the second part describes the RRAM model. The selector model is developed with two current components representing the *G_tun* and *G_hop* voltage-controlled voltage source (VCCS), as discussed in the previous session. The magnitude of the

TABLE 9.2
Parameters Used in the LTSPICE Model of Ta_2O_5/TaO_x RRAM [48]

Symbol	Value	Unit	Description
φ_t	2.3 [48]	eV	Trap energy
m_{ox}	0.3m0 [49]	Kg	The effective mass of electrons in metal oxide
m_{no}	0.1 m0 [50]	Kg	The effective mass of electrons in the metal
m0	9.1×10^{-31}	Kg	Mass of free electron
A	9×10^{-9}	m^{-2}	Area
D	4 [8]	nm	Metal oxide thickness
Ea	1.3 [51]	eV	Activation energy
E_0	0.2 [50]	eV	Total energy of an electron

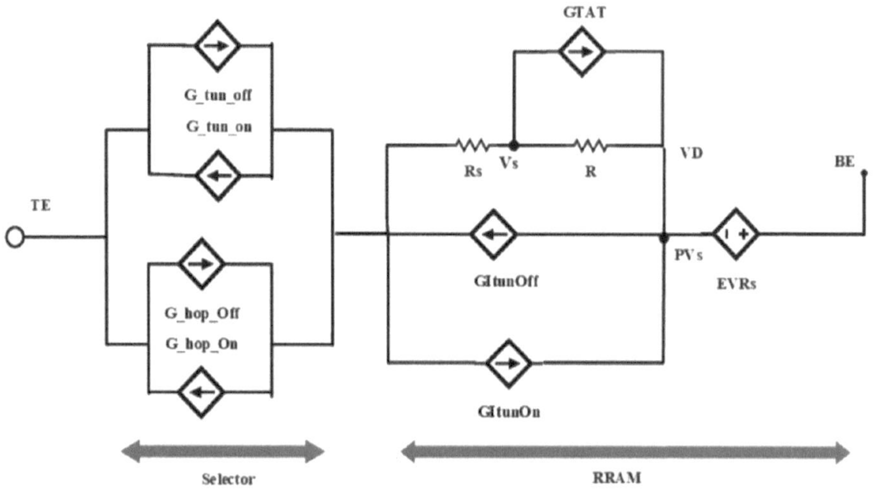

FIGURE 9.19 Electrical model of 1S-1R configuration marking all the associated current sources.

current changes with the input polarity due to asymmetry in the device structure; hence, both forward as well as the reverse bias current components are depicted in Figure 9.19, where *G_tun _On* and *G_hop _On* represent the forward current and *G_tun _Off* and *G_hop _Off* represent reverse current. In the second part of the combined circuit, three VCCS are used to represent the current path subcircuit of the RRAM, where the first element GTAT is the VCCS that represents the electron tunneling current (TAT) driven by the electric field across the dielectric layer at the positive bias of the input voltage.

Similarly, the sources *GItunON* and *GItunOFF* represent the contribution of the Schottky barrier tunneling current through the conduction filament (CF) activated at an applied voltage of $V < 0$ and $V > 0$, respectively. These current sources deliver charge to the internal node (PVs) between the VCCSs and voltage source eVs, as shown in Figure 9.19. The voltage-controlled voltage source (eVs) is connected in series with the current sources representing the voltage drop on the bulk TaO_x layer (IRB). The remaining voltage ($V–V_{sel}–IRB$) drops across the VCCS, representing the sum of the potential drop across the Schottky interface (V_s), as well as the Ta_2O_5 layer (V_D) that appears across the current sources of *GItunOFF* and *GItunON*. Thus, the 1S-1R model is described and the total current through the device is the sum of Schottky barrier tunneling as well as the TAT current, which contributes charge to PV nodes connected in series with eVRs; both current and voltage sources are controlled by the dependent voltage sources (VCVS). Based on Kirchhoff's voltage law, the total input voltage applied is distributed across the selector and RRAM circuit models, as shown in Figure 9.20(a), indicating that most of the input voltage (V_{tot}) from ~ –1.0 V to 1.5 V appears across the selector. Therefore, the voltage developed across the RRAM is negligible when changing its current memory state; thereby, the leakage current can be eliminated to an extent. At V_{tot} > –0.85 V, the voltage across the RRAM starts to build up due to the

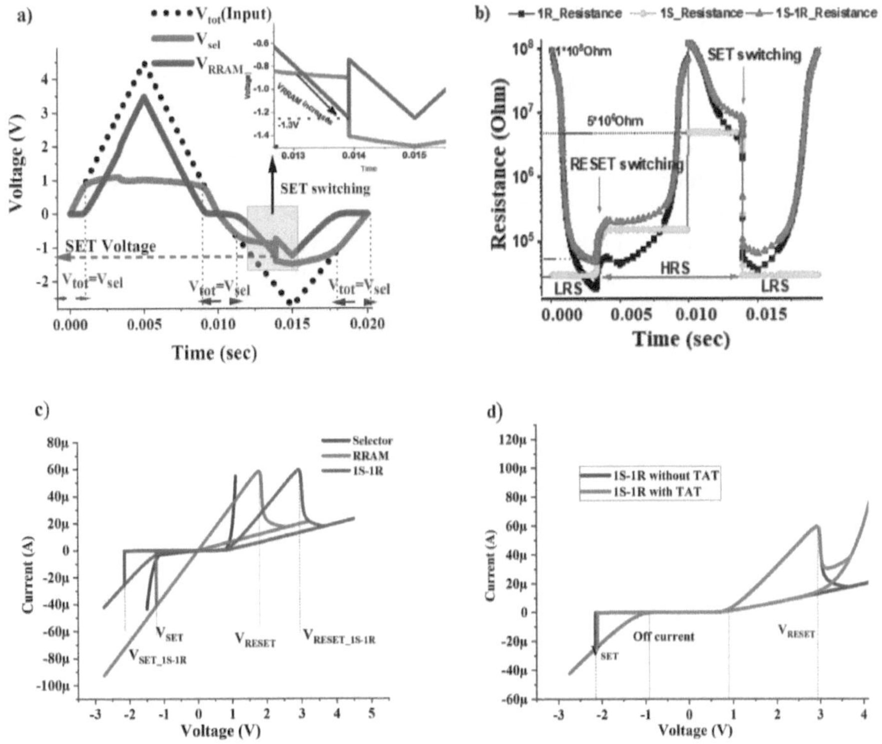

FIGURE 9.20 Comparison of related parameters in 1S-1R, 1S, and 1R.

formation of the conduction filament; when it reaches –1.3 V, the device switches to a LRS; therefore, the voltage drop across it reduces abruptly back to –0.7 V, as given in inset of Figure 9.20(a).

After that, a further increase in voltage leads to a rise in potential across both models, with more voltage drops occurring at the selector. Figure 9.20(b) illustrates the comparison of the resistance across 1S, 1R, and 1S-1R combinations. The circuit of the 1S-1R model shows comparatively higher resistance than 1S and 1R for the entire cycle of testing, owing to the effect of series resistance. However, during ON-state (LRS), the resistance of the combined circuit is in the same order as the resistance that appears in the RRAM model. In contrast, in low-voltage regions the total resistance of 1S-1R becomes ~ 10 GΩ which is high enough to prevent a sneak-path current. Figure 9.20(c) is plotted on a linear scale, showing the proposed selector, RRAM response, and both models together. As expected, the simulated model of 1S-1R demonstrates the characteristics of the memory window, but there is also a change in SET and RESET voltages of 1S-1R compared to 1R. Similarly, the combination of the proposed selector model with the RRAM model incorporating the current conduction beyond the RESET phase is also found to be compatible with the selector model, as shown in Figure 9.20(d). From the I-V characteristics, the preferable range of read/write voltages required for both selected and unselected bit/word lines to avoid the excess of sneak-path current can be predicted.

Moreover, the optimum input voltage asserted to attain SET and RESET voltage in the 1S-1R cell can also be predicted in advance. For instance, the input voltage required for SET switching in the RRAM is ~ –1.2 V, whereas, in an integrated circuit, this parameter becomes ~ 2.7–3.6 V. Likewise, the voltage needed for RRAM and 1S-1R to drive into the RESET-state is found to be ~ 1.9 V and ~ 4.5 V, respectively. Even though the nonlinearity of RRAM is ~ 2, the combined circuit acquired a nonlinearity factor of ~ 50, which was inherited from the selector model. Similarly, the off-current in 1S-1R is as low as 0.04 μA to facilitate a reduction in sneak-path current. The analysis shows that the low off-current present in the selector is directly reflected in the 1S-1R setup. However, the R_{off}/R_{on} ratio of the combined circuit is ~ 160, inherited from the RRAM when calculated at the read voltage of both circuits. To compare the simulated characteristics of RRAM and 1S-1R models, extracted parameters are listed in Table 9.3.

9.4.1 Analysis and Results

The read voltage required to supply the selected RRAM device can be predicted from the above analysis. In general, the read voltage for 1S-1R can be assigned in the range of $|V_{thr}| < |V_{sel-read}| < |V_{write}/2|$, where V_{sel_read} is the read voltage to be applied to the selected cell, Vthr is the threshold voltage measured from the selector, $V_{write}/2$ is half of the write voltage applied to the bit line of the selected RRAM cell. The read bias of unselected cells can be allotted below the threshold voltage of the selector so that the selector drives to a high resistance state

TABLE 9.3
Comparison of Parameters Derived From RRAM and Proposed 1S-1R Model

Parameters	1R	1S-1R
R_{off}/R_{on} ratio	~ 150 (at 1.0 V)	~ 160 at -2 V
Nonlinearity (calculated at $V_{read}/2$) Vread)	~ 2 (at 0.5&1 V)	50 (at 0.7 & 1.4 V)
OFF-current (*A*) (at 0.5V)	3 μA	0.04 μA
ON-state current (*A*)	60 μA	60 μA
RESET voltage	1.9 V	> = 4.5 V
SET voltage	~ –1.2 V	–2.7 to –3.6 V

and effectively blocks the sneak-path current from the unselected cells, whereas, beyond the threshold, the selector model enables the RRAM to function write or read.

Next, the response of the integrated (1S-1R) model to pulse input is tested and given in Figure 9.21, where Figures 9.21(a) and 9.21(b) show the negative and positive input pulses applied, Figure 9.21(c) shows the corresponding hysteresis characteristics. While simulating with pulse input, the SET current increased up to 36 μA in response to the negative input pulse of amplitude –2.7 V, a pulse width of 180 μs with a fall delay time of 80 μs. Similarly, to achieve RESET switching, a positive pulse with an amplitude of 4.5 V and a pulse width of 180 μs, and a transition delay of 80 μs (*T_rise*) was applied. The result showed the hysteresis curve with negligible off-current (10 nA) at an input of –1 V to 1 V and followed the RRAM current characteristics beyond that (~ –1.5 V to –3.5 V and ~ 1.5 V to 4.5 V). While analyzing, we noticed that the pulse transition time (rise/fall) and the amplitudes of positive and negative pulses affected the switching characteristics.

When considering the 1S-1R circuit from the perspective of sneak current, it is evident that the proposed 1S-1R circuit is beneficial to use in a crossbar array as its off-current within –1 to 1 V settles at nano-amperes and the low/high resistances are distinguishable [19]. Moreover, the nonlinearity exhibited by the 1S-1R model enables the choice of a different voltage range to bias both selected and unselected cells. This study predicts the read and write voltage required to supply trails at cross points of selected and unselected cells. An input voltage ranging from –2.6 V to –3.6 V is required to drive the device to the ON-state, and at least 4.4 V is needed to drive the device to the OFF-state, whereas to perform the read operation with a selected cell, the range of voltages from 1.3 to 1.8 V is preferred.

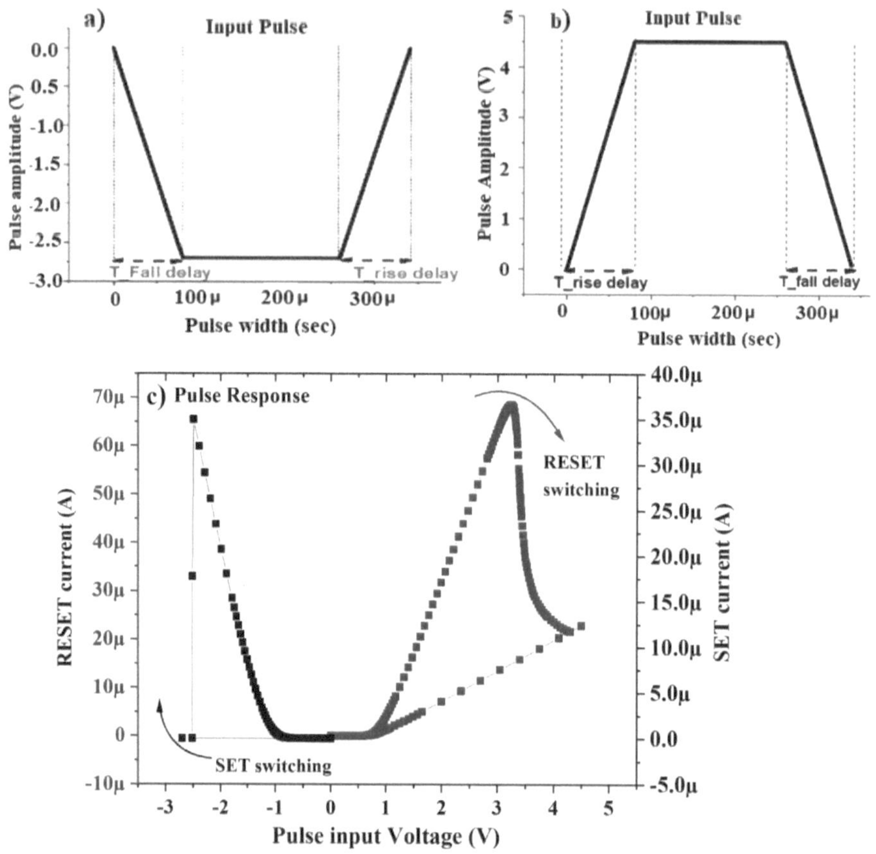

FIGURE 9.21 Response of the 1S-1R model to pulse input: (a) negative pulse input, (b) positive pulse input, (c) 1S-1R I-V characteristics with the above-mentioned applied input pulses.

9.5 FORMATION AND SNEAK-PATH CURRENT MEASUREMENT OF 2 × 2 CBA OF 1S-1R WITH A Pt/Ta_2O_5/TaOx/TiO_2/Pt SELECTOR AND Pt/Ta_2O_5/TaOx/Pt RRAM

CBAs of 2 × 2 were created in a Pt/Ta_2O_5/TaO_x/Pt RRAM model (1R) and Pt/Ta_2O_5/TaO_x/TiO_2/Pt selector model to investigate the performance of 1S-1R in reducing the sneak-path current. The sneak-path current problem in CBA becomes critical while performing the read operation; therefore, in this study, read is performed in worst-case conditions by keeping the selected cell in HRS and all neighboring unselected cells in LRS [19]. After that, the sneak-path current is measured during the read process. The read/write scheme followed in the proposed design is the *V*/3 scheme with the selected word line supplied with V_{dd}, bit line with the ground (GND), word lines of all unselected cells are biased with *V*/3 potential, and their bit lines with 2*V*/3 bias. In order to compare the reduction in sneak-path current, two 2 × 2 crossbar

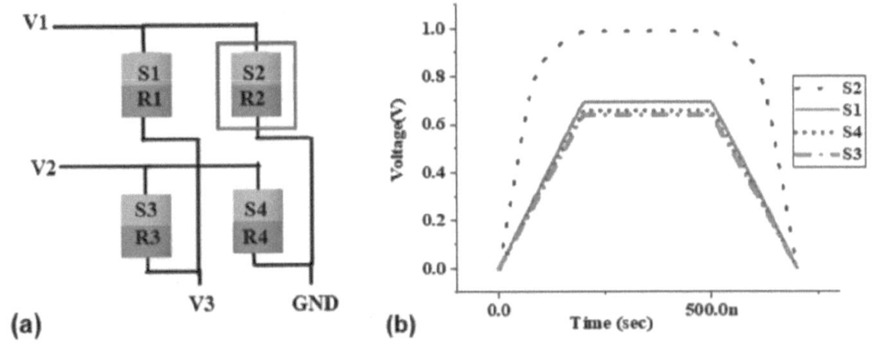

FIGURE 9.22 (a) The layout of a 2 × 2 crossbar array, where S2-R2 is the selected cell marked with a red square. (b) The voltage distribution across the selectors S1, S2, S3, and S4.

arrays were developed; one is an array of RRAMs only (not shown), and the other is with 1S-1R cells.

A 1S-1R 2 × 2 crossbar array is shown in Figure 9.22(a), where S2-R2 is the selected cell, and all remaining cells are unselected. The read process is initiated after performing the write.

As discussed in the previous session, read is performed in a RRAM array with a positive pulse amplitude of ~ 1 V, whereas, considering the drop across the selector, the read voltage selected for 1S-1R is 1.8 V. To read the data from the RRAM array, an input pulse of 1 V with 300 ns pulse width is applied for a transient of 0.7 μs and the measured sneak-path current is 10 μA, whereas, in the 1S-1R array, the worst-case read process is invoked with 1.8 V, 300 ns pulse with rise and fall time delay of 200 ns each, and the corresponding sneak-path current calculated as 56 nA. Hence, the observation is that 98% of the sneak current is reduced in the 1S-1R array compared to the 1R array. Moreover, it has been noticed that the voltage drop across the unselected 1S-1R cell is ~ 0.6 V, whereas the drop across the selected 1S-1R is ~ 1.9 V. Also, the voltage distribution across the selector in the selected 1S-1R cell during read is 0.98V and at the RRAM cell it is ~ 1 V. Figure 9.22 (a) shows the layout of the 1S-1R array and Figure 9.22(b) shows the voltage drop across the selectors in the array. In the figure, S2 is the selector in the selected 1S-1R cell and the remaining selectors, such as S1, S3, and S4, are unselected. As the voltage drop across the selector in the selected cell is ~ 1 V, the current across the model increases nonlinearly, facilitating more voltage to drop across the RRAM cell. The voltage across the unselected selectors is in the range of 0.6 V to 0.7 V, which provides a comparatively lower current to drive the RRAM device. Hence the unselected memory device remains switched off to prevent a sneak current during the read.

REFERENCES

1. J. J. Yang, D. B. Strukov, and D. R. Stewart, "Memristive devices for computing," *Nat. Nanotechnol.*, vol. 8, no. 1, pp. 13–24, 2012, doi: 10.1038/nnano.2012.240.

2. S. Venkatesan and M. Aoulaiche, "Overview of 3D NAND technologies and outlook invited paper," in 2018 Nonvolatile Memory Technology Symposium (NVMTS), 2018, pp. 1–5, doi: 10.1109/NVMTS.2018.8603104.
3. J. You, J. Song, and C. Kim, "A 2-Gb/s/ch data-dependent swing-limited on-chip signaling for single-ended global I/O in SDRAM," *IEEE Trans. Circuits Syst. II Express Briefs*, vol. 64, no. 10, pp. 1207–1211, 2017, doi: 10.1109/TCSII.2015.2483158.
4. *International Technology Roadmap for Semiconductors (ITRS) Report, 2.0*, 2020.
5. J. Meena, S. Sze, U. Chand, and T.-Y. Tseng, "Overview of emerging non-volatile memory technologies," *Nanoscale Res. Lett.*, vol. 9, no. 1, p. 526, 2014, doi: 10.1186/1556-276X-9-526.
6. S. Kvatinsky, E. G. Friedman, A. Kolodny, and U. C. Weiser, "The desired memristor for circuit designers," *IEEE Circuits Syst. Mag.*, vol. 13, no. 2, pp. 17–22, 2013, doi: 10.1109/MCAS.2013.2256257.
7. D. S. Jeong et al., "Emerging memories: Resistive switching mechanisms and current status," *Reports Prog. Phys.*, vol. 75, no. 7, p. 076502, 2012, doi: 10.1088/0034-4885/75/7/076502.
8. https://slideplayer.com/slide/13852296/.
9. L. O. Chua, "The fourth element," *Proc. IEEE*, vol. 100, no. 6, pp. 1920–1927, 2012, doi: 10.1109/JPROC.2012.2190814.
10. D. B. Strukov, G. S. Snider, D. R. Stewart, and R. S. Williams, "The missing memristor found," vol. 453, no. May, pp. 80–84, 2008, doi: 10.1038/nature06932.
11. J. Joshua Yang et al., "The mechanism of electroforming of metal oxide memristive switches," *Nanotechnology*, vol. 20, no. 21, p. 215201, 2009, doi: 10.1088/0957-4484/20/21/215201.
12. Z. Wei et al., "Switching mechanism of TaOx ReRAM," in International Conference on Solid State Devices and Materials, Sendai, 2009, pp. 1202–1203.
13. F. O. Hatem, P. W. C. Ho, T. N. Kumar, and H. A. F. Almurib, "Modeling of bipolar resistive switching of a nonlinear MISM memristor," *Semicond. Sci. Technol.*, vol. 30, no. 11, p. 115009, Nov. 2015, doi: 10.1088/0268-1242/30/11/115009.
14. P. Sarkar and A. Roy, "Improvement of retentivity in TiOx/HfOx bilayer structure for low power resistive switching memory applications," *Surf. Rev. Lett.*, vol. 22, p. 150201225054003, Feb. 2015, doi: 10.1142/S0218625X15500316.
15. J. A. Lekshmi, T. Nandha Kumar, and K. Jinesh, "Multilevel non-volatile memory based on Al_2O_3/ZnO bilayer device," *Micro Nano Lett.*, vol. 15, no. 13, pp. 910–914, Nov. 2020, doi: https://doi.org/10.1049/mnl.2020.0335.
16. Z. Fang, H. Y. Yu, X. Li, N. Singh, G. Q. Lo, and D. L. Kwong, "HfOx/TiOx/HfOx/TiOxmultilayer-based forming-free RRAM devices with excellent uniformity," *IEEE Electron Device Lett.*, vol. 32, no. 4, pp. 566–568, 2011, doi: 10.1109/LED.2011.2109033.
17. D. Ielmini, "Resistive switching memories based on metal oxides: mechanisms, reliability and scaling," *Semicond. Sci. Technol.*, vol. 31, no. 6, p. 063002, 2016, doi: 10.1088/0268-1242/31/6/063002.
18. E. Linn, R. Rosezin, C. Kügeler, and R. Waser, "Complementary resistive switches for passive nanocrossbar memories," *Nat. Mater.*, vol. 9, no. 5, pp. 403–406, 2010, doi: 10.1038/nmat2748.
19. L. Jagath and T. N. Kumar, "A comparative study on the performance of 1S-1R and Complementary resistive switching models," in 2020 IEEE International Conference on Semiconductor Electronics (ICSE), 2020, pp. 9–12, doi: 10.1109/ICSE49846.2020.9166874.
20. Y. Li et al., "Bipolar one diode–one resistor integration for high-density resistive memory applications," *Nanoscale*, vol. 5, no. 11, pp. 4785–4789, 2013, doi: 10.1039/C3NR33370A.

21. G. Niu et al., "Material insights of HfO 2-based integrated 1-transistor-1-resistor resistive random access memory devices processed by batch atomic layer deposition," *Nat. Publ. Gr.*, 2016, doi: 10.1038/srep28155.
22. Y. S. Chen et al., "Highly scalable hafnium oxide memory with improvements of resistive distribution and read disturb immunity," in 2009 IEEE International Electron Devices Meeting (IEDM), Kyoto, Japan, 2009, pp. 1–4, doi: 10.1109/IEDM.2009.5424411.
23. J. Woo et al., "Multilayer tunnel barrier (Ta2O5/TaOx/TiO2) engineering for bipolar RRAM selector applications," in 2013 Symposium on VLSI Technology, Kyoto, Japan, 2013, pp. T168–T169.
24. B. J. Choi et al., Trilayer tunnel selectors for memristor memory cells, *Adv. Mater.*, 2016, 28, 356–362, 2015.
25. W. Lee et al., "High current density and nonlinearity combination of selection device based on TaO x / TiO 2 / TaO x structure for one selector À one resistor arrays," no. 9, pp. 8166–8172, 2012.
26. B. Govoreanu, P. Blomme, M. Rosmeulen, J. Van Houdt, and K. De Meyer, "VARIOT: A novel multilayer tunnel barrier concept for low-voltage non-volatile memory devices," *IEEE Electron Device Lett.*, vol. 24, no. 2, pp. 99–101, 2003, doi: 10.1109/LED.2002.807694.
27. Y. Sun et al., "Performance-enhancing selector via symmetrical multilayer design," *Adv. Funct. Mater.*, vol. 29, Jan. 2019, doi: 10.1002/adfm.201808376.
28. Y. Cassuto, S. Kvatinsky and E. Yaakobi, "Sneak-path constraints in memristor crossbar arrays," in 2013 IEEE International Symposium on Information Theory, 2013, pp. 156–160, doi: 10.1109/ISIT.2013.6620207.
29. T. N. Kumar, H. A. F. Almurib, and F. Lombardi, "Design of a memristor-based look-up table (LUT) for low-energy operation of FPGAs," *Integration*, vol. 55, pp. 1–11, 2016, doi: https://doi.org/10.1016/j.vlsi.2016.02.005.
30. A. Chen, "A highly efficient and scalable model for crossbar arrays with nonlinear selectors," in 2018 IEEE International Electron Devices Meeting (IEDM), 2018, pp. 37.2.1–37.2.4, doi: 10.1109/IEDM.2018.8614505.
31. J. Choi et al., "Trilayer tunnel selectors for memristor memory cells," *Adv. Mater.*, vol. 28, no. 2, pp. 356–362, Jan. 2016, doi: 10.1002/adma.201503604.
32. S. H. Jo, T. Kumar, S. Narayanan, and H. Nazarian, "Cross-point resistive RAM based on field-assisted superlinear threshold selector," *IEEE Trans. Electron Devices*, vol. 62, no. 11, pp. 3477–3481, 2015, doi: 10.1109/TED.2015.2426717.
33. W. Lee et al., "High current density and nonlinearity combination of selection device based on TaOx/TiO2/TaOx structure for one selector: One resistor arrays," *ACS Nano*, vol. 6, no. 9, pp. 8166–8172, Sep. 2012, doi: 10.1021/nn3028776.
34. J. Woo, D. Lee, E. Cha, S. Lee, S. Park, and H. Hwang, "Multilayer-oxide-based bidirectional cell selector device for cross-point resistive memory applications," *Appl. Phys. Lett.*, vol. 103, no. 20, p. 202113, Nov. 2013, doi: 10.1063/1.4831680.
35. J. Woo, D. Lee, E. Cha, S. Lee, S. Park, and H. Hwang, "Vertically stacked ReRAM composed of a bidirectional selector and CB-RAM for cross-point array applications," *IEEE Electron Device Lett.*, vol. 34, no. 12, pp. 1512–1514, 2013, doi: 10.1109/LED.2013.2285583.
36. J. Huang, Y.-M. Tseng, W.-C. Luo, C.-W. Hsu, and T. Hou, "One selector-one resistor (1S1R) crossbar array for high-density flexible memory applications," in 2011 International Electron Devices Meeting, Washington, DC, USA, 2011, pp. 31.7.1–31.7.4. doi: 10.1109/IEDM.2011.6131653.
37. Q. Luo et al., "Highly uniform and nonlinear selection device based on trapezoidal band structure for high density nano-crossbar memory array ," *Nano Res.*, vol. 10, pp. 3295–3302, 2017. https://doi.org/10.1007/s12274-017-1542-2.

38. S. H. Jo, T. Kumar, S. Narayanan, W. D. Lu, and H. Nazarian, "3D stackable cross-bar resistive memory based on field assisted superlinear threshold (FAST) selector," in 2014 IEEE International Electron Devices Meeting, 2014, pp. 6.7.1–6.7.4, doi: 10.1109/IEDM.2014.7046999.
39. L. Jagath, T. N. Kumar, H. A. Almurib, and K. B. P. Jinesh, "Analytical modelling of tantalum/titanium oxide-based multilayer selector to eliminate sneak-path current in RRAM arrays," *IET Circuits, Devices Syst.*, vol. 14, no. 7, pp. 1092–1098, Oct. 2020, doi: https://doi.org/10.1049/iet-cds.2019.0480.
40. L. Jagath, T. N. Kumar, and H. A. F. Almurib, "Modeling of current conduction during RESET phase of Pt/Ta2O5/TaOx/Pt bipolar resistive RAM devices," in 2018 IEEE 7th Nonvolatile Memory Systems and Applications Symposium (NVMSA), 2018, pp. 55–60, doi: 10.1109/NVMSA.2018.00014.
41. L. Jagath, C. Hock Leong, T. N. Kumar, and H. F. Almurib, "Insight into physics-based RRAM models: review," *J. Eng.*, vol. 2019, no. 7, pp. 4644–4652, Jul. 2019, doi: https://doi.org/10.1049/joe.2018.5234.
42. K. M. Kim, S. R. Lee, S. Kim, M. Chang, and C. S. Hwang, "Self-limited switching in Ta2 O5 /TaOx memristors exhibiting uniform multilevel changes in resistance," *Adv. Funct. Mater.*, vol. 25, no. 10, pp. 1527–1534, 2015, doi: 10.1002/adfm.201403621.
43. S. Ruffell, P. Kurunczi, J. England, Y. Erokhin, J. Hautala, and R. G. G. Elliman, "Formation and characterization of Ta2O5/TaOx films formed by O ion implantation," *Nucl. Instruments Methods Phys. Res. Sect. B Beam Interact. with Mater. Atoms*, vol. 307, no. 0, p. 491, 2013, doi: 10.1016/j.nimb.2012.11.092.
44. J. H. Hur, M. J. Lee, C. B. Lee, Y. B. Kim, and C. J. Kim, "Modeling for bipolar resistive memory switching in transition-metal oxides," *Phys. Rev. B - Condens. Matter Mater. Phys.*, vol. 82, no. 15, pp. 1–5, 2010, doi: 10.1103/PhysRevB.82.155321.
45. J. Choi et al., "Trilayer Tunnel Selectors for Memristor Memory Cells," *IEEE Electron Device Lett.*, vol. 28, no. 5, pp. 261–267, Sep. 2016, doi: 10.1002/adma.201503604.
46. F. O. Hatem, T. N. Kumar, and H. Almurib, "A SPICE model of the Ta2O5/TaOx bi-layered RRAM," *IEEE Trans. Circuits Syst.*, vol. 63, no. 9, pp. 1487–1498, 2016, doi: 10.1109/TCSI.2016.2579503.
47. L. Jagath, T. Nandha Kumar, and H. A. F. Almurib, "Electrical model of Ta2O5/TaOx RRAM device with current conduction beyond RESET phase," 2019 IEEE 9th International Nanoelectronics Conferences (INEC), 2019, pp. 1–5, Kuching, Malaysia, doi: 10.1109/INEC.2019.8853845.
48. R. Ramprasad et al., "Oxygen vacancy defects in tantalum pentoxide: A density functional study," *Microelectron. Eng.*, vol. 69, no. 2–4, pp. 190–194, 2003, doi: 10.1016/S0167-9317(03)00296-X.
49. M. Houssa et al., "Trap-assisted tunneling in high permittivity gate dielectric stacks," *J. Appl. Phys.*, vol. 87, no. 12, pp. 8615–8620, May 2000, doi: 10.1063/1.373587.
50. S. Fleischer, P. T. Lai, and Y. C. Cheng, "Trap-assisted conduction in nitride-oxide and re-oxidized nitrided-oxide n-channel metal-oxide-semiconductor field-effect transistors," *J. Appl. Phys.*, vol. 73, p. 8353, 1993.
51. H. L. Chee, T. Nandha Kumar and H. A. Almurib, "Multifilamentary conduction modelling of bipolar Ta2O5/TaOx bi-layered RRAM," in 2018 IEEE 7th Non-Volatile Memory Systems and Applications Symposium (NVMSA), 2018, pp. 113–114, doi: 10.1109/NVMSA.2018.00029.

10 SRAM
An Essential Part of Integrated Circuits

Vishal Sharma and Shivani Godha

CONTENTS

10.1 INTRODUCTION

Static Random-Access Memory (SRAM) is a volatile memory that retains its data as long as the power is connected. It has been an indispensable part of almost all processing units by effectively being used as an on-chip cache memory, where the data is required to match the processor's speed. SRAM is preferred over other available memory solutions due to its faster response. Also, it does not need to be refreshed

DOI: 10.1201/9781003155751-10

repeatedly and thus consumes less standby power. In addition, the economic fabrication of SRAM through regular CMOS processing without any additional processing cost, unlike the other counterpart memories, such as DRAM, Flash, or advanced Resistive RAM (ReRAM) memories, further makes it a preferred on-chip cache over the other available memories.

Therefore, due to the ever-increasing complexity and higher bandwidth requirement of modern system-on-chip (SoC) designs, almost all the integrated circuits (ICs) necessitate on-chip SRAM for their complex processing units. The application area of such SoCs ranges widely from the battery-operated wireless sensing units to the artificial intelligence (AI) based image/voice processing units. Based on the *International Technology Roadmap for Semiconductors (ITRS) Report*, SRAM occupies more than 90% of the chip area of recent SoC designs and is expected to increase further for the upcoming more complex IC designs, due to technology advancements and the never-ending demands of electronic consumers.

Since on-chip SRAM occupies a major portion of the system area, the power consumed by *SRAM* also dominates the overall system power. It has thus been necessary to design area and power-efficient SRAM as this has been identified as one of the core barriers to the effective implementation of recent SoCs. Moreover, with technological advancements, the design challenges of SRAM become more critical at a scaled supply voltage, since the susceptibility toward PVT variation increases due to the reduced voltage swing values. This is because the inter/intra-die variations are inversely proportional to the channel area and thus continuously increase with the technology's downscaling. These process variations for on-chip memory are more crucial than any other subblocks of SoCs, due to the use of minimum-sized devices for SRAM cells. For SRAM memory, such variations majorly degrade the parameters, including read current (I_{Read}) and the read and write static noise margins (RSNM and WSNM, respectively) of SRAM cells at a low supply voltage. Also, read performance deteriorates due to the large variability of device ON-current (I_{ON}) caused by the (V_{TH}) fluctuations. Apart from the regular data stability challenges in modern nanotechnology nodes, SRAM half-select issues faced at a reduced supply voltage restricts the use of bit-interleaving array architecture and make it prone to the soft error occurrence.

Looking into the above-mentioned design challenges of SRAM, the modern integrated circuit market requires a lot of design input at both the bit cell and architecture levels of SRAM. This is due to the demand for low-power consumption for battery longevity, reliable data storage, and faster performance to support the accuracy and speed of advanced machine learning-based algorithms.

This chapter discusses the various applications of SRAM and its critical design parameters for respective applications. Then, the standard SRAM bit-cell architectures, their design metrics, and the major challenges will also be covered. The simulation-based results and data included in this chapter were generated through the Cadence Virtuoso design environment and ADEXL simulator. This chapter also covers the various possible design strategies for the SRAM of next-generation IC designs at both the bit cell and architecture levels. This book chapter focuses on the memory design solutions required for modern Internet-of-Things (IoT) edge devices and highly accurate AI edge devices.

10.2 APPLICATION WINDOWS

SRAM comprises a significant percentage of the total chip area for different SoC designs. For this reason, the design of SRAMs has been a very challenging task over the past few decades in the semiconductor market, especially at the advanced technology nodes. However, the required design performance of SRAM, like low power, high speed, or high density varies based on its targeted application; hence, the designers must carefully select the design strategies for the given application. Based on its applications, SRAM can be categorized into three main performance categories, as illustrated in Table 10.1. The given table also presents the possible design strategies that can be adopted while targeting a particular SRAM application category.

10.2.1 IoT-Enabled Wireless Sensor Networks

With the revolutionary growth of the internet, IoT has become ubiquitous in human life by connecting the many digital devices, and this is expected to increase in the foreseeable future, as shown in Figure 10.1 [1–3]. The applicability of IoT-enabled wireless sensor networks ranges widely from smart homes and smart cities to smart automobiles and also smart health monitoring systems. This plethora of connected things using wireless sensor networks (WSNs) is based on the functioning of multiple wireless sensor nodes. A typical IoT-based wireless sensor network and the basic architecture of a sensor node are provided in Figure 10.2. The role of sensor nodes in such networks is to sense the intended activities of targeted objects, collect the relevant data, process it, and then transfer it to the base station (gateway) [5]. During this process, the data is temporarily stored within battery-operated sensor nodes for

TABLE 10.1
Application-Dependent SRAM Design Strategies

Performance Category	Low Power	High Density	High Performance
Application Area	IoT edge devices, wireless sensor networks, wearable biomedical implants	Mobile phones, laptops, other multimedia gadgets	Image processing, high-end server machines
Design Strategies	VDD scaling, small device sizes, high-threshold devices, hierarchical bit line array architectures	Small device sizes, long bit line array architectures	Larger device sizes, low-threshold devices, hierarchical bit line array architectures

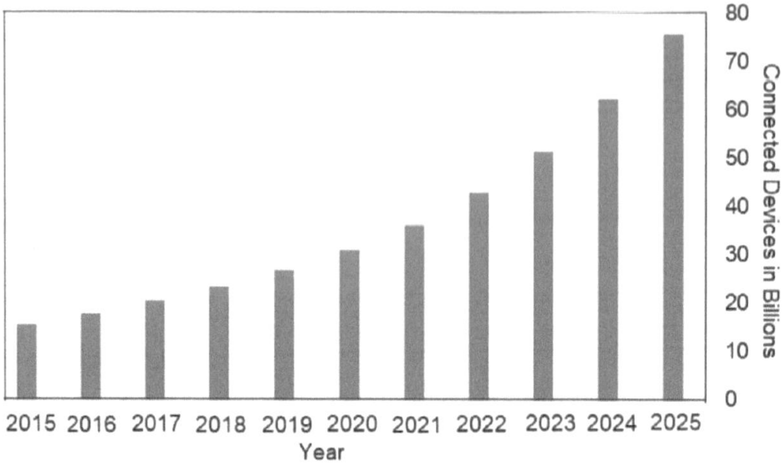

FIGURE 10.1 Growth of IoT devices [1].

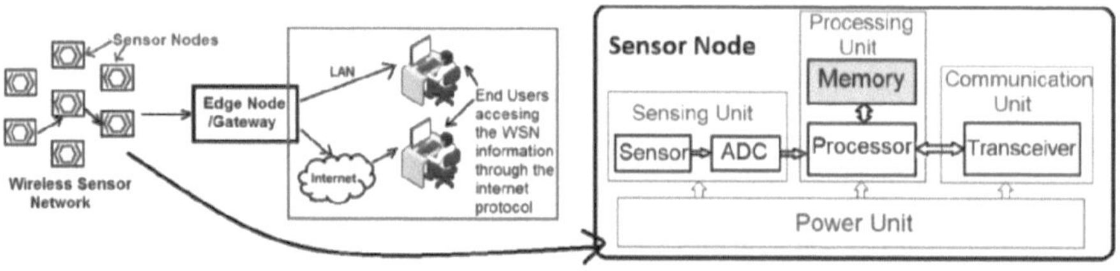

FIGURE 10.2 A typical IoT-based wireless sensor network and sensor node architecture [4].

processing, and hence independent of any applications there is a common trait for each of the sensor nodes, that low-power memory is required to extend system functions under limited battery-driven power resources [6].

Since SRAM is always preferred for such memory requirements due to its faster operation that supports the high data transfer rate of the system, a large percentage of the total area of each sensor node is occupied by an on-chip SRAM to support the complex run-time detection data. Therefore, due to the compact size of the sensor nodes with the limited battery-driven power budget for wearable IoT applications, a low-power and high-density SRAM is needed.

10.2.2 Multimedia Applications

Multimedia applications include voice, image or video processing and this is implemented by compressing the mobile data at the transmission end (content provider) by the encoder before communicating with the server. Then at the receiving end, whether it may be a smartphone or laptop unit, the signal processing unit of the decoder extracts the original data, such as the motion estimation. As shown in Figure 10.3, to

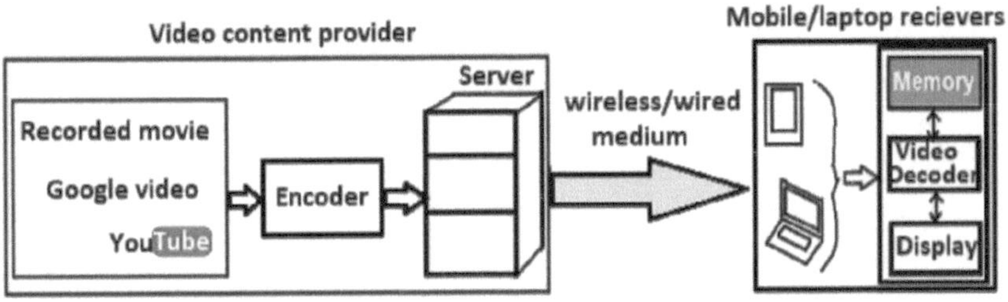

FIGURE 10.3 Memory needed for computation in multimedia applications [7].

extract the information the processing unit requires a significant amount of on-chip computation and thus demands a large, embedded memory [7]. The IoT processors of such high-end devices require fast read–write performance to support the mobile data while maintaining the low processing power [8]. Voltage scaling seems the best choice to reduce the power dissipation of the high-density on-chip memory. However, it may increase the bit-error-rate (BER) for different pixels of the video/image data and lead to a degraded quality of service in terms of blurred image quality [9]. Moreover, susceptibility to parametric failures (read and write operation) caused by the manufacturing and intrinsic device variations in SRAM increases at lower voltages [10]. Therefore, a low-power, high-stability, and high-density memory with a fast read/write operation is required for multimedia applications.

10.2.3 High-Performance Processors

The operating speed of new-generation machine learning-based complex computing processors and high-speed server machines is much faster than the speed achieved by on-chip cache memory, and, based on user demand, this gap continues to increase [11]. The semiconductor industry is trying to cope with this challenge by increasing cache sizes. This is the reason the on-chip cache keeps increasing from 256 KB in Intel's Pentium III Coppermine with 180nm technology to 8 MB in the 6th generation Intel Core i7-6700k processor with 14nm technology node. This cache size is larger for recently evolved multi-core architectures to assist with their large bandwidth requirements [12]. Memory designers see the possibility to achieve faster cache memory (SRAM) by reducing device threshold voltage (V_{TH}) or using larger device sizes. However, these solutions lead to increased standby memory power. Therefore, additional design strategies are needed to solve these issues.

10.3 CONVENTIONAL SRAM CELL

As the name suggests, SRAM is said to be static as the data stored can be retained indefinitely (as long as a supply voltage is provided), without any need for a periodic refresh operation, and it can access any of the given memory bits randomly within the same access time. The basic schematic architecture of a conventional 6T SRAM cell and its layout design using a standard 65nm technology node are shown in Figure 10.4. A 6T SRAM bit cell invariably consists of a simple latch made of two cross-coupled inverters, PM1-NM1 and PM0-NM0, as shown in Figure 10.4, which have two stable operating points (states) defined by the complementary storage nodes Q and QB. Based on the preserved stable state of the latch, the data stored in the bit cell is said to store logic 1 and 0 when the complementary storage nodes Q and QB store "10" and "01," respectively. To access the cell storage nodes for write or read operation, there are two NMOS access switches, NM3 and NM2, controlled by the row selecting word line (WL), those connecting Q and QB with the column-wise selected complementary bit lines BL and BLB, respectively. During standby mode, WL remains 0, and the bit lines remain isolated from storage nodes Q and QB. In this condition, the inherent feedback loop of the inverter pair maintains its latching property and preserves the stored data.

For the write operation, complementary bit lines are driven by the write data and then connected to the storage nodes Q and QB by enabling WL at logic 1. During the write operation of the 6T SRAM bit cell, logic 0 is stored first in one of its complementary storage nodes, which then assists in storing 1 on another node. For instance, suppose a cell holds logic 1 (Q = 1 and QB = 0), and a write 0 operation is needed to flip the stored data. Accordingly, BL and BLB are driven by logic 0 and 1, respectively, for write 0, allowing node Q to discharge through turned-on access device NM2. However, initially stored logic 0 at QB opposes the discharging of node Q due to the turned-on pull-up device PM0, which keeps drawing a charging current from VDD. Therefore, a successful write operation mandates that the access device must be stronger than the pull-up device. Once node Q is discharged, it completely turns off pull-down device NM1 and turns on PM1, which helps to attain logic 1 at node QB. This is how the write operation is completed.

For the read 0 operations of the SRAM cell, the access devices are turned on by activating the row selecting signal WL at logic 1, and the bit lines are precharged at VDD. Hence, logic 0 stored at node Q provides a discharge path to the corresponding bit line BL and creates a voltage difference between the complementary bit lines BL and BLB, which is then sensed by the sense amplifier to generate the read output as logic 0. For the read 1 operation, the polarity of the generated voltage difference between complementary bit lines is opposite to that of read 0, and the sense amplifier detects this as a logic 1 read output.

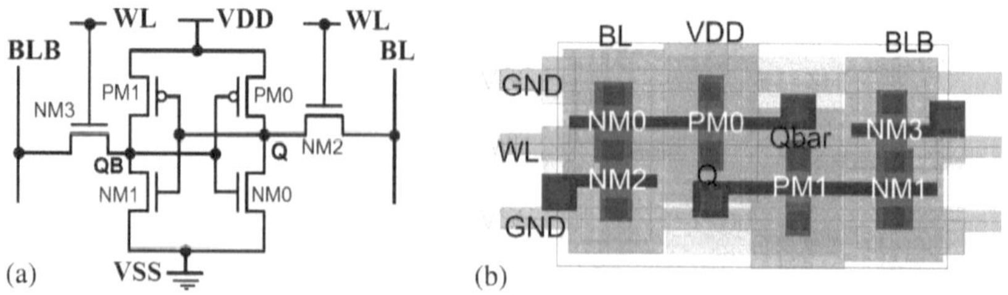

FIGURE 10.4 Conventional 6T SRAM bit cell and its layout using standard 65nm CMOS technology.

However, in both read operations (read 1 or 0) there is always potential divider formation through the access and pull-down devices of the SRAM bit cell (when BL discharges through the path formed by NM2 and NM0 in the case of read 0), and thus creates a non-zero potential at node Q. If the access device (NM2) is stronger than the pull-down device (NM0) in this situation, there may be a voltage rise at Q higher than the switching threshold voltage of the opposite inverter formed by PM1 and NM1. It may cause the stored data to flip, which is known as read failure. To ignore such read failures in the bit cells, the SRAM bit cell requires that the access devices are weaker than the pull-down devices.

10.3.1 Device Sizing Strategy for Functional 6T SRAM Cell

As described above, there are two main design constraints for device sizing of a typical 6T SRAM cell: (i) the access devices should be stronger than the pull-up devices to allow for successful write operation, and (ii) access devices are required to be weaker than pull-down devices for a successful read without read failure. This identifies a read/write conflict as both operations are performed through the access devices. For improved read stability and reduced risk of read failure, the cell access ratio (also called βratio), defined by the pull-down to access device size ratio, should be high and thus requires smaller cell access devices [13]. On the other hand, an improved write ability is achieved through a lower pull-up ratio (also termed γ_{ratio}) defined by the pull-up to access device size ratio, and this needs larger cell access devices. Therefore, conclusive design sizing for the acceptable read stability and write ability of the bit cell may be written as:

Pull-up device size < Access device size < Pull-down device size

10.3.2 Memory Architecture

A basic memory architecture for m-row address bits and n-column address bits is shown in Figure 10.5. Based on the given row address, the row decoder is used to select a particular WL out of 2^m lines, while the column decoder sets a pair of column switches to connect one of the 2^n bit line pairs with the read/write peripheral circuits. The column control circuitry of memory consists of a precharge circuit, column multiplexer (MUX), sense amplifier, output latch, and write amplifier, as seen in the figure.

The read operation of this memory starts by inserting the corresponding address into the decoder circuit. A short-duration precharge-enabled (PE) pulse is generated to detect each address transition, enabling the precharge circuit. The precharge circuit equalizes both bit lines (BL and BLB) and initializes at a particular voltage level (V_{PRE}). Once the bit lines are precharged, the row decoder activates the WL signal. This WL signal is asserted just after the initialization of the bit. Hence, the internal node of the bit cell that holds logic 0 pulls its bit line low through the access device, while the other bit line preserves its precharged voltage value. The bit cell is

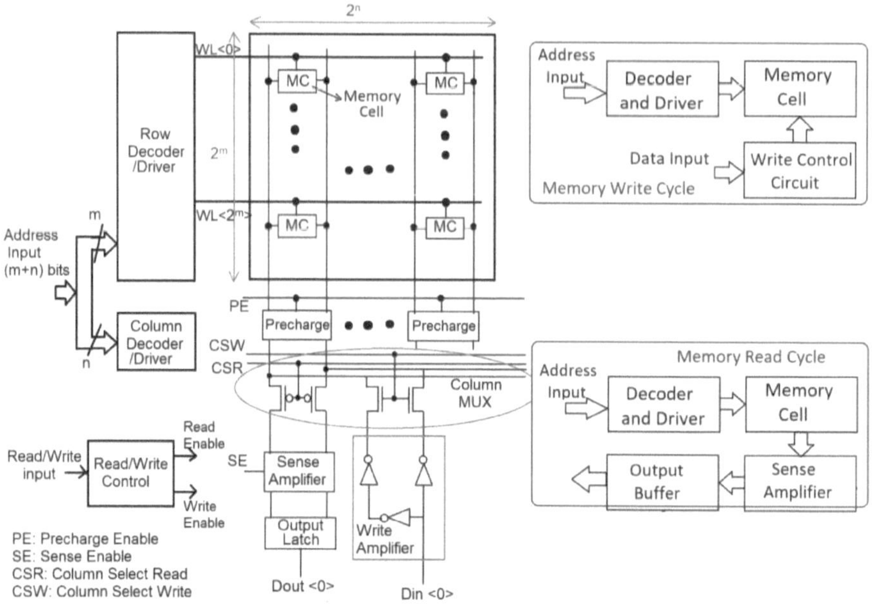

FIGURE 10.5 Memory array structure and write/read cycle.

optimized for a small layout area, so its small read current results in a slow bit line discharge rate. To speed up the read operation, sense amplifiers are used that sense the small voltage difference in the bit line pair (V_{PRE} and $V_{PRE}/\Delta\ V_{PRE}$) after a small interval from when the WL was activated. This small voltage difference of $\Delta\ V_{PRE}$ is converted into full swing output by the sense amplifier only when the column switch (pass transistors forming column MUX) and the sense amplifier are turned on through the column select read (CSR) and sense enable (SE) signals, respectively. Then the amplified data is finally stored in the output latch, where it remains available as output data (D_{out}).

The write operation of the memory is activated by sending the input data to be written to the write amplifiers, which generates complementary signals for BL and BLB. Then after the row address decoder activates the required WL signal, the column decoder activates the column switches to connect the complementary data of the write amplifier to the desired column with the help of the column select write signal (CSW). Then the bit cell is written using the bit line voltage values based on the input data.

10.4 MAJOR DESIGN CHALLENGES OF SRAM MEMORY FOR MODERN INTEGRATED CIRCUITS

Since SRAM dominates the power consumption of modern SoC devices, as an important building block of almost all integrated circuits, it plays a major role in operating speed, power consumption, supply voltage, and chip size of the overall system [14, 15]. Therefore, while designing SRAMs it is crucial to achieve optimized power, increased bandwidth, high density, and reliable functionality as per the target application. However, due to the minimum-sized devices, the memory is the most vulnerable part of any system for the ever-increasing process variations in aggressive technology downscaling. This SRAM vulnerability appears because of the increased V_{TH} variations of the reduced device geometry in modern scaled technologies, and hence this increases the memory bit cells failure probability due to the difficulty in maintaining an acceptable sizing ratio [16, 17].

Moreover, voltage scaling is always the preferred power-reducing very-large-scale integration (VLSI) design technique as it offers an exponential reduction in leakage current and a quadratic reduction in its dynamic power dissipation. This scaled voltage further challenges bit-cell stability, due to the smaller voltage swing, to tolerate the effect of noise voltage. The situation of low-voltage operation gets worse in the presence of increased process variations at the lower technology nodes.

10.4.1 Process Variation Effects

Process variation effects are classified into two categories:

1. Inter-die variations (also known as global variations)
2. Intra-die variations (also known as local or mismatch variations)

Inter-die variations are identified as the variations in the average value of different device parameters, such as device threshold voltage, dielectric thickness, poly width, etc., for multiple dies. All devices on a single die are affected in a similar manner due to such global variations, which appear due to the systematic processing changes that affect the individual dies.

On the contrary, intra-die variations, also termed mismatch variations, are the differences in parameter values among the various devices on the same die. Such local variations occur due to poly-line edge roughness, lithographic loss, difference in the number of NMOS/PMOS channel-adjust doping ions, as well as the transient effects, such as negative bias temperature instability (NBTI) [18].

Since the local mismatch variations are inversely proportional to the channel area and thus increase with the technology downscaling, these variations become a more serious design challenge than the global variations of the advanced technology nodes used in recent ICs [19–21].

It is observed that PMOS devices are more vulnerable to mismatch than NMOS devices at lower technology nodes and thus a lower V_{TH} value due to such variations may make the PMOS device stronger than the NMOS. In the case of memory, this situation may degrade the write ability of the SRAM bit cell as a better write ability demands that the PMOS devices of the bit cell are weaker than in the NMOS devices [6]. The read performance of bit cells also gets degraded by a large variability in device ON-current (I_{ON}) caused by V_{TH} fluctuation. Moreover, the sense amplifier offset voltage disturbed by the mismatch variations limits the minimum bit line swing. However, this swing is required to be low for a higher read speed and lower read power associated with the charging and discharging of highly capacitive bit lines during the read operation. Although upsizing the sense amplifier may overcome this offset issue caused by the mismatch, it increases the layout area and the read power due to larger device sizes. However, for recent memory designs at lower technology nodes, the sense amplifier power must be minimized as it is the major power-consuming part of the memory, and modern IC designs can't afford that level of power consumption [21].

Aside from all the above-discussed issues, the RSNM and write static noise margin (WSNM) also get severely affected by the process variations. RSNM and WSNM are the two most common design parameters of SRAM memory that are used to determine the SRAM cell's read and write ability, respectively [22]. RSNM is the maximum noise voltage that can be tolerated during read operation without flipping the stored data, and it is estimated as the side length of the largest square that can be inscribed into the smaller lobe of the read buttery curve (combined voltage transfer curve of bit-cell inverter pairs) [23]. RSNM is required to be as large as possible for better read stability of SRAM. On the other hand, WSNM is the ability of the memory cell to pull down the node, storing 1, to a value below the switching threshold voltage of another inverter, storing 0, so that it can successfully flip the cell stored data [12]. This is determined from the write buttery curve, which is the side length of the smallest square inscribed in the write buttery curve [23]. A larger WSNM represents a better write ability of SRAM memory cells. Figure 10.6 shows 2000 Monte Carlo (MC) simulation results for the RSNM and WSNM of a 6T SRAM cell, including both the local and global variations, and it can be observed that the process variations significantly degrade the read/write ability of the bit cell.

10.4.2 Temperature Variation Effect on Read and Write Ability

Considering the global process variations, the combination of fast NMOS and slow PMOS (FS corner) is identified as the worst corner for read stability, while the slow

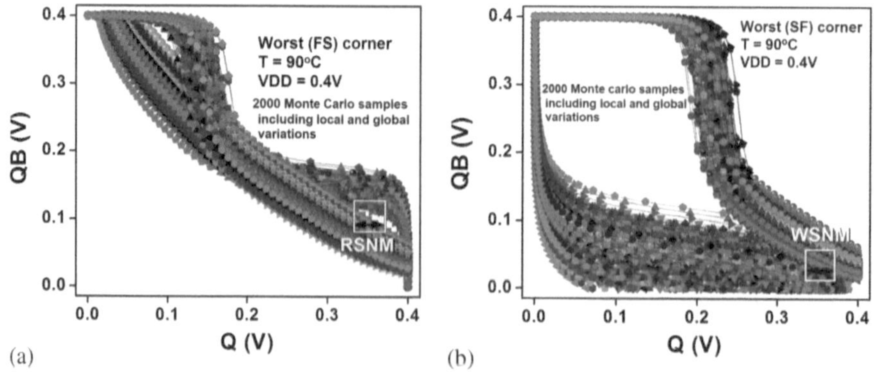

FIGURE 10.6 Simulation results of a 6T SRAM cell with 2000 Monte Carlo samples, including local and global variations for (a) RSNM and (b) WSNM using 65nm CMOS technology node [24].

NMOS and fast PMOS (SF corner) cause a worst-case write ability scenario for the conventional SRAM cells. Since the NMOS V_{TH} reduces with increasing temperature values, the fast pull-down and access NMOS transistors result in a reduced cell read stability and improved write ability at high temperatures. With a similar effect, stronger NMOS devices with increased V_{TH} at a low-temperature range leads to an improved read stability and poor write ability [25, 26]. Dependent on the temperature variations, the read current of the SRAM cell also becomes vulnerable (increasing/decreasing) with high/low-temperature values, and thus ultimately the read performance gets disturbed.

Therefore, temperature tolerance is an important design requirement for a SRAM cell using minimum-sized devices, so that the memory read and write failure probabilities can be reduced, and the overall yield can be improved for modern integrated circuits and systems operating at a low supply voltage.

10.4.3 Read/Write Conflict and Read-Decoupled 8T SRAM Cell

As discussed in Section 10.3.1, due to device sizing restrictions of the read and write operation trade-off, it is difficult to simultaneously improve both the read and write abilities of the conventional 6T SRAM bit cell. Read stability can be improved using high-V_{TH} access devices, but this degrades the write ability. On the other side, low-V_{TH} devices may improve write ability at the cost of poor read stability [25].

To remove such read/write trade-offs, a read decoupling approach has been widely used for the various SRAM cells proposed in the literature, in which the storage nodes are isolated from the read path. One of the most common read-decoupled 8T SRAM cells is illustrated in Figure 10.7(a) [27]. This cell demonstrated significant improvement in its read stability (RSNM) equivalent to that of its hold static noise margin (HSNM) value for the selected row and column due to the isolated read path, as shown in Figure 10.7(b). From the figure, it can be observed that the 8T cell shows

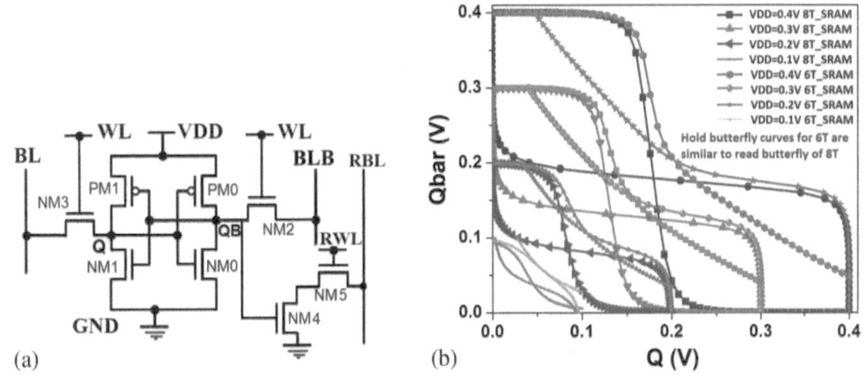

FIGURE 10.7 (a) Read-decoupled 8T SRAM cell [27] and simulated (b) read butterfly curves at different V_{DD} for 6T and 8T SRAM cells using 65nm CMOS technology node.

read stability even at 0.1V, while the 6T cell is totally read unstable at 0.1V. However, despite the improved read stability of the selected 8T SRAM cell, it is observed that the RSNM for row half-selected 8T cell (for the selected row and unselected column) remains equivalent to that of RSNM for the selected conventional 6T SRAM cell. Therefore, this 8T cell is also susceptible to read disturbance at a low supply voltage and needs other design strategies to solve the issue.

10.4.4 Read Current to Leakage Current Ratio (I_{READ}/I_{LEAK})

Aside from the read/write trade-off issue, the reduced read current to leakage current ratio (I_{Read}/I_{leak}) is another big issue for the conventional SRAM cells of recent low-power integrated circuits operating at a scaled supply voltage. A lower I_{Read}/I_{leak} ratio limits the maximum number of bit cells that can be connected in a single column of the memory array, and thus the memory density is restricted. To better understand this, let us assume that a read operation is performed for the selected 6T cell with storage node Q at logic 0, as shown in Figure 10.8. Then a bit line (BL0) discharging a read current I_{Read} flows from BL0 to the ground through the corresponding read path (access and pull-down NMOS devices). This read current (I_{Read}) develops a minimum potential difference between BL0 and BLB0 so that it can be sensed by the sense amplifier at the output. Now, if the selected cell stores logic 1 and all the column half-selected cells (unselected cells in a selected column) store logic 0 at their corresponding storage node Q, then the leakage current I_{leak} of all those half-selected

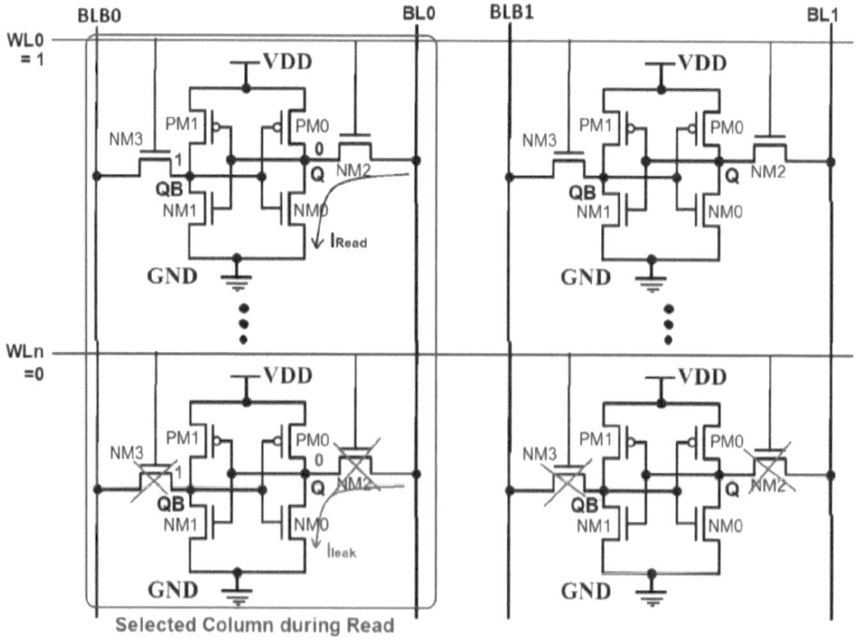

FIGURE 10.8 Memory array showing read current (I_{Read}) and leakage current (I_{Leak}).

cells get combined. The worst-case I_{leak} through a memory column must be lower than the I_{Read} value so that it will not be able to generate the required potential difference equivalent to the actual read 0 operations. Therefore, the correct read functionality of the memory array demands the condition $\sum I_{Leak} < I_{Read}$. However, connecting a larger number of bit cells in a memory column may lead $\sum I_{Leak} > I_{Read}$ and cause the false read operation [28, 29]. Therefore, to implement a reliable read operation the I_{Read}/I_{Leak} ratio of an individual bit cell must be as high as possible. It will also allow the connection of a larger number of bit cells in a single column, increasing the array density.

10.4.5 Performance and Stability Trade-off

Since the enhanced read stability of the SRAM bit cell to reduce the risk of memory read failure demands a large β_{ratio}, this is achieved by lowering the access device size [13]. However, this deteriorates the read performance by suppressing the read current (I_{Read}). Therefore, the conventional 6T SRAM cell structure has an inherent disadvantage in the trade-off between the read performance (I_{Read}) and stability (RSNM).

Moreover, supply voltage scaling has been widely adopted to quadratically reduce the switching power of modern SoC devices. However, the reduced driving current at a lower supply voltage weakens the operating frequency by several orders of magnitude. Therefore, the leakage power at such slow clock cycles starts to dominate the overall system power and thus creates a boundary for supply voltage scaling in power-limited applications [29].

10.4.6 Half-Select Issue

Writing in a particular bit cell of the memory array is performed by driving the complementary bit lines of the selected column through the required data and activating the word line of the selected row with high logic, as shown in Figure 10.9. However, apart from the bit cell of that selected row, other cells connected in the same column and unselected rows with logic low WL may also get affected by the bit line voltages due to current leakage through the bit lines to cell access devices. If the operating supply voltage is very small, as in the modern low-power SoC devices, then this leakage may be sufficient to flip the stored data of the unselected cells in the worst process-temperature conditions. The issue is known as the column half-select issue.

Also, the high logic value on selected WL activates all the bit cells of a row, while the bit line pairs of unselected columns in that row remain floating. This logic is high on WL, and the floating bit line voltage of the unselected columns has a high probability of potentially disturbing the stored data with any noise occurrence on the bit lines (as shown by the positive/negative glitches in Figure 10.9). This is known as the row half-select issue.

Therefore, the low-voltage operation of SRAM is very sensitive to these half-select issues and mandatorily needs a half-select free SRAM bit cell.

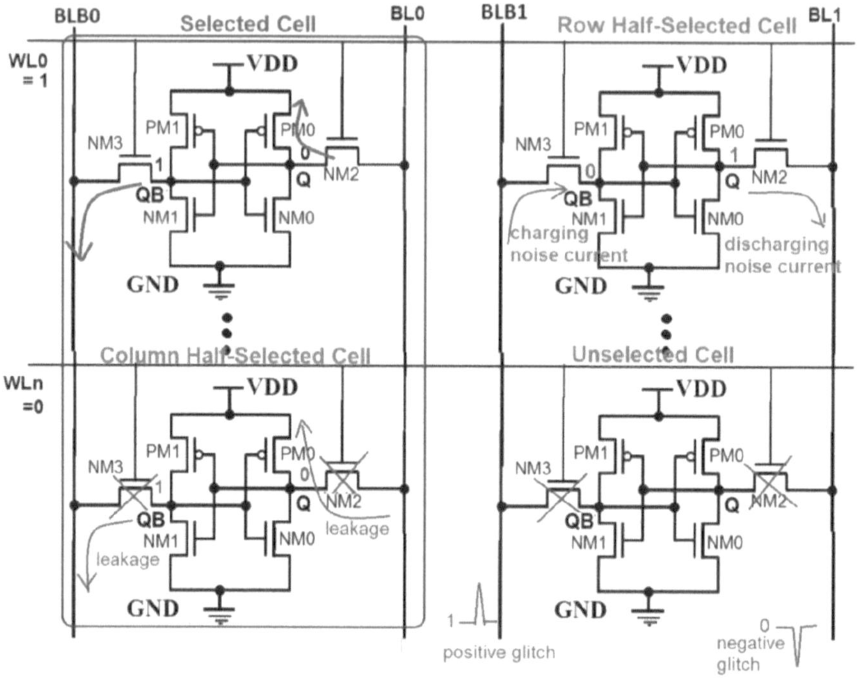

FIGURE 10.9 Half-selected cells in a 6T SRAM array during selected write '1' operation.

In a summary, the major challenges identified for SRAM memory of modern low-voltage IC designs using the conventional 6T bit cell at advanced technology nodes are:

- Sensitivity to PVT variations
- Read/write conflict
- Low I_{Read}/I_{leak} ratio due to increasing leakage
- Degraded cell stability
- Half-select issues

10.5 VARIOUS APPROACHES TO SOLVE THE DESIGN ISSUES OF CONVENTIONAL BIT CELLS

Looking into the design challenges of conventional 6T SRAM bit cells, different research groups have proposed several design solutions to improve the read/write ability of SRAM cells and reduce their power consumption. Some of the most popular design strategies include:

- multi-V_{TH} devices [29–31],
- single-ended operation [32–34],
- read decoupling [11, 35],

- dynamic feedback cutting [36–38],
- Schmitt trigger-based design [39, 40],
- stacking effect [11, 35, 41].

Using one or more of the above-mentioned approaches, SRAM designers keep exploring the implementation of efficient SRAM bit cells, and as a result different state-of-the-art SRAM cells have been developed. To reduce the power consumption of the memory, voltage scaling has been observed as the most effective solution [28]. However, this low-voltage operation in sub-nanometer technologies has already been a severe headache for SRAM designers, due to the loss of static noise margin, intensive current variations due to the random dopant fluctuations (RDFs), and limitations on the number of cells connected to a single bit line [42, 43]. Also, the low-voltage operation of the SRAM cell is prone to half-select issues and soft errors, which ultimately leads to diminishing cell stability [41]. Wang et al. [30] proposed the use of multi-V_{TH} devices where the cross-coupled inverters employ high-V_{TH} devices to improve the leakage power, while low-V_{TH} devices are used in the read access path to improve read performance. However, high-V_{TH} devices of inverter pairs slow down the write speed and thus nullify the improved energy of the cell. Moreover, the multi-V_{TH} devices of bit cells demand additional processing steps during fabrication. To improve energy efficiency, researchers have proposed using single-ended SRAM bit cells [32–34]. But the single-ended cells lack write ability compared with that of a differential architecture and hence are not suitable for the low-voltage operation of modern ICs [35]. Moreover, the low sense margin during the read operation is another critical issue for single-ended SRAM cells [44]. To improve write ability, a data-dependent feedback cutting strategy is an effective approach, as suggested by Chang et al. [38]. Using this feedback cutting approach, a 9T SRAM cell was proposed in [38] to independently control the read and write operations. The leakage power was also reduced for this 9T cell due to the stacking devices in the inverter pair. However, the cell has limitations in terms of overall area and a high power overhead due to the additional data-dependent control signal-generating circuitry. Saeidi et al. [36] proposed a differential 8T memory cell in which the feedback cutting approach was used for the read path to enhance its read margin. However, the cell required wider access devices to improve the write ability and thus resulted in an area overhead. A PPN 10T cell proposed in [35] explored the stacking effect of PMOS devices in the inverter pair to obtain a low cell leakage and high immunity to data-dependent bit line leakage. However, the feasibility of this cell was limited by its slower write speed.

Therefore, the design of low-power and highly stable SRAM memory has been a necessity with the increasing demand for advanced IoT-enabled wearable electronic devices and mobile health monitoring systems. However, it is clear from the above discussion that the research is still going on into SRAM cells, which may effectively suppress the rising leakage power of power-hungry applications with high read and write abilities. It is also important to note that the common cross-coupled inverter pair topology of SRAM cells, like the standard 6T SRAM cell, is less immune to process variations at lower supply voltage.

10.6 LOW-VOLTAGE HIGHLY STABLE 12T SRAM CELL DESIGN FOR WSNs

Since the reliable operation of a wireless system is highly dependent on the reliable storage of detected data and then the ability to read out the processed data without any noise inference, so that it can be transferred to the gateway (base station), as illustrated in Figure 10.2. Hence the improved read and write abilities of SRAM cells and half-select removal are essential tasks for the memories used in wireless sensor nodes while operating at a low supply voltage of modern SoCs. The operation of such memories should be process-tolerant and should not be affected by environmental conditions to reliably monitor the physical data. As the frequency of such sensing data lies in the range of a few tenths of KHz [45], the speed of the memory is not a major concern when reducing the power consumption and improving the read/write ability.

This section describes an efficient cross-point selection-based 12T SRAM memory cell suitable for wearable wireless sensor nodes. The basic design strategies and the striking features of the cell can be summarized as follows:

- Cross-point selection (both row- and column-wise activated write access paths) through additional NMOS in the write path completely solves the half-select issues of the write operation. Half-select removal allows using this bit cell within a bit-interleaving structure, which is considered as a single-bit error-tolerant structure.
- The stacking effect used in the pull-down networks of the inverter pair, controlled by the bit lines, efficiently enhances the write ability of the SRAM cell.
- A decoupled read path in the cell removes the probability of any noise occurrence and therefore offers read stability (RSNM) equivalent to that of a hold SNM (HSNM). The stacking of NMOS further improves the static noise margin of the cell.
- The stacking effect in cross-coupled inverters increases the equivalent path resistance and thus assists in reducing the leakage power.
- Therefore, improved read/write ability and half-select removal make this cell suitable for modern low-voltage ICs.

The basic architecture of the 12T SRAM bit cell and the orientation of various control signal routing is illustrated in Figure 10.10. While the status of the different control signals during write, read, and hold operations is given in Table 10.2. Since the write operation of this 12T SRAM cell is performed by activating the row-based WL and column-based write word line (WWL), a row and column-based selection enables a cross-point write structure. In addition, cascaded NMOS access devices (M7, M9, and M8, M10) increase the resistance of the write path and thus help to restrict the write leakage current of half-selected/unselected cells. Therefore, it can be stated that the half-select issues are absent in the given 12T cell.

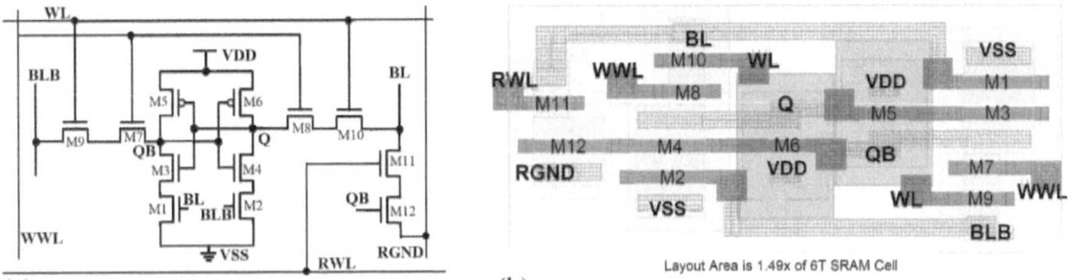

(a) (b)

FIGURE 10.10 (a) Schematic and (b) layout design of a half-select free, highly stable 12T SRAM cell [24].

TABLE 10.2
Truth Table for Control Signal of 12T Bit Cell

	WL	WWL	RWL	RGND	BL	BLB
Write (1/0)	1	1	0	1	(1/0)	(0/1)
Read	0	0	1	0	Precharged at 1	1
Hold	0	0	0	1	1	1

According to Table 10.2, the write 0 operation is activated by enabling WL and WWL at logic 1, and driving the bit lines BL and BLB at logic 0 and 1, respectively. During the operation, RWL and RGND are disabled at logic 0 and 1, respectively. As logic 0 at bit line BL turns off device M1, QB gets disconnected from the ground. It offers BLB to quickly charge up node QB to above the threshold voltage of device M4 through the write access devices M7 and M9. Therefore, a direct discharge path is formed for node Q to V_{SS} through M4 and the BLB-driven NMOS M2. The high voltage at QB completely turns off device M6 and thus helps to immediately discharge node Q to zero through the pull-down path (M4, M2) in addition to the write access path (M8, M10). This shows how the data-dependent pull-down path improves write ability by cutting off the feedback of the cross-coupled inverter pair and successfully writes 0 into the selected cell. A symmetric write structure was used for 12T SRAM cell implementation. Therefore, a similar write 1 operation can also be explained in a similar manner with the feedback cutting through M2.

Based on simulation results using 2000 Monte Carlo samples at the worst process corner and 65nm node, shown in Figure 10.11(a), the write stability (WSNM) is observed as 9x higher than that of a conventional 6T cell.

As the read path in a 12T SARM is completely isolated from the inverter pair, the read operation does not affect the stored data. Therefore, the cell read stability represented by a RSNM value tends to be equivalent to that of the hold stability, which is always much higher than the read stability of a normal 6T cell structure. Apart from the read stability improvement due to the decoupled read path, the stacked pull-down NMOS devices of the inverter pair of this 12T bit cell also improve its RSNM value. The RSNM of a SRAM cell is defined by the side length of the largest square

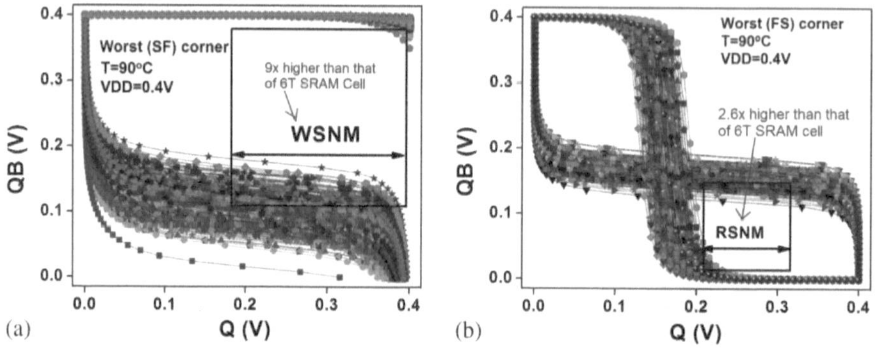

FIGURE 10.11 Simulation results of a 12T cell using 2000 Monte Carlo samples at the worst process corner, 90°C and 0.4 V supply using standard 65nm technology node for (a) WSNM, and (b) RSNM [24].

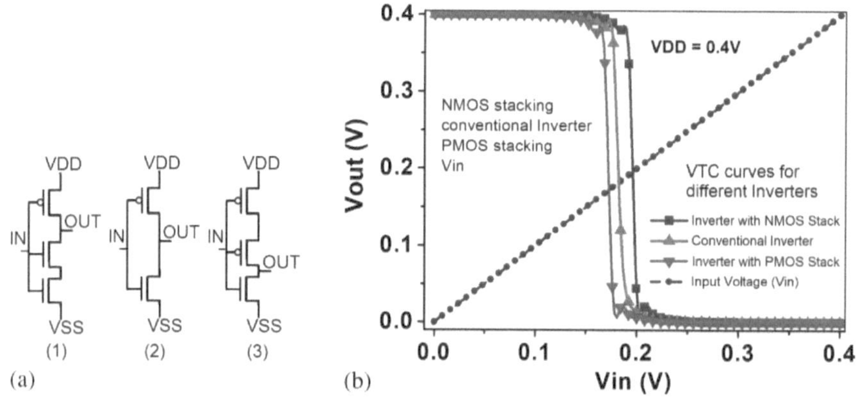

FIGURE 10.12 (a) Three different inverter configurations and (b) the corresponding VTC curve.

that can best fit into the smaller lobe of the buttery curve [23]. Figure 10.12(a) illustrates three different configurations of inverter, and Figure 10.12(b) demonstrates the comparison of their corresponding VTC curves. As the cascading of NMOS devices increases the resistance of the pull-down path, the switching voltage of the inverter is lowered, and the VTC curve is shifted toward the right, as shown in the figure. On the contrary, the cascading of PMOS devices weakens the pull-up path resistance and results in the VTC curve shifting to the left. As a result, it can be analyzed that the given 12T SRAM cell has a higher switching voltage for its inverter pair, resulting in larger lobe formation in the read buttery curves. Therefore, the read stability of the cell has been improved. Monte Carlo simulation with 2000 samples, including both the local and global variations at the 65nm node, was performed as shown in Figure 10.11(b). Based on the simulation results, the RSNM of the 12T cell is 2.6 times higher than that of a conventional 6T SRAM cell.

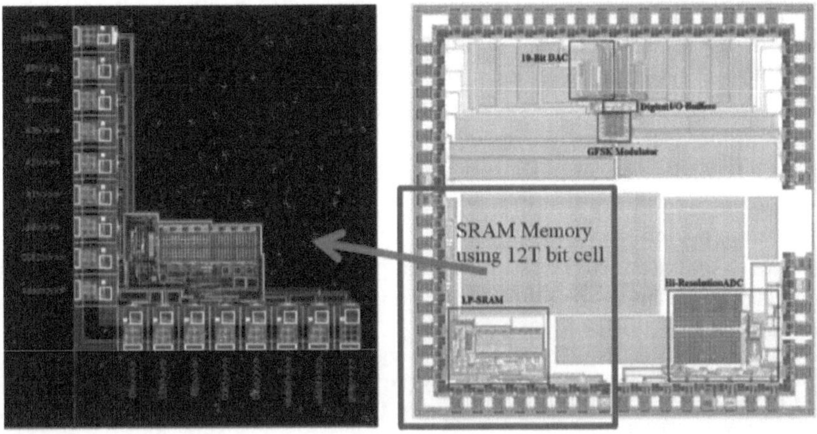

FIGURE 10.13 Layout of a small and customized 256-bit SRAM memory using a standard 180 nm node [24].

Also, unlike the conventional read-decoupled 8T SRAM cell, as illustrated in Figure 10.7(a) [27], this 12T cell does not use an extra read bit line (RBL) and hence avoids leakage through an additional bit line.

Looking into the attractive features of 12T bit cells, a customized memory was designed as part of an international research project entitled "Aalto-IITI Cooperation for Skill Developments of IoT-based Implementation," funded by the Centre for International Mobility, Finland (CIMO grant no: Intia-1- 2016-03). The top-level memory layout and its placement in the overall chip of the project can be seen in Figure 10.13.

10.7 IN-MEMORY COMPUTING FOR THE MODERN AI EDGE DEVICES AND ITS CHALLENGES

Advanced SoC systems, including artificial intelligence (AI), have led to explosive growth in the semiconductor market. Moreover, the growth of the internet and competitive digital IT services in every area of human life have developed an era of big data. Hence, the most widely existing data-driven application is machine learning in big data storage systems, where the artificial neural networks (ANNs) based algorithms find the implicit pattern of data to perform real-time speech, image, or object recognition. Such pattern or object recognition would require receiving and immediately processing a large amount of complex data. For example, in an image recognition system, which may have millions of possible images, feature extraction of all the images just to process a single image will lead to significant congestion at the data bus. This is due to communicating the data between memory and processor. Therefore, most of the energy and time of such processing is consumed in the data movement between the processing unit and the storage memory, rather than the actual computing of data. As a result, the ongoing Von Neumann-based processing units struggle to match the required processing speed of real time data, even with advanced processors. These issues restrict the continuous performance improvement of the overall systems. In addition, the large volume of on-chip memory used for holding the matched data results in significant power leakage.

To improve the energy and speed performance of such ANN-based algorithms, there is a recent widely used approach called "in-memory computing" [46, 47]. The basic idea behind this is that instead of feeding the processor all the raw data, it is beneficial to pre-process the data inside the memory itself and then provide the processor only the intermediate result. This approach suppresses the required memory bandwidth by reducing the required data communication between the memory and the processor and saves a lot of energy.

10.7.1 Key Challenges of In-Memory Computing

In-memory computing (IMC) has been identified as a solution for modern AI edge devices for reducing the data traffic between the memory and processor. This is possible by providing the basic computing features of multiply and accumulate (MAC) inside the memory itself. To effectively serve the purpose and be feasible for the improved accuracy of AI-based algorithms, the memory design needs the following requirements [47, 48]:

- High precision of input, weight, and output of memory to provide better accuracy.
- Short access time to provide a faster response for the AI edge devices.
- Better energy efficiency to support battery-operated applications.

As high precision memory provides better accuracy, there have been various approaches suggested for such implementation; however, multi-bit MAC implementation faces the following challenges based on the used approaches [48]:

- Approach: Simultaneously turning ON multiple WLs of memory
  Challenge:
  - High BL current density due to the multiple bit cells turned on in a single column.
  - Wide range of BL current leads to inaccurate BL clamping voltage as the large voltage variation of BL voltage changes the operating region of the corresponding MOS devices.
  - Approach: Applying the various inputs at different WLs and turning on a single WL at a time.
  Challenge:
- Multi-bit input needs multiple cycles, which results in a long access time.

10.8 CONCLUSION

This chapter has highlighted the role and necessity of SRAM for modern integrated circuit designs. The various application-dependent design strategies have also been discussed to help the effective design of SRAM memory for a given application. The major design challenges of conventional memory, which are the key barriers for the existing SRAM used in modern IC designs, have been explained. The various possible solutions for the design challenges of SRAM bit cells, available in the literature, have been incorporated, which may be very helpful for the SRAM designer. A low-voltage, highly stable 12T SRAM cell design has also been incorporated as a potential candidate for the modern battery-operated IoT-enabled WSN nodes. The final section of this chapter described the unavoidable role of memory in advanced AI edge devices. The use of the most recent memory design strategies known as "in-memory computing" and its main design challenges for AI applications were also included in this section.

REFERENCES

1. Statista Research Department. 2016. Internet of Things: Number of connected devices worldwide 2015–2025. https://www.statista.com/statistics/471264/iot-number-of-connected-devices-worldwide/
2. Patrick, Gordon, and A. Gattani. "Memory plays a vital role in building the connected world." *Electronic Design: Hong Kong, China* (2015). https://www.electronicdesign.com/technologies/iot/article/21800623/memory-plays-a-vital-role-in-building-the-connected-world.
3. Rich Blomseth. 2016. Developing IoT devices with future flexibility in mind. http://blog.echelon.com/echelon_blog/2016/03/developing-iot-devices-with-future-exibility-in-mind.html. (Accessed Apr 30, 2016).
4. Sharma, Vishal, Maisagalla Gopal, Pooran Singh, and Santosh Kumar Vishvakarma. "A 220 mV robust read-decoupled partial feedback cutting based low-*Leakage* 9T *SRAM* for *Internet of Things* (IoT) applications." *AEU-International Journal of Electronics and Communications* 87 (2018): 144–157
5. Hodge, Victoria J., Simon 'Keefe, Michael Weeks, and Anthony Moulds. "Wireless sensor networks for condition monitoring in the railway industry: A survey." *IEEE Transactions on Intelligent Transportation Systems* 16, no. 3 (2014): 1088–1106.
6. Chiu, Yi-Wei, Yu-Hao Hu, Ming-Hsien Tu, Jun-Kai Zhao, Yuan-Hua Chu, Shyh-Jye Jou, and Ching-Te Chuang. "40 nm *bit-interleaving* 12T subthreshold *SRAM* with data-aware write-assist." *IEEE Transactions on Circuits and Systems I: Regular Papers* 61, no. 9 (2014): 2578–2585.
7. Gong, Na, Seyed Alireza Pourbakhsh, Xiaowei Chen, Xin Wang, Dongliang Chen, and Jinhui Wang. "Spider: sizing-priority-based application-driven memory for mobile video applications." *IEEE* Transactions *on Very Large Scale Integration (VLSI) Systems* 25, no. 9 (2017): 2625–2634.
8. Synopsis. 2017. DesignWare IP for IoT. https://www.synopsys.com/designware-ip/ip-market-segments/internet-of-things.html.
9. Cho, Minki, Jason Schlessman, Wayne Wolf, and Saibal Mukhopadhyay. "Reconfigurable *SRAM* architecture with spatial *Voltage scaling* for *low-power* mobile multimedia applications." *IEEE Transactions on Very Large Scale Integration (VLSI) Systems* 19, no. 1 (2009): 161–165.
10. Mukhopadhyay, Saibal, Hamid Mahmoodi, and Kaushik Roy. "Modeling of failure probability and statistical design of *SRAM* array for yield enhancement in nanoscaled CMOS." *IEEE Transactions on Computer-Aided Design of Integrated Circuits and Systems* 24, no. 12 (2005): 1859–1880.

11. Pal, Soumitra, and Aminul Islam. "9-T *SRAM* cell for reliable ultralow-power applications and solving multi-bit soft-error issue." *IEEE Transactions on Device and Materials Reliability* 16, no. 2 (2016): 172–182.
12. Islam, Aminul, and Mohd Hasan. "Variability aware low *Leakage* reliable *SRAM* cell design technique." *Microelectronics Reliability* 52, no. 6 (2012): 1247–1252.
13. Zhang, Kevin, Fatih Hamzaoglu, and Yih Wang. "Low-power *SRAMs* in nanoscale CMOS technologies." *IEEE Transactions on Electron Devices* 55, no. 1 (2007): 145–151.
14. Galib, Md Mehedi Hassan, Ik Joon Chang, and Jinsang Kim. "Supply voltage decision methodology to minimize SRAM standby power under radiation environment." *IEEE* Transactions *on Nuclear Science* 62, no. 3 (2015): 1349–1356.
15. Kumar, Animesh, Jan Rabaey, and Kannan Ramchandran. "*SRAM* supply *voltage scaling*: A reliability perspective." In 2009 10th International Symposium on Quality Electronic Design, pp. 782–787. IEEE, 2009.
16. Ahmad, Sayeed, Naushad Alam, and Mohd Hasan. "Pseudo differential multi-cell upset immune robust *SRAM* cell for ultra-*low power* applications." *AEU-International Journal of Electronics and Communications* 83 (2018): 366–375.
17. Karl, Eric, Yih Wang, Yong-Gee Ng, Zheng Guo, Fatih Hamzaoglu, Uddalak Bhattacharya, Kevin Zhang, Kaizad Mistry, and Mark Bohr. "A 4.6 GHz 162Mb *SRAM* design in 22nm tri-gate CMOS technology with integrated active V MIN-enhancing assist circuitry." In 2012 IEEE International Solid-State Circuits Conference, pp. 230–232. IEEE, 2012.
18. Karl, Eric, Zheng Guo, James W. Conary, Jeffrey L. Miller, Yong-Gee Ng, Satyanand Nalam, Daeyeon Kim, John Keane, Uddalak Bhattacharya, and Kevin Zhang. "17.1 A 0.6 V 1.5 GHz 84mb *SRAM* design in 14nm ~~finfet~~ FinFET CMOS technology." In 2015 IEEE International Solid-State Circuits Conference-(ISSCC) Digest of Technical Papers, pp. 1–3. IEEE, 2015.
19. Goud, A. Arun, Rangharajan Venkatesan, Anand Raghunathan, and Kaushik Roy. "Asymmetric underlapped FinFET based robust *SRAM* design at 7nm node." In 2015 Design, Automation & Test in Europe Conference & Exhibition (DATE), pp. 659–664. IEEE, 2015.
20. Shyu, J-B., Gabor C. Temes, and Kung Yao. "Random errors in MOS capacitors." *IEEE Journal of Solid-State Circuits* 17, no. 6 (1982): 1070–1076.
21. Pelgrom, Marcel JM, Aad C.J. Duinmaijer, and Anton P.G. Welbers. "Matching properties of MOS transistors." *IEEE Journal of Solid-State Circuits* 24, no. 5 (1989): 1433–1439.
22. Sharma, Vishal, Maisagalla Gopal, Pooran Singh, Santosh Kumar Vishvakarma, and Shailesh Singh Chouhan. "A robust, ultra low-power, data-dependent-power-supplied 11T *SRAM* cell with expanded read/write stabilities for internet-of-things applications." *Analog Integrated Circuits and Signal Processing* 98, no. 2 (2019): 331–346.
23. Seevinck, Evert, Frans J. List, and Jan Lohstroh. "Static-noise margin analysis of MOS *SRAM* cells." *IEEE Journal of Solid-State Circuits* 22, no. 5 (1987): 748–754.
24. Sharma, Vishal, Santosh Vishvakarma, Shailesh Singh Chouhan, and Kari Halonen. "A write-improved low-power 12T *SRAM* cell for wearable wireless sensor nodes." *International Journal of Circuit Theory and Applications* 46, no. 12 (2018): 2314–2333.
25. Sharma, Vibhu, Francky Catthoor, and Wim Dehaene. *SRAM Design for Wireless Sensor Networks: Energy Efficient and Variability Resilient Techniques*. Springer Science & Business Media, 2012.
26. Goetz, Jay. "Sensors that can take the heat, part 1: Opening the high-temperature toolbox." *Sensors-the Journal of Applied Sensing Technology* 17, no. 6 (2000): 20–39.

27. Chang, Leland, Robert K. Montoye, Yutaka Nakamura, Kevin A. Batson, Richard J. Eickemeyer, Robert H. Dennard, Wilfried Haensch, and Damir Jamsek. "An 8T-*SRAM* for variability tolerance and *low-voltage* operation in high-performance *Caches*." *IEEE Journal of Solid-State Circuits* 43, no. 4 (2008): 956–963.
28. Duan, Chuhong, Andreas J. Gotterba, Mahmut E. Sinangil, and Anantha P. Chandrakasan. "Energy-efficient reconfigurable *SRAM*: Reducing read power through data statistics." *IEEE Journal of Solid-State Circuits* 52, no. 10 (2017): 2703–2711.
29. Wang, Bo, Truc Quynh Nguyen, Anh Tuan Do, Jun Zhou, Minkyu Je, and Tony Tae-Hyoung Kim. "Design of an ultra-low voltage 9T *SRAM* with equalized bitline *Leakage* and CAM-assisted energy efficiency improvement." *IEEE Transactions on Circuits and Systems I: Regular Papers* 62, no. 2 (2014): 441–448.
30. Wang, Bo, Jun Zhou, and Tony Tae-Hyoung Kim. "*SRAM* devices and circuits optimization toward energy efficiency in multi-Vth CMOS." *Microelectronics Journal* 46, no. 3 (2015): 265–272.
31. Upadhyay, Prashant, Rajib Kar, Durbadal Mandal, and Sakti Prasad Ghoshal. "A design of low swing and multi threshold voltage based *low power* 12T *SRAM* cell." *Computers & Electrical Engineering* 45 (2015): 108–121.
32. Kushwah, C. B., Santosh Kumar Vishvakarma, and Devesh Dwivedi. "Single-ended boost-less (SE-BL) 7T process tolerant *SRAM* design in sub-threshold regime for ultra-low-power applications." *Circuits, Systems, and Signal Processing* 35, no. 2 (2016): 385–407.
33. Ahmad, Sayeed, Mohit Kumar Gupta, Naushad Alam, and Mohd Hasan. "Low *Leakage* single bitline 9Tt (SB9T) static random access memory." *Microelectronics Journal* 62 (2017): 1–11.
34. Tu, Ming-Hsien, Jihi-Yu Lin, Ming-Chien Tsai, Chien-Yu Lu, Yuh-Jiun Lin, Meng-Hsueh Wang, Huan-Shun Huang et al. "A single-ended disturb-free 9T subthreshold *SRAM* with cross-point data-aware write word-line structure, negative bit line, and adaptive read operation timing tracing." *IEEE Journal of Solid-State Circuits* 47, no. 6 (2012): 1469–1482.
35. Lo, Cheng-Hung, and Shi-Yu Huang. "PPN based 10T *SRAM* cell for low-*Leakage* and resilient subthreshold operation." *IEEE Journal of Solid-State Circuits* 46, no. 3 (2011): 695–704.
36. Saeidi, Roghayeh, Mohammad Sharifkhani, and Khosrow Hajsadeghi. "A subthreshold symmetric *SRAM* cell with high read stability." *IEEE Transactions on Circuits and Systems II: Express Briefs* 61, no. 1 (2014): 26–30.
37. Kushwah, C. B., and Santosh Kumar Vishvakarma. "A single-ended with dynamic feedback control 8T subthreshold *SRAM* cell." *IEEE Transactions on Very Large Scale Integration (VLSI) Systems* 24, no. 1 (2015): 373–377.
38. Chang, Meng-Fan, Shi-Wei Chang, Po-Wei Chou, and Wei-Cheng Wu. "A 130 mV *SRAM* with expanded write and read margins for subthreshold applications." *IEEE Journal of Solid-State Circuits* 46, no. 2 (2010): 520–529.
39. Ahmad, Sayeed, Mohit Kumar Gupta, Naushad Alam, and Mohd Hasan. "Single-ended Schmitt-trigger-based robust low-power *SRAM* cell." *IEEE Transactions on Very Large Scale Integration (VLSI) Systems* 24, no. 8 (2016): 2634–2642.
40. Kulkarni, Jaydeep P., and Kaushik Roy. "Ultra*low-voltage* process-variation-tolerant Schmitt-trigger-based *SRAM* design." *IEEE Transactions on Very Large Scale Integration (VLSI) Systems* 20, no. 2 (2011): 319–332.
41. Pal, Soumitra, and Aminul Islam. "Variation tolerant differential 8T *SRAM* cell for ultra*low power* applications." *IEEE Transactions on Computer-Aided Design of Integrated Circuits and Systems* 35, no. 4 (2015): 549–558.

42. Bhavnagarwala, Azeez J., Xinghai Tang, and James D. Meindl. "The impact of intrinsic device fluctuations on CMOS *SRAM* cell stability." *IEEE Journal of Solid-State Circuits* 36, no. 4 (2001): 658–665.
43. Kim, Tae-Hyoung, Jason Liu, and Chris H. Kim. "A voltage scalable 0.26 V, 64 kb 8T *SRAM* with Vmin lowering techniques and deep sleep mode." *IEEE Journal of Solid-State Circuits* 44, no. 6 (2009): 1785–1795.
44. Chang, Ik Joon, Jae-Joon Kim, Sang Phill Park, and Kaushik Roy. "A 32 kb 10T subthreshold *SRAM* array with *bit-interleaving* and differential read scheme in 90 nm CMOS." *IEEE Journal of Solid-State Circuits* 44, no. 2 (2009): 650–658.
45. Shriram, Revati, Asmita Wakankar, Nivedita Daimiwal, and Dipali Ramdasi. "Continuous cuffless blood pressure monitoring based on PTT." In 2010 International Conference on Bioinformatics and Biomedical Technology, pp. 51–55. IEEE, 2010.
46. Zhang, Jintao, Zhuo Wang, and Naveen Verma. "In-memory computation of a machine-learning classifier in a standard 6T *SRAM* array." *IEEE Journal of Solid-State Circuits* 52, no. 4 (2017): 915–924.
47. Biswas, Avishek, and Anantha P. Chandrakasan. "Conv-RAM: An energy-efficient *SRAM* with embedded convolution computation for low-power CNN-based machine learning applications." In 2018 IEEE International Solid-State Circuits Conference-(ISSCC), pp. 488–490. IEEE, 2018.
48. Su, Jian-Wei, Xin Si, Yen-Chi Chou, Ting-Wei Chang, Wei-Hsing Huang, Yung-Ning Tu, Ruhui Liu et al. "15.2 a 28nm 64kb inference-training two-way transpose multi-bit 6t *SRAM* compute-in-memory macro for ai edge chips." In 2020 IEEE International Solid-State Circuits Conference-(ISSCC), pp. 240–242. IEEE, 2020.

11 Implementation of 512-bit SRAM Tile Using the Lector Technique for Leakage Power Reduction

Rajeevan Chandel, K Madhu Kiran, and Rohit Dhiman

CONTENTS

11.1 INTRODUCTION

In modern complementary metal-oxide semiconductor (CMOS) very large scale integrated (VLSI) circuits, chip complexity and operating speeds have increased by manifolds. This causes an increase in processing capacity per chip and thereby power dissipation. In battery-powered applications utilizing CMOS ICs, the higher the power consumption, the lower the battery life. Because of high power consumption, cooling costs, reliability, and packaging are much affected. The main constituents of power dissipation [1] are: i) Dynamic power dissipation due to charging and discharging load capacitance. ii) Short circuit power dissipation because of the conducting path between the power supply and the ground. This power dissipation due to short circuit appears for a small interval when transition happens in a logic

DOI: 10.1201/9781003155751-11

gate [2]. iii) Static power dissipation occurs as a leakage current mainly comprises parasitic diode reverse-saturation and subthreshold currents. Because of the stored voltage between the drain and bulk activity, a transistor reverse bias diode current occurs, whereas a subthreshold current occurs due to carrier diffusion between the source terminal and drain terminal of the OFF transistors. Those circuits which have an equal rise and fall edge for both input and output can reduce the power dissipated due to the short circuit of up to 10% of the overall dissipated power [1–3]. The miniaturization in feature size degrades the supply voltage. If the power supply voltage and ground ratio are in an order that does not affect the operation of the CMOS circuit, it can give improved noise margins and also help to escape the hot carrier effect in short-channel devices. But the reduction of threshold voltage pays the penalty for the increase in the subthreshold leakage current. In order to achieve low power dissipation, leakage power reduction techniques play an important role in designing deep submicron and nanometer regime VLSI circuits [3–12]. Many portable e-gadgets and multimedia applications consist of dual modes of performance, namely an active mode and a standby mode. Current mobile phones have a low activity factor, meaning their idle time is greater than their active time. Even in an idle state, many portable devices are prone to power leakage, which further diminishes the utility of the battery. In the present chapter, a leakage power reduction technique called lector [10–12] is utilized to analyze the performance of an SRAM array.

The rest of the chapter is formulated as follows. Section 11.2 provides a brief literature review. Section 11.3 illustrates SRAM operation. Section 11.4 introduces the leakage reduction technique for SRAM design. Section 11.5 presents the results and provides a discussion of the different designs under consideration. Finally, a conclusion is drawn in Section 11.6.

11.2 LITERATURE REVIEW

Low power dissipation and high performance are the main requirements of ICs in the present era. For this, different techniques have been presented by researchers in the literature [1–4]. Further CMOS leakage reduction, mitigation, and control have received significant attention from the researchers for performance efficient reliable designs [5–12]. Gu et al. [4] developed an IGFET model and identified the working of a FET under different threshold voltages. As an example, an SRAM cell was implemented. This architecture enhances the power-delay product by varying the upper and lower bounds of the supply voltage. Johnson et al. [5] minimized the leakage levels to between 35% to 90% less than conventional ones. A leakage control option was given showing an efficient use of transistor stacks in a single-threshold CMOS. Narendra et al. [6] reported scaling of the stack effect for leakage reduction. Sirichotiyakul et al. [7] proposed an accurate leakage estimation and optimization tool for dual-V circuits. Choi et al. [8] reported an extreme UV scatter-resistant SRAM designed using 7 nm FinFET technology. Though the architecture encounters hot carrier injection, bias temperature stability, and dielectric breakdown, these problems do not affect SRAM functionality. Therefore, these can be accessed in the future for logic production with a high reliability. Qin et al. [9] describe SRAM leakage suppression by minimizing standby supply voltage.

Hanchate et al. [10] designed an effective power leakage minimizer known as a lector. The leakage current is cut down in this design with no increment in dynamic power dissipation. For this, the gate terminal was controlled by the source of another gate. This happens because the path resistance was increased, and decay in current was observed; 79.4% of leakage reduction can be obtained using the lector technique. Gaonkar et al. [11] designed a CMOS inverter using the lector technique to reduce leakage. Alekhya et al. [12] reported a 64-bit SRAM using the lector technique for low leakage power with read and write enabled.

The literature shows that various attempts have been made by researchers to improve the performance of SRAM cells for efficient memory design. 6T, 7T, and 8T SRAM cell designs have been designed and analyzed by various researchers [1, 13–24]. Some researchers have used a higher number of transistors in SRAM cell design [25–30]. An overview of these SRAM cells is further provided in this section.

Torrens et al. [13] proposed a 6T SRAM cell for faster operations and minimization of the soft error rate (SER). The widths of NMOS and PMOS are adjusted for nominal voltages. But the results vary considerably according to the input voltages. It works well but only under low-voltage conditions. Sah et al. [14] presented a 6T SRAM cell performance analysis in 180 nm CMOS technology. Khare et al. [15] discuss VLSI design and analysis of low-power 6T SRAM cells using the Cadence tool. Surana and Mekie [16] designed an energy-efficient 6T SRAM for large memory accessing applications such as multimedia with a better minimized bit error rate and minimized power (28%) and delay (5%) than conventional memories. This design was compared with hybrid and heterogeneous approximate memories to justify the performance and it passed all the tests. Wu et al. [17] developed a 256Kb mini-array with 6T SRAM using 28 nm technology with low power utilization in sleep mode and a faster wakeup mode. It enhanced the write ability using Vtrip-tracking, which minimized the power supply to the sub-array using cell inverter trip voltage.

Mohammadi et al. [18] designed a 128Kb SRAM with 7T architecture single boost using 28 nm with retention voltage logic. This approach boosted the memory operation of SRAM and limited the power, for example, it was mentioned that read energy was 8.54 fJ/bit at 90MHz with an operating voltage of 0.3 V. Here the memory included a Built-In Self-Test (BIST) approach. Gupta et al. [19] proposed a 7T architecture subthreshold-based operation for read assistance using 32 nm technology. These read operation characteristics were compared with the 6T architecture to improve the performance with local and global variations and use the subthreshold effect to minimize the read and write conflict in the previous architecture. This architecture operated at 0.4 V, which improved the dynamic memory write ability.

Kumar and Ubhi [20] evaluated the performance of 6T, 7T, and 8T SRAM for 180 nm technology nodes. Kushwah and Vishvakarma [21] developed 8T SRAM to improve data stability using subthreshold operation. This architecture used dynamic feedback control on a single side, enhancing the static noise margin for ultra-low-power operation. The write time reduced by 71% compared to the 5T and 6T architectures. Huang and Chiou [22] developed an asynchronous mode bit-interleaved ultra-low-voltage single bit-line 8T SRAM cell for increasing the read stability of the system. This architecture was implemented on 90 nm technology with operating voltage of 360mV and power consumption of 2.68 μW/bit. To obtain stability in read, write, and hold operation, no additional peripherals were accessed. Pal and Islam [23] presented variation-tolerant differential 8T SRAM cells suitable for ultra-low-power applications.

An area and power-efficient 9T SRAM with one-sided Schmitt trigger operation, which aids low-power operations, high read stability and write ability, and a hold stability with no write-back scheme due to the bit-interleaving structure of the cell, was designed by Cho et al. [24]. Wen et al. [25] proposed a 10T SRAM with high power utility reduction at lower voltages and non-destructive column selection. It was implemented in 65 nm technology, and a differential supply voltage technique improves the write ability of the SRAM, reduces dynamic power consumption, and was 3.3 times more effective in write operation compared to conventional SRAMs. Maroof and Kong [26] designed a 10T architecture dynamically controlled virtual rail and reduced the read bit line (RBL) leakage power. This architecture was implemented using 65 nm and helped to minimize the power dissipation by 50% compared with 6T and 8T architectures.

Jiang et al. [27] designed 10T and 12T SRAM bit cells for 130 nm technology nodes using quadruple cross couple storage cells. The comparison of these two architectures was performed with 6T architectures and resulted in more effective power utilization by minimizing the read time to 50% when referred to 8T and 6T architectures.

Sharma and Shrivastava [28] reported a low-power and high speed 11T SRAM cell with bit-interleaving capability. Yadav et al. [29] designed a 12T SRAM for space applications with bit-interleaving configuration that decreased the effect of upsets and increased stability and reliability during read and write operations. Nobakht and Niaraki [30] implemented a subthreshold region-based interleaving capability using 32 nm technology and compared power consumption and delay with 6T, 7T, 9T, 10T, and 12T SRAM architectures, and finally it was inferred that 7T provided the best results and more effective utilization compared to the other architectures.

Different design techniques and technologies have been reported for low-power, process-variation tolerant, and efficient SRAM designs [31–44]. Pal et al. [31] reported a reliable write-assist low-power SRAM cell useful for wireless sensor network applications. Boumcheda et al. [32] developed a single rail two-port SRAM for 28 nm SOI technology for synchronous and asynchronous accesses to Internet-of-Things (IoT) applications. It could be operated between 0.25 V to 1.25 V and minimizes static power consumption compared to nominal SRAM units. At 27°C this SRAM was operated with 0.25 V with a leakage power of 25 pW/bit. Sharma and Chandel [33] analyzed various SRAM cell designs for low-power applications. Raine et al. [34] designed a nanowire FET-based SRAM with ultra-thin memory material effectively. 3D-SRAM cells were designed with Monte Carlo simulations using computer-aided design tools to demonstrate the sensitivity effect on the memory chip.

A SRAM architecture cell design using single-gated feedback field-effect transistors was given by Cho et al. [35]. The write speed was ~ 0.1 ns, and retention time was 3600 s. The advantages of this architecture were high performance for low-power memory applications and high-density architecture with low standby power (0.24 nW/bit). Kobayashi et al. [36] fabricated a 65 nm technology SRAM that limits the sensitivity issues in on-chip memories. In fact, every on-chip system varies in power supply voltage in its cross-sections, resulting in total dose changes. This system overcomes the negative bias temperature instability effect of the system. Sharma and Chandel [37] designed variation-tolerant low-power SRAM cells using FGMOS and conducted a simulative stability analysis. Wang et al. [38] presented a hybrid (in-near-memory) computational SRAM, which was accessible for all arithmetic operations and small IoT processors, verified with various bit functionality, and developed in 28 nm technology.

Giordano et al. [39] developed a redundant memory utilizing SRAM, which made logic operation faster. It was demonstrated using the Xilinx Kintex-7 FPGA unit. The architecture comprised a reconfigurable nature, and the repetitive use of the memory is another advantage. Without accessing any other external peripherals for this operation, it is effective in terms of power-delay products compared to other groups of FPGA. Sun et al. [40] designed a single-level cell with reduced power up to 53.97% and 62.61% to restore minimization in energy efficiency. This architecture is built for energy efficiency with precharge and sense amplifier updating. Ahmad et al. [41] designed low-leakage fully half-select-free robust SRAM cells and presented a BTI reliability analysis. Ueyoshi et al. [42] showcased an external software program based on SRAM optimization with 3D representation for 40 nm technology that uses inductive coupling for the write stage to minimize the power. Dave and Lekshmi [43] developed a two-level adiabatic logic-based SRAM to minimize the effect of stability and also minimize leakage power. The conventional SRAM was compared with two-level adiabatic logic-based SRAM using the power dissipation factor. Two-level adiabatic logic is known as the zero-CMOS model. The resulting power dissipation in a two-level adiabatic logic SRAM is ten times lower than the conventional SRAM. An energy-efficient SRAM design using FinFETs and potential alteration topology schemes was presented by Samarth and Chandel [44].

Saini et al. [45] designed a latch-based sense amplifier for SRAM using 6T architecture implemented using 45 nm CMOS technology with an input voltage of 1.2 V. This enhances the optimization of static power consumption. The power consumption is reduced by 243.8% compared to conventional sense amplifiers. Yang and Kim [46] designed a low-power SRAM using a hierarchical bit line and local sense amplifiers. Jeong et al. [47] demonstrated a switching pMOS sense amplifier for a high-density low-voltage single-ended SRAM. Further, Vijayan and Kodavanti [48] designed and implemented an 8 × 8 SRAM cell array for low-power applications. Kulkarni et al. [49] developed 8T SRAM arrays using 14 nm FinFET technology to minimize operating voltage to 560 mV with an asymmetric sense amplifier to access memory with high speed. It exhibits a low power dissipation of 24%.

Different technologies and architectures for SRAM cell design are discussed. The important performance metrics for SRAMs include the number of transistors in a cell, area, power, and delay. However, a complete memory tile or SRAM array design needs more detail. Hence an attempt has been made in this chapter to design and analyze an SRAM tile. The methods reviewed in this section help with the design of a memory array with effective power-delay-product reduction and comparative analysis with some of the conventional SRAM designs.

11.3 STATIC RANDOM-ACCESS MEMORY (SRAM) – ARCHITECTURE AND OPERATION

The architecture and working operation of a conventional six-transistor (6T) SRAM cell are presented briefly in this section.

11.3.1 SRAM Architecture

The basic SRAM cell comprises two cross-coupled CMOS inverters and two access/pass transistors used to access the data. These two access/pass transistors further connect to the bit line and bit line bar, i.e. BL and BLB, respectively, to perform read, and write operations. The conventional design of a six-transistor (6T) one-bit SRAM cell is shown in Figure 11.1 [1]. BL and BLB are connected to the access transistors whose gate is controlled by a word line. The word line should be high to perform any operation in the cell. The memory or storage cell preserves one of its two possible stable cases, only if the power supply is given to the cell. The read from or write to storage circuits are controlled by the access devices depending on the data fed to those devices. To achieve the perfect read and write operations in the memory cell, the width to length ratio (W/L) of the transistors of the SRAM cell play a crucial role. The two essential conditions of the W/L ratio of the memory cell are, namely, i) the data read operation should not damage the data placed in the memory cell and ii) the memory cell should allow for changes during the write operation [1].

11.3.2 Read Operation

The transistors are turned on and off during read operations based on the stored logic or value in the memory cell, as shown in Figure 11.2. Assume that logic 0 is already stored in the SRAM memory cell. The node voltages of the memory cell at the initial stage of read operation are shown in Figure 11.2. For the assumed logic 0 stored at node1, the transistors M_2 and M_5 are in an off state (shown in the lighter color), while transistor M_1 and transistor M_6 (shown in the darker color) continue to conduct in the linear region. Thus, voltages at nodes 1 and 2 are $V_1 = 0V$ and $V_2 = V_{DD}$, respectively, before M_3 and M_4 transistors are turned on using bit lines. Read and write operations

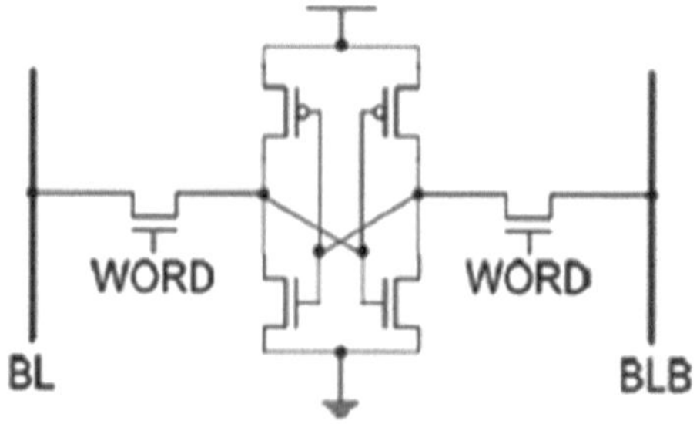

FIGURE 11.1 Conventional 6T (one-bit) SRAM cell.

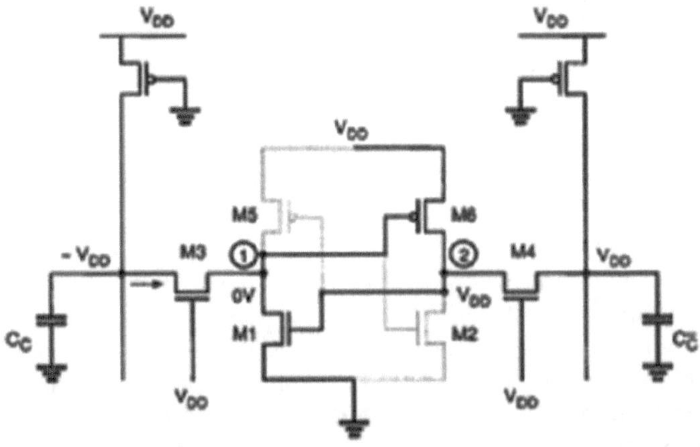

FIGURE 11.2 Read operation for a one-bit memory cell (6T) [1].

occur only when access transistors M_3 and M_4 are turned on; otherwise, the memory continues to be in the hold state. Thereafter, when the row selection circuitry is turned on, i.e. the access transistors, there is no variation in the voltage level at $\overline{C}$ since there is no flow of current in transistor M_4. However, transistors M_1 and M_3 start to conduct a non-zero current and the column voltage decreases. On the other hand, transistor M_1 and transistor M_3 start discharging the capacitance C_C and increasing the voltage at node V_1 from its initial value.

If the W/L ratio of the pass/access transistor M_3 is larger than that of transistor M_1, there is a chance of voltage at V_1 increasing from the threshold voltage of transistor M_2. The key design for the read operation to be perfect is that transistor M_2 should remain turned off, which means voltage V_1 should not rise above the threshold voltage of transistor M_2 during read operation.

$$V_1 \max \leq V_{T,2} \tag{11.1}$$

When the access or pass transistors start to conduct we can assume that column voltage V_C is nearly the same as V_{DD}. This makes transistor M_3 conduct in saturation and M_1 in the linear region, thus leading to Eq. (11.2).

$$\frac{k_{n,3}}{2}\left(V_{DD} - V_1 - V_{T,n}\right)^2 = \frac{k_{n,1}}{2}\left(2\left(V_{DD} - V_{T,n}\right)V_1 - V_1^2\right) \tag{11.2}$$

Using the condition given in Eqs. (11.1) and (11.2) leads to,

$$\frac{k_{n,3}}{k_{n,1}} = \frac{\left(\frac{W}{L}\right)_3}{\left(\frac{W}{L}\right)_1} < \frac{2\left(V_{DD} - 1.5v_{T,n}\right)V_{T,n}}{\left(V_{DD} - 2V_{T,n}\right)^2} \tag{11.3}$$

Eq. (11.3) must be fulfilled in order to make the correct read operation.

11.3.3 Write Operation

Consider a write operation for 0 by estimating that logic 1 is put into the memory cell, as shown in Figure 11.3. The node voltages at the initial stage of the write operation are also shown in Figure 11.3. The transistors M_1 and M_6 are off depending on the value already put into the memory, and the transistors M_5 and M_2 operate in linear mode. Initially, the node voltages are $V_1 = V_{DD}$ and $V_2 = 0V$ before the access or pass transistors start to conduct. The voltage V_C is forced to logic 0 with the help of write circuitry. It can be estimated that voltage V_C is nearly equal to 0.

When the access/pass transistors are turned on using the row decoder, we estimate that the node voltage at V_2 is lower than the threshold voltage of transistor M_1. To write or change the data stored in the memory cell, the voltage at V_2 is forced to V_{DD} and V_1 to 0V. It has to be ensured that the transistor M_2 should not conduct by keeping the node voltage below the threshold voltage of M_2 during the data write phase. During this phase, M_3 conducts in the linear region and M_5 in the saturation region, and M_2 in the cut-off region, so the change in the data takes place. This leads to Eq. (11.4), and using the condition that M_2 remains in cut-off gives Eqs. (11.5) and (11.6).

$$\frac{k_{P,5}}{2}\left(0-V_{DD}-V_{T,P}\right)^2=\frac{k_{n,3}}{2}\left(2\left(V_{DD}-V_{T,n}\right)V_{T,n}-V_{T,n}^2\right) \tag{11.4}$$

$$\frac{K_{P,5}}{K_{n,3}}<\frac{2\left(V_{DD}-1.5V_{T,n}\right)V_{T,n}}{\left(V_{DD}+2V_{T,p}\right)^2} \tag{11.5}$$

$$\frac{\left(\frac{W}{L}\right)_5}{\left(\frac{W}{L}\right)_3}<\frac{\mu_n}{\mu_p}\cdot\frac{2\left(V_{DD}-1.5V_{T,n}\right)V_{T,n}}{\left(V_{DD}+2V_{T,P}\right)^2} \tag{11.6}$$

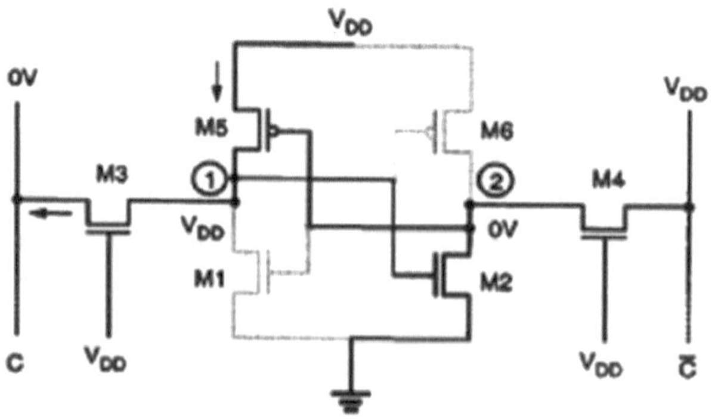

FIGURE 11.3 Write operation for a one-bit memory cell (6T) [1].

Thus, to allow modifications in the stored value in the SRAM cell, the write operation must fulfill Eq. (11.6).

11.4 THE LECTOR TECHNIQUE

The key concept for leakage power reduction is transistor stacking in the path of the supply voltage to the ground [10]. Generally, the lector is referred to as a leakage control transistor (LCT), which means the transistor can control leakage current in CMOS circuits with no rise in dynamic power dissipation [10–12]. This technique has a similar structure to conventional CMOS with the addition of two LCTs whose gate is controlled by the source of another gate. By introducing these LCTs between the power supply and ground, one of these two LCTs will always conduct near to its cut-off mode for any input combination. The main principle behind this technique is that a circuit with more than one transistor is off between the power supply and ground and is less leaky compared to in the case where only one transistor is off between the power supply and ground [10–12]. An inverter using the lector technique is shown in Figure 11.4.

The two leakage-controlled transistors are introduced between the nodes *N_1* and *N_2*, whose source terminals are connected to transistors M_1 and M_2 of the CMOS gate. There is no need for an external circuit since LCTs are self-controlled. When the input applied is low, the pull-down transistor M_2 and LCT_1 get turned off; hence, stacking occurs between the two series-connected transistors. On the other hand, when the input applied is high. The pull-up transistors M_1 and LCT_2 are turned off,

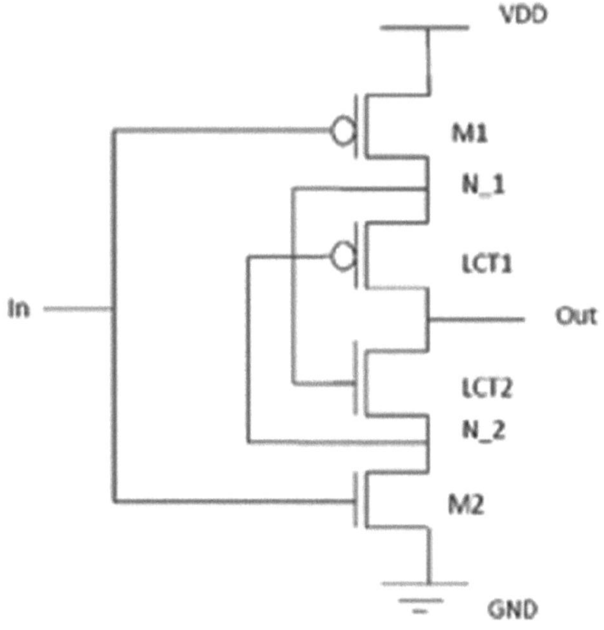

FIGURE 11.4 CMOS inverter based on a lector.

TABLE 11.1
State Matrix of a Lector-Based CMOS Inverter

Transistor reference	Input 0	Input 1
M_1	ON	OFF
M_2	OFF	ON
LCT_1	Near cut-off	ON
LCT_2	ON	Near cut-off

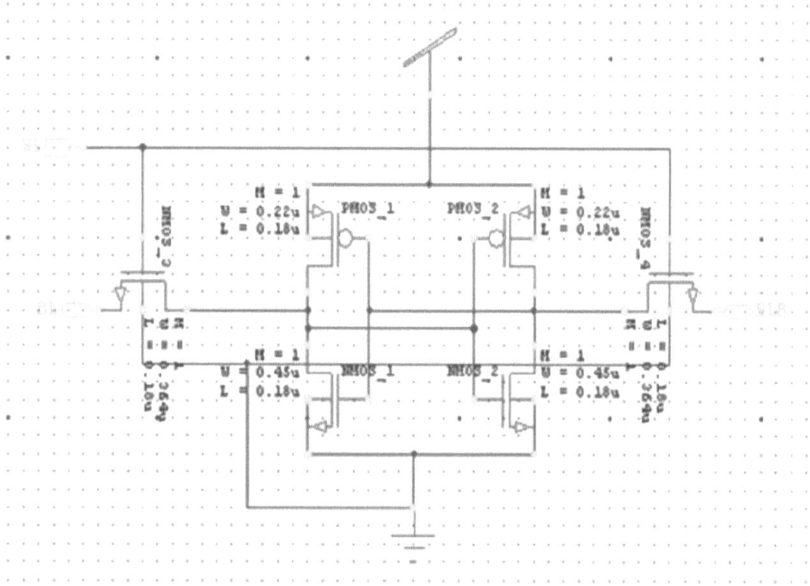

FIGURE 11.5 SRAM without lector.

leading to increased off resistance of the supply to the ground. Table 11.1 shows the state matrix of a lector-based CMOS inverter.

11.4.1 Applying the Lector Technique to SRAM

As the SRAM cell is made of two cross-coupled CMOS inverters, the lector technique is applied for both the inverters and the peripheral circuits of a SRAM [10–12]. Figures 11.5 and 11.6 show the SRAM without the lector and with the lector technique, respectively. The blocks in Figure 11.6 are implemented using lector inverters.

11.5 RESULTS AND DISCUSSION

The static random-access memory consists of many peripheral circuits, such as using write circuitry to rewrite the data into the cell. Decoders are used to pass the data to

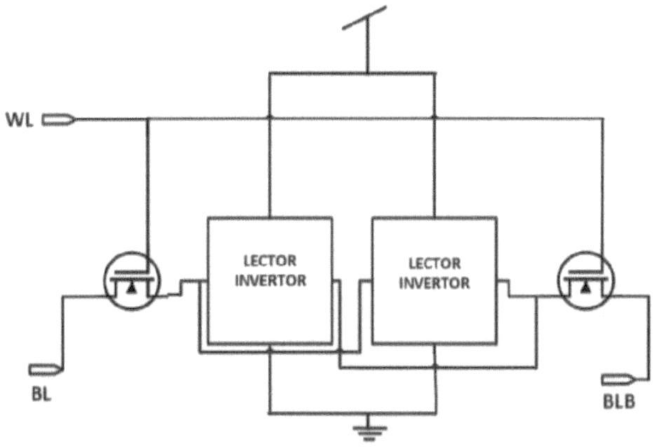

FIGURE 11.6 SRAM with lector.

TABLE 11.2
Sizing Used in SRAM

S.No.	Transistor	Width (μm)
1	Pull-up transistor M_5 and M_6	0.22
2	Pull-down transistor M_1 and M_2	0.45
3	Access transistors M_3 and M_4	0.364

the cell through a column circuit. A sense amplifier facilitates sensing the data and computing the delay for input and output. All the results in this chapter are implemented using Tanner EDA tools [50], using a 180 nm technology node. Table 11.2 presents the dimensions of the transistors computed for the SRAM cell design.

Figure 11.7 gives the schematic of the precharge circuit design. Figure 11.8 provides the sense amplifier circuit design schematic. Figure 11.9 is the schematic of the write circuitry. Figure 11.10 shows the schematic of the output buffer circuitry of the SRAM.

In Figure 11.11, a 4 × 16 lector decoder is seen. Waveforms of the 4 × 16 lector decoder are presented in Figure 11.12. Figure 11.13 gives the schematic of the 5 × 32 lector decoder. Further, Figure 11.14 shows the waveforms of the lector 5 × 32 decoder. Figure 11.15 shows a 16 × 16 SRAM memory array, where each block contains a single-bit SRAM cell.

Finally, Figure 11.16 provides the schematic of desired 32 × 16 (512-bit) SRAM array. Figure 11.17 shows the waveforms of the 32 × 16 SRAM array for data read and write phases of the 6T configuration. Figure 11.18 gives the waveforms for data read and write phases of a 32 × 16 SRAM array for 8T design. Furthermore, Figure 11.19 presents the read and write operation of a 32 × 16 SRAM array with the lector technique.

Table 11.3 presents a comparative analysis of power dissipation in three 32 × 16 (512-bit) SRAM array designs. Figure 11.20 is the graphical representation of the

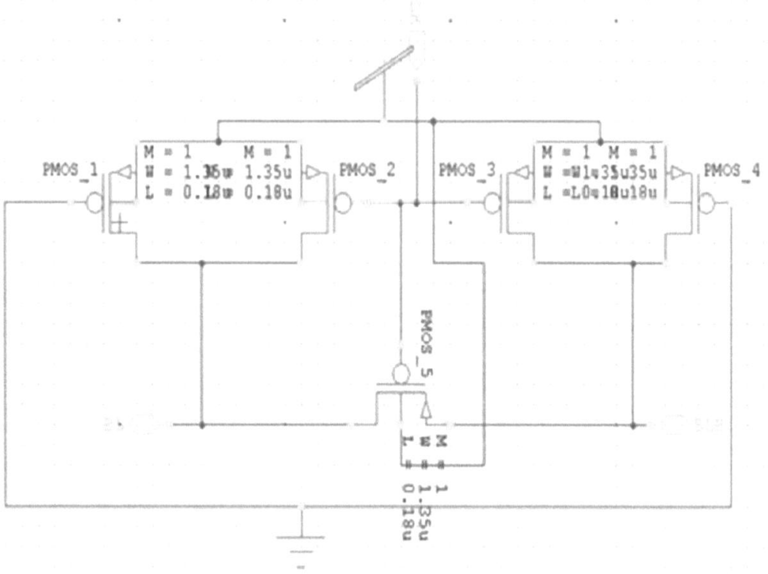

FIGURE 11.7 Design of a precharge circuit.

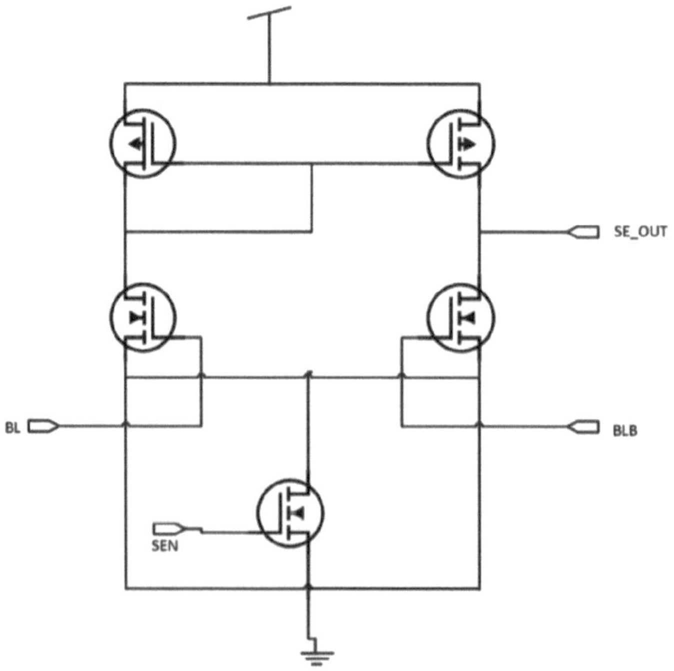

FIGURE 11.8 Design of a sense amplifier.

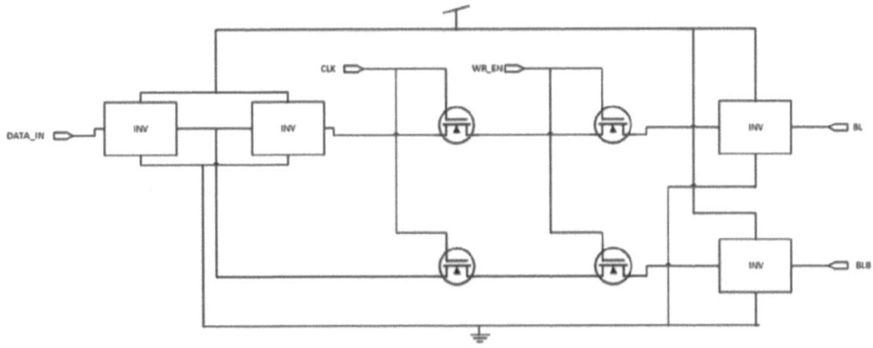

FIGURE 11.9 Design of write circuitry.

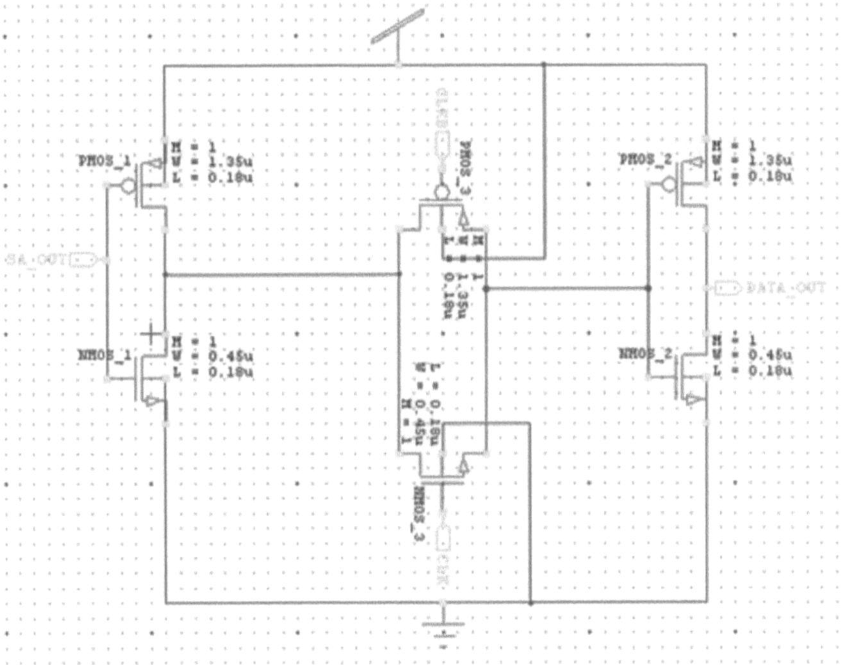

FIGURE 11.10 Schematic of output buffer circuitry.

power dissipation in the three SRAM array designs under consideration. Table 11.3 and Figure 11.20 show that the lector technique leads to lesser power dissipation than the 6T and 8T SRAM designs. The lector technique provides a 24.49% decrease in power consumption compared to the 6T SRAM array design; it has 8.77% lesser power dissipation than the 8T SRAM array design.

512-bit SRAM tiles were designed and their read and write operations were verified. The SRAM tile performance metrics, viz., read and write access time or delay,

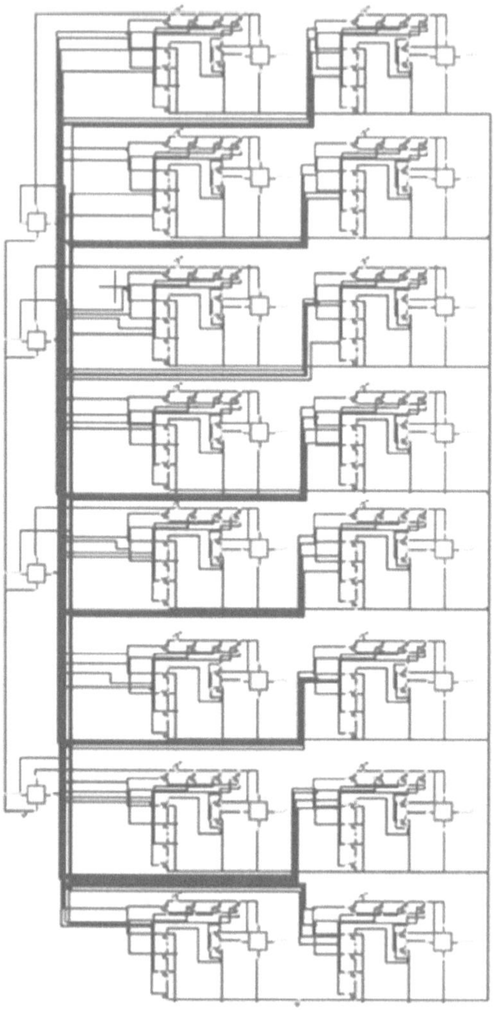

FIGURE 11.11 The designed 4 × 16 lector decoder.

power dissipation, and energy, were determined as shown in Table 11.4. It is seen from the table that the SRAM tile designed using the lector technique has the least read access time of 1.25 ns. The write access time is the minimum for the 6T tile, followed by the lector tile, which provides 1.92 ns, and 8T has the maximum write delay of 2.13 ns. Though the lector technique SRAM cell uses ten transistors, it has an advantage due to a low power dissipation of 37.624 μW over the complete tile. Furthermore, a lower power-delay-product implies lower energy dissipation and, consequently, better-integrated circuit performance. It can also be seen from Table 11.4 that the average SRAM tile power-delay-product is the least among the

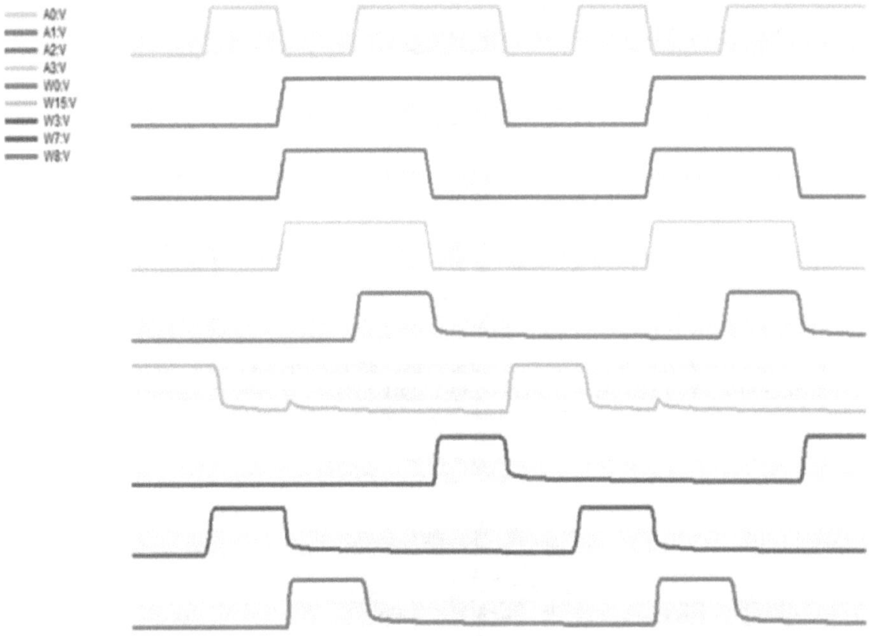

FIGURE 11.12 Waveform of a 4 × 16 lector decoder.

three cases in the case of the lector-based SRAM tile. Energy dissipation is 32.43% lesser than 6T and 23.19% lesser than the 8T designs for the designed 512-bit SRAM array. Thus, the lector-based SRAM tile design has better and higher performance for both power and energy dissipation criteria.

11.6 CONCLUSION

Leakage reduction plays a crucial role in VLSI circuit design. The lector technique is found to be significantly effective in reducing the leakage current. The simulative analysis presented above inferred that the lector technique leads to lesser power dissipation when compared to 6T and 8T SRAM array designs. The lector technique gives a 24.49% and 8.77% decrease in power consumption compared to 6T and 8T SRAM array designs, respectively. Further, the technique also provides an advantage in terms of read access time for the 512-bit SRAM tiles. As the technology nodes get miniaturized there is a huge demand for less leaky circuit designs. The lector technique benefits from leakage reduction without much affecting the dynamic power dissipation; hence, it is one of the best solutions. It is analyzed that the lector technique is extremely appropriate for power and energy-centric MOS memory designs. The current work shall be highly useful for VLSI designers and researchers.

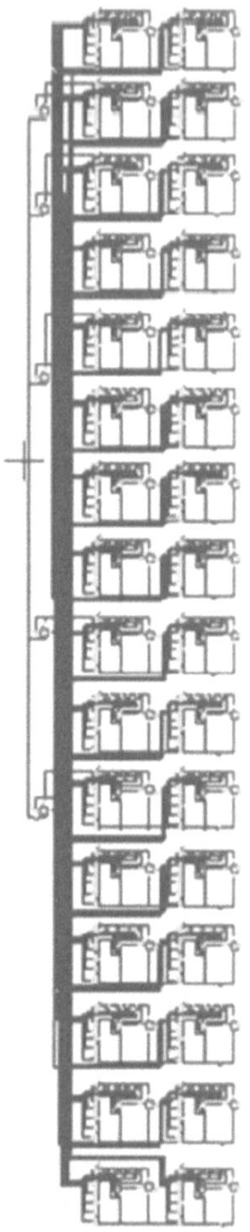

FIGURE 11.13 The designed 5 × 32 lector decoder.

ACKNOWLEDGMENT

The technical support of the SMDP-C2SD project of the Ministry of Electronics and Information Technology (MeitY), GoI New Delhi, awarded to the Department of ECE, NIT Hamirpur (HP), is duly acknowledged. This work in part is also the outcome of the technical and financial support received under the Core Research Grant (Ref. No. CRG/2021/000780) sponsored by the Science and Engineering Research Board (SERB), Department of Science and Technology, Government of India. The work is also the outcome of the research and development work undertaken in the project under the Visvesvaraya PhD Scheme of MeitY, Government of India [Ref. No: MeitY-PhD-3186], being implemented by the Digital India Corporation (formerly Media Lab Asia).

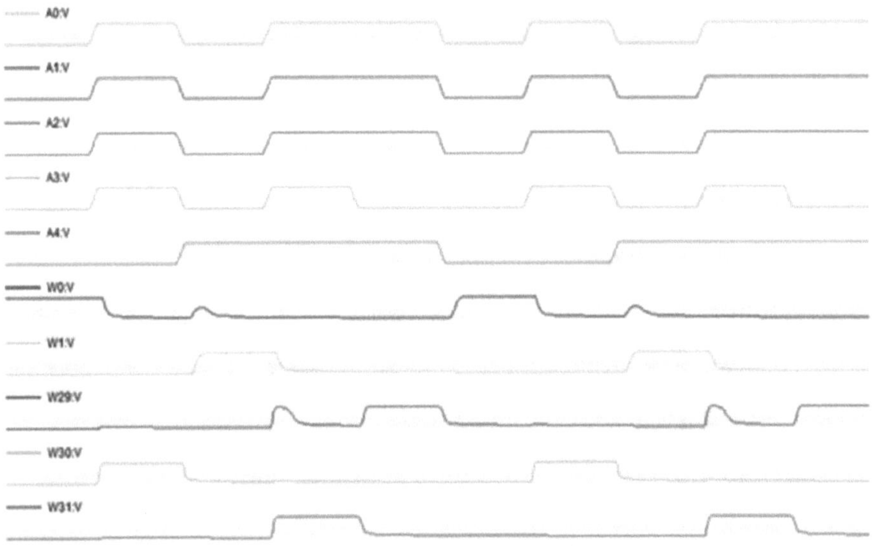

FIGURE 11.14 Waveform of a lector 5 × 32 decoder.

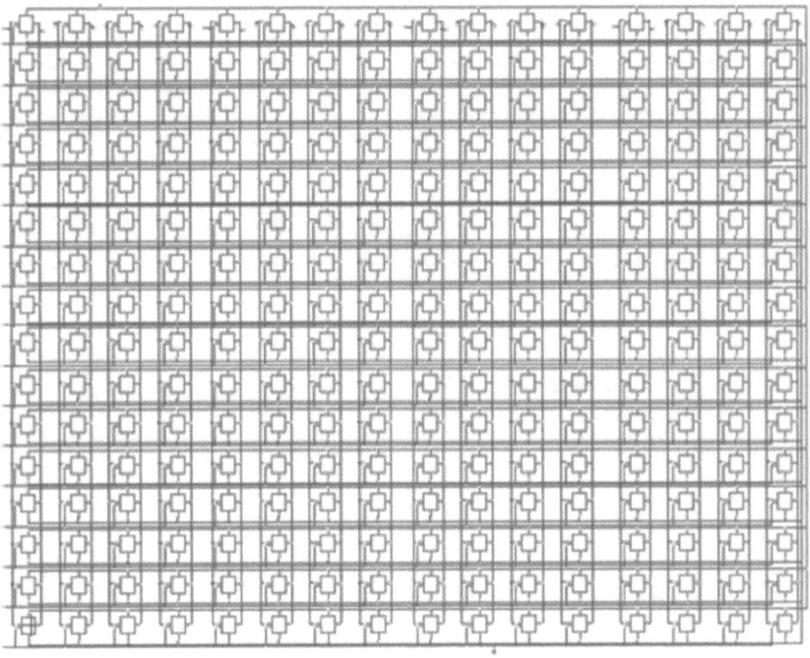

FIGURE 11.15 16×16 memory array (each block contains a single bit cell).

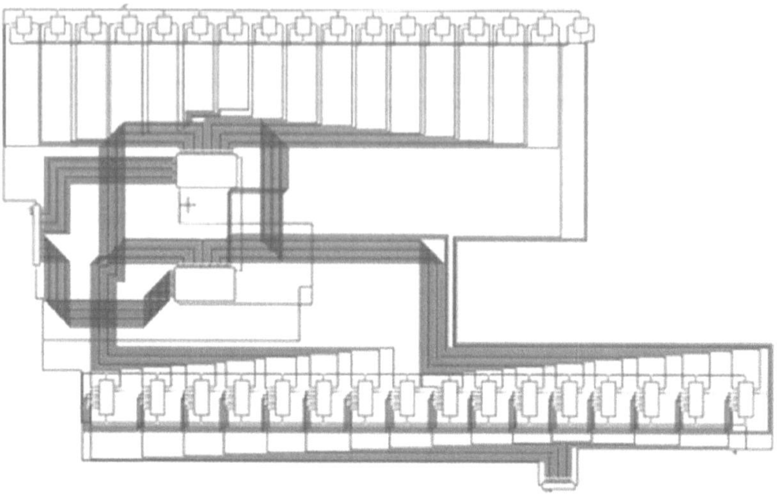

FIGURE 11.16 Schematic of the designed 32 × 16 (512bit) SRAM array.

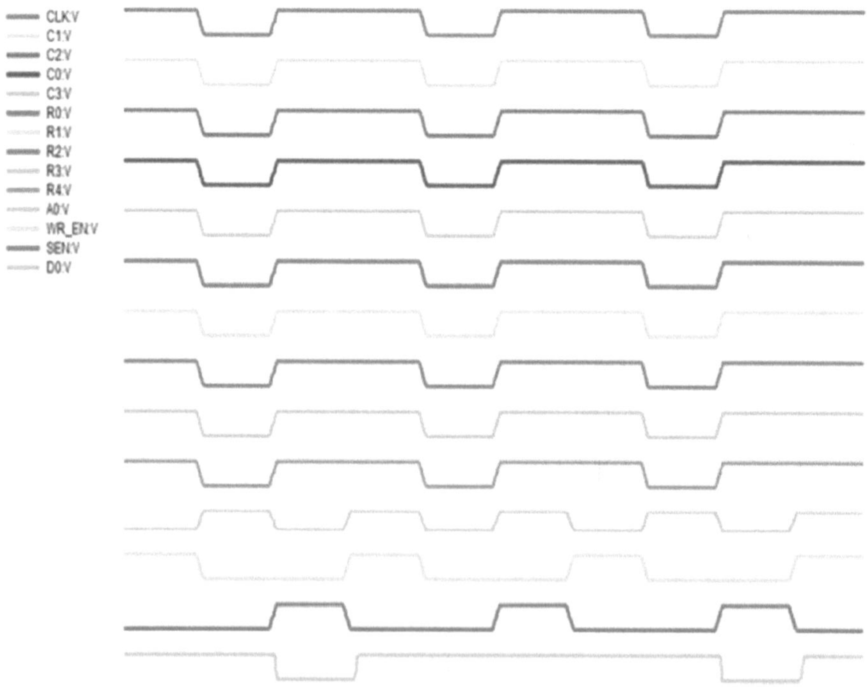

FIGURE 11.17 Waveform of a 32 × 16 array for data read and write phase (6T).

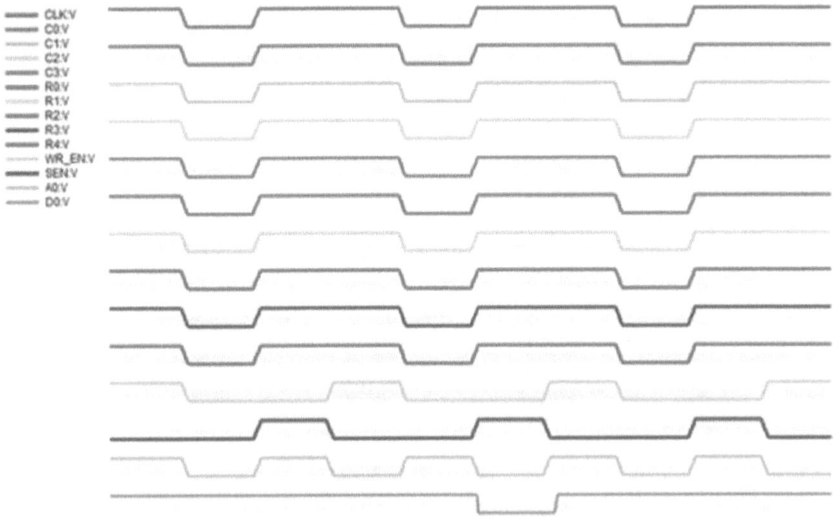

FIGURE 11.18 Waveform of a 32 × 16 SRAM array for data read and write phase (8T).

FIGURE 11.19 Read–write operation of a 32 × 16 SRAM array with lector technique.

TABLE 11.3
Power Dissipation in 32×16 (512 bit) SRAM Arrays

512Bit SRAM Array Design using	Average power consumption (microwatts)
6T	49.831
8T	40.925
Lector	37.624

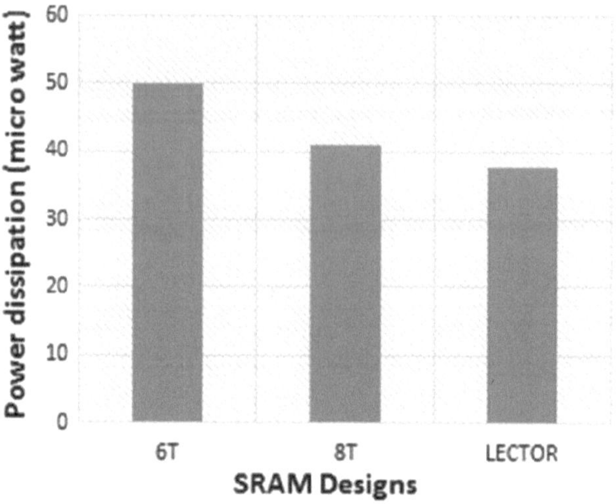

FIGURE 11.20 Bar graph indicating the power dissipation in the three SRAM designs.

TABLE 11.4
Performance Analysis of the Three Differently Designed 512 Bit SRAM Tiles

	SRAM Tile design technique		
Performance metrics	**6T**	**8T**	**Lector**
Number of transistors per SRAM cell	6	8	10
Tile Read access time (ns)	1.79	1.46	1.25
Tile Write access time (ns)	1.38	2.13	1.92
Tile Power dissipation (μW)	49.831	40.925	**37.624**
Tile Energy dissipation (wrt read time) (aJ)	89.19	59.75	47.03
Tile Energy dissipation (wrt write time) (aJ)	68.76	87.17	72.23
Average Energy dissipation (aJ)	78.97	73.46	**59.63**

REFERENCES

1. S. M. Kang, and Y. Leblebici, *CMOS Digital Integrated Circuits*, Tata McGraw-Hill Edition, New Delhi, 2003.
2. H. J. M. Veendrick, "Short circuit dissipation of static CMOS circuitry and its impact on the design of buffer circuits," *IEEE J. Solid-State Circuits*, vol. SC-19, no. 4, pp. 468–473, Aug. 1984.
3. V. De, and S. Borkar, "Technology and design challenges for low power and high performance," ACM ISLPED99, San Diogo, CA, pp. 163–168, Aug. 1999.

4. R. X. Gu, and M. I. Elmasry, "Power dissipation analysis and optimization for deep submicron CMOS digital circuits," *IEEE J. Solid-State Circuits*, vol. 31, no 5, pp. 707–713, May. 1999.
5. M. C. Johnson, D. Somasekhar, L. Y. Chiou, and K. Roy, "Leakage control with efficient use of transistor stacks in single threshold CMOS," *IEEE Trans. VLSI Syst.*, vol. 10, no. 1, pp. 1–5, Feb. 2002.
6. S. Narendra, S. Borkar, V. De, D. Antoniadis, and A. P. Chandrakasan, "Scaling of stack effect and its application for leakage reduction," *Proc. IEEE*, IEEE International Symposium on Low Power Electronics and Design (ISLPED), Huntington Beach, Califomia, USA, pp. 195–200, Aug. 2001.
7. S. Sirichotiyakul, T. Edwards, C. Oh, R. Panda, and D. Blaauw, "Duet: An accurate leakage estimation and optimization tool for dual-V circuits," *IEEE Trans. VLSI Syst.*, vol. 10, no. 2, pp. 79–90, Apr. 2002.
8. K. Choi, H. Shim, et al., "Enhanced reliability of 7nm process technology featuring EUV," *IEEE Trans. Electron Devices*, vol. 66, no. 12, pp. 5399–5403, June. 2019.
9. H. Qin, Y. Cao, D. Markovic, A. Vladimirescu, and J. Rabaey, "SRAM leakage suppression by minimizing standby supply voltage," IEEE International Symposium on Signals, Circuits and Systems (SCS), San Jose, CA, USA, pp. 55–60, 2004.
10. N. Hanchate, and N. Ranganathan, "LECTOR: A technique for leakage reduction in CMOS circuits," *IEEE Trans. VLSI Syst.*, vol. 12, no. 2, pp. 196–205, Feb., 2004.
11. S. Gaonkar, and C. Engineering, "Design of CMOS inverter using lector technique to reduce the leakage," *IEEE Trans. Very Large Scale Integr.*, vol. 31, no. 31, pp. 231–233, Sept. 2015.
12. Y. Alekhya, and J. Sudhakar, "Design of 64 bit SRAM using lector technique for low leakage power with read and write enable," *IOSR J. VLSI Signal Process.*, vol. 7, no. 2, pp. 10–19, April. 2017.
13. G. Torrens, A. Alheyasat, B. Alorda, S. Barcelo, J. Segura, and S. A. Bota, "Transistor width effect on the power supply voltage dependence of α-SER in CMOS 6T SRAM," *IEEE Trans. Nucl. Sci.*, vol. 67, no. 5, pp. 811–817, March. 2020.
14. R. K. Sah, I. Hussain, and M. Kumar, "Performance analysis of a 6T SRAM cell in 180nm CMOS technology," *IOSR J. VLSI Signal Process.*, vol. 5, no. 2, pp. 20–22, Oct. 2015.
15. K. Khare, N. Khare, V. K. Kulhade, and P. Deshpande, "VLSI design and analysis of low power 6T SRAM cell using Cadence tool," IEEE International Conference on Semiconductor Electronics (ICSE), Johor Bahru, Malaysia, pp. 117–119, Nov. 2008.
16. N. Surana, and J. Mekie, "Energy efficient single-ended 6T SRAM for multimedia applications," *IEEE Trans. Circuits Syst. II Express Briefs*, vol. 66, no. 6, pp. 1023–1027, June. 2019.
17. S. L. Wu, K.Y. Li, P.T. Huang, W. Hwang, M.H. Tu, S.C. Lung, W.S. Peng, H.S. Huang, K.D. Lee, Y.S. Kao, and C.T. Chuang, "A 0.5-V 28-nm 256-kb mini-array based 6T SRAM with Vtrip-tracking write-assist," *IEEE Trans. Circuits Syst. I Regul.*, vol. 64, no. 7, pp. 1791–1802, July. 2017.
18. B. Mohammadi, O. Andersson, J. Nguyen, L. Ciampolini, A. Cathelin, and J. N. Rodrigues, "A 128 kb 7T SRAM using a single-cycle boosting mechanism in 28-nm FD-SOI," *IEEE Trans. Circuits Syst. I Regul.*, vol. 65, no. 4, pp. 1257–1268, April. 2018.
19. S. Gupta, K. Gupta, and N. Pandey, "A 32-nm subthreshold 7T SRAM bit cell with read assist," *IEEE Trans. Very Large Scale Integr. Syst.*, vol. 25, no. 12, pp. 3473–3483, Dec. 2017.
20. M. Kumar, and J. S. Ubhi, "Performance evaluation of 6T, 7T & 8T SRAM at 180 nm technology," 8th International Conference on Computer Communication Network Technology, ICCCNT 2017, Delhi, India, pp. 2–7, July. 2017.

21. C. B. Kushwah, and S. K. Vishvakarma, "A single-ended with dynamic feedback control 8T subthreshold SRAM cell," *IEEE Trans. Very Large Scale Integr. Syst.*, vol. 24, no. 1, pp. 373–377, Jan. 2016.
22. C. Huang, and L. Chiou, "Single bit-line 8T SRAM cell with asynchronous dual word-line control for bit-interleaved ultra-low voltage operation," *IET Circuits, Devices Syst.*, vol. 12, no. 6, pp. 713–719, Nov. 2018.
23. S. Pal, and A. Islam, "Variation tolerant differential 8T SRAM cell for ultralow power applications," *IEEE Trans. Comput.-Aided Des. Integr. Circuits Syst.*, vol. 35, no. 4, pp. 549–558, April. 2016.
24. K. Cho, J. Park, T. W. Oh, and S. Jung, "One-sided Schmitt-trigger-based 9T SRAM cell for near-threshold operation," *IEEE Trans. Circuits Syst. I: Regul. Papers*, vol. 67, no.5, pp. 1551–1561, May. 2020.
25. L. Wen, Y. Zhang, and X. Zeng, "Column-selection-enabled 10T SRAM utilizing shared diff-Vdd write and dropped-Vdd read for power reduction," *IEEE Trans. Very Large Scale Integr. Syst.*, vol. 27, no. 6, pp. 1470–1474, June. 2019.
26. N. Maroof, and B. Kong, "10T SRAM using half- V_{DD} precharge and row-wise dynamically powered read port for low switching power and ultralow RBL leakage," *IEEE Trans. Very Large Scale Integr. Syst.*, vol. 25, no. 4, pp. 1193–1203, April. 2017.
27. J. Jiang, Y. Xu, W. Zhu, J. Xiao, and S. Zou, "Quadruple cross-coupled latch-based 10T and 12T SRAM Bit-cell designs for highly reliable terrestrial applications," *IEEE Trans. Circuits Syst. I Regul.* vol. 66, no. 3, pp. 967–977, March. 2019.
28. S. K. Sharma, and B. P. Shrivastava, "Low power and high speed 11T SRAM cell with bit-interleaving capability," *Int. J. Comput. Appl.*, vol. 141, no. 11, pp. 21–24, May. 2016.
29. N. Yadav, A. P. Shah, and S. K. Vishvakarma, "Stable, reliable, and bit-interleaving 12T SRAM for space applications: A device circuit co-design," *IEEE Trans. Semicond. Manuf.*, vol. 30, no. 3, pp. 276–284, Aug. 2017.
30. M. Nobakht, and R. Niaraki, "A new 7T SRAM cell in sub-threshold region with a high performance and small area with bit interleaving capability," *IET Circuits, Devices Syst. IET Circuits*, vol. 13, no. 6, pp. 873–878, Sept. 2019.
31. S. Pal, S. Bose, W. Ki, and A. Islam, "Reliable write assist low power SRAM cell for wireless sensor network applications," *IET Circuits, Devices Syst.*, vol. 14, no. 2, pp. 137–147, March. 2020.
32. R. Boumchedda, A. Makosiej, J.P. Noel, B. Giraud, J.F. Christmann, I.M. Panades, L. Ciampolini, P. Royer, C. Mounet, D. Turgis, and E. Beigne, "1.45-fJ/bit access two-port SRAM interfacing a synchronous/asynchronous IoT platform for energy-efficient normally off applications," *IEEE Solid-State Circuits Lett.*, vol. 1, no. 9, pp. 186–189, 2018.
33. C. K. Sharma, and R. Chandel, "Analysis of SRAM cell designs for low power applications," IEEE International Conference for Convergence of Technology (I2CT-2014), Pune, India, pp. 1–4, 2014.
34. M. Raine, M. Gaillardin, T. Lagutere, O. Duhamel, and P. Paillet, "Estimation of the single-event upset sensitivity of advanced SOI SRAMs," *IEEE Trans. Nucl. Sci.*, vol. 65, no. 1, pp. 339–345, Jan. 2018.
35. J. Cho, D. Lim, S. Woo, K. Cho, and S. Kim, "Static random access memory characteristics of single-gated feedback field-effect transistors," *IEEE Trans. Electron Devices*, vol. 66, no. 1, pp. 413–419, Jan. 2019.
36. D. Kobayashi et al., "Process variation aware analysis of SRAM SEU cross sections using data retention voltage," *IEEE Trans. Nucl. Sci.*, vol. 66, no. 1, pp. 155–162, Jan. 2019.
37. N. Sharma, and R. Chandel, "Variation tolerant and stability simulation of low power SRAM cell analysis using FGMOS," *Int.J. Model., Simul. Sci. Comput.*, (online), https://doi.org/10.1142/S179396232150029X, March. 2021.

38. J. Wang et al., "A 28-nm compute SRAM with bit-serial logic/arithmetic operations for programmable in-memory vector computing," *IEEE J. Solid-State Circuits*, vol. 55, no. 1, pp. 76–86, Jan. 2020.
39. R. Giordano, S. Perrella, V. Izzo, G. Milluzzo, and A. Aloisio, "Redundant-configuration scrubbing of SRAM-based FPGAs," *IEEE Trans. Nucl. Sci.*, vol. 64, no. 9, pp. 2497–2504, Sept. 2017.
40. Y. Sun et al., "Energy-efficient nonvolatile SRAM design based on resistive switching multi-level cells," *IEEE Trans. Circuits Syst. II Express Briefs*, vol. 66, no. 5, pp. 753–757,May. 2019.
41. S. Ahmad, B. Iqbal, N. Alam, and M. Hasan, "Low leakage fully half-select-free robust SRAM cells with BTI reliability analysis," *IEEE Trans. Device Mater. Reliab.*, vol. 18, no. 3, pp. 337–349, Sept. 2018.
42. K. Ueyoshi et al., "QUEST: Multi-purpose log-quantized DNN inference engine stacked on 96-MB 3D SRAM using inductive coupling technology in 40-nm CMOS," *IEEE J. Solid-State Circuits*, vol. 54, no. 1, pp. 186–196, Feb. 2019.
43. A. Dave, and C. A. Lekshmi, "A novel adiabatic SRAM design using two level adiabatic logic," 15th Biennial Baltic Electronics Conference (BEC), pp. 51–54, Nov. 2016.
44. S. Agarwal, and R. Chandel, "Energy efficient SRAM design using FinFETs and potential alteration topology schemes," 10th International Symposium *on* Embedded Computing *and* System Design (ISED-2021), DA-IICT, Gandhinagar, Gujarat, 16–18 July, 2021.
45. A. Saini, P. K. Gupta, and R. Gupta, "Analysis of low power SRAM sense amplifier," International Conference on Electrical, Electronics and computer Engineering (UPCON-2019), pp. 1–6, Nov. 2019.
46. Yang, and L. Kim, "A low-power SRAM using hierarchical bit line and local sense amplifiers," *IEEE J. Solid-State Circuits*, vol. 40, no. 6, pp. 1366–1376, June. 2005.
47. H. Jeong, T. Kim, K. Kang, T. Song, G. Kim, H. S. Won, and S. O. Jung. "Switching pMOS sense amplifier for high-density low-voltage single-ended SRAM,'. *IEEE Transactions on Circuits and Systems I: Regular Papers*, *62*(6), 1555–1563, June. 2015. [7112585]. https://doi.org/10.1109/TCSI.2015.2415171
48. S. Vijayan, and P. Kodavanti, "Novel design and implementation of 8x8 SRAM cell array for low power applications," *International Journal of Engineering Research and Technology*, vol. 3, no.7, pp. 1355–1359, July. 2014.
49. J. P. Kulkarni, J. Keane, K. H. Koo, S. Nalam, Z. Guo, E. Karl, and K. Zhang, "5.6 Mb/mm^2 1R1W 8T SRAM arrays operating down to 560 mV utilizing small-signal sensing with charge shared bitline and asymmetric sense amplifier in 14 nm FinFET CMOS technology," *IEEE J. Solid-State Circuits*, vol. 52, no. 1, pp. 229–239, Jan. 2017.
50. *Tanner EDA Tools* Version 16.3, released June 2015.

12 Characterization of Stochastic Process Variability Effects on Nano-Scale Analog Circuits

Soumya Pandit

CONTENTS

DOI: 10.1201/9781003155751-12

12.1 INTRODUCTION

The traditional approach to designing an integrated circuit at the transistor level of abstraction consists of determining the transistor level design parameters while the targeted design specifications are being met [1]. A fundamental assumption for this deterministic design methodology approach is that the semiconductor fabrication process is well characterized, and the circuit design parameters are accurately fabricated [2]. This assumption is acceptably true for large geometry transistor processes, where the random effects of manufacturability errors are not significant compared to the deterministic values of the performance parameters [3]. However, with the scaling of CMOS technology on the nano-scale dimension, the variability of the process parameters has become critically high, so the fundamental assumption of the deterministic VLSI design methodology fails [4, 5]. The performance parameters of an integrated circuit need to be formulated as random or stochastic variables. The set of values that can be taken by a performance parameter can lie anywhere between a range, and the chance of taking each value is characterized by a probability measure. There will always be a non-zero probability that the value taken by a stochastic performance variable will not lie within the acceptable specification range. The increasing amount of local/random or intra-die process variabilities on the yield of nano-scale integrated circuits has imposed a serious challenge to the conventional deterministic approach to circuit design. The global/systematic or inter-die process variabilities, although more prominent in effect, are nowadays resolved through special treatments. However, the former needs a paradigm shift in design methodology, leading to a new design methodology, referred to as the statistical design methodology. In such a design methodology, modeling process variability and accurate characterization of performance variability are critical tasks for the designers.

The rest of this chapter is organized as follows. Section 12.2 provides the necessary background concepts. Section 12.3 gives a detailed description of the statistical characterization methodology for stochastic process variabilities. The implementation of the methodology is given in Section 12..4. An application example illustrating the use of the methodology is given in Section 12.5. Finally, Section 12.6 concludes the chapter.

12.2 BACKGROUND CONCEPTS

This section introduces important concepts, definitions, and mathematical formulations of the statistical characterization methodology.

Definition 1 (Circuit Parameters): A circuit is characterized by a set of parameters that determine the circuit's performance. These are referred to as circuit parameters.

Definition 2 (Circuit Design Parameters): The type of circuit parameters that are used by circuit designers for optimizing the performance of the circuit are referred to as the circuit design parameters.

Definition 3 (Process Parameters): The type of circuit parameters that depend upon the device's physical characteristics used for implementing the circuit are known as process parameters.

Definition 4 (Circuit Operating Parameters): The performance of a circuit is dependent on a set of circuit parameters that characterize operating conditions. These parameters are referred to as circuit operating parameters. They may be supply voltage values or temperature.

Definition 5 (Circuit Performance Parameters): The performance of a circuit is characterized by a set of parameters referred to as circuit performance parameters. They form a part of the circuit design specifications.

The mathematical relationship between these various parameters may be stated as follows. Let Φ be the set of circuit parameters and Φ_d be the set of circuit design parameters; let Φ_p be the set of process parameters, and Φ_o be the set of operating parameters. The relationship between these four sets is

$$\Phi = \Phi_d \cup \Phi_p \cup \Phi_o \tag{12.1}$$

The relationship between the set of circuit performance parameters and the set of circuit parameters is written as

$$P = f(\Phi) \tag{12.2}$$

These definitions may be explained with the help of a simple example. Consider the design of a simple CMOS inverter: The circuit design parameters include the channel widths and lengths of the n-channel and p-channel MOS transistors. The process parameters include the threshold voltages of the transistors, their oxide thickness, surface mobility of electrons and holes, and so on. On the other hand, the circuit performance parameters include the switching threshold voltage of the circuit, propagation delay, etc. The functional form f as specified in Eq. (12.2) is evaluated through the circuit simulation mechanism [1] in the design procedure. This is illustrated in Figure 12.1, a detailed discussion on which may be found in [1].

12.2.1 Randomness of Circuit Parameters and Circuit Performance Parameters

Due to non-idealities and practical limitations, an inevitable fact is that actual values of the circuit parameters assumed in the design procedure will not be achieved after the fabrication of the circuit. Therefore, the actual values of the circuit parameters

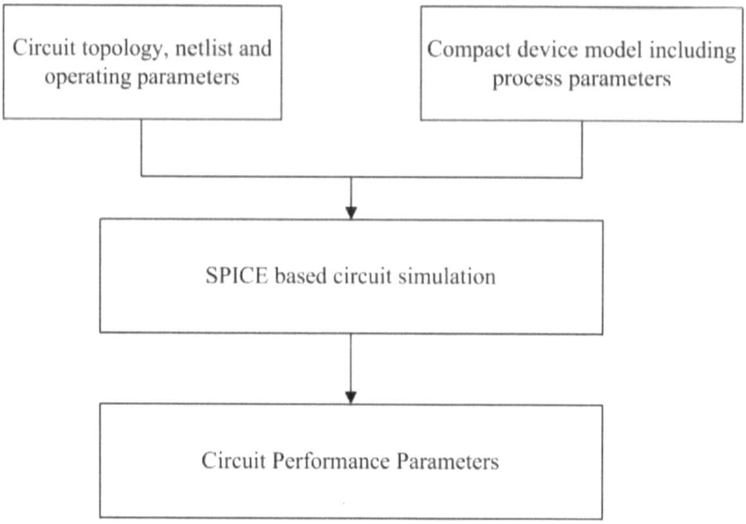

FIGURE 12.1 Idea of circuit simulation.

will be different from their designed values. The impact is that the circuit performance parameters' actual values will also differ from their target values. This true practical aspect of circuit design is mathematically represented by adding an uncontrollable random varying component to the deterministic values of each circuit parameter. For a single design parameter, this is written as

$$\Phi_i = \overline{\Phi}_{id} + \tilde{\Phi}_{is} \tag{12.3}$$

where $\overline{\Phi}_{id}$ is the deterministic or nominal component of the circuit parameter and $\Phi\tilde{\Phi}_{is}$ is the uncontrollable random/stochastic component of the circuit parameter. It is because of the presence of this uncontrollable stochastic component that a circuit design parameter Φ_i becomes a stochastic variable. The various values taken by this variable will lie within a certain range. The probability of occurrence of each of these values will be non-zero and finite. Therefore, the randomness of each circuit design parameter will be characterized by a probabilistic description of each of the values taken by the random circuit design parameter variable. For example, the actual channel width W of the MOS transistor may be decomposed into a deterministic component $\overline{W}$ and a stochastic component $\tilde{W}$. A similar formulation can be made for the channel length of a MOS transistor.

Since a performance parameter of a circuit depends on the values of each element of the circuit parameter set and each such element is a stochastic variable, a performance parameter also becomes a stochastic variable, which is defined below.

Definition 6 (Stochastic Performance Parameters): Let Φ be a stochastic variable defined on a certain event space and let $f(.)$ be a function such that $P=f(\Phi)$ is also a stochastic variable defined on the same event space. The probability description of Φ determines the probability description of *P*.

12.2.2 Major Sources of Stochastic Process Variability

The two major sources of front-end process variabilities are (i) systematic or global or inter-die process variability and (ii) stochastic or local or intra-die process variability [5, 6].

Definition 7 (Systematic or Global Process Variability): The process variability between identical devices that spread across one die to another, one wafer to another, and one lot to another is referred to as systematic or global process variability. Such variability causes a shift in the mean value of circuit parameters like channel length, channel width, gate oxide thickness, resistivity, doping concentration, etc.

Definition 8 (Stochastic or Local Process Variability): The process variability between identical devices placed within the same die is referred to as random/stochastic or local process variability. Such variability causes parametric fluctuations or mismatch between a pair of identical devices and therefore causes variation around a mean.

The distinction between systematic and random/stochastic types of front-end process variabilities is illustrated in Figure 12.2. It may be observed that the systematic variation involves a shift of the mean of the distribution from one value to another, whereas the random variation is identified by the variance around the mean value. A comprehensive discussion on the various sources of systematic and random process variations is provided in [6, 7]. In this section, we provide the definitions of the major sources of random/stochastic process variabilities.

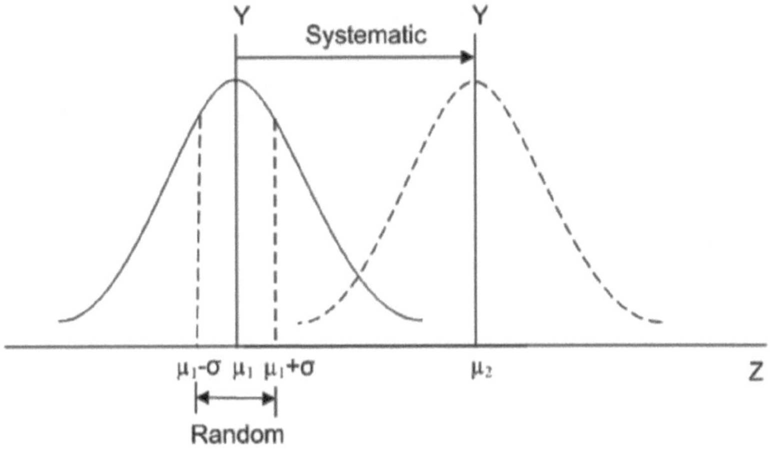

FIGURE 12.2 Distinction between random and systematic types of front-end process variabilities.

12.2.2.1 Random Discrete Dopant Effect

In a MOS transistor, the channel region is doped with dopant atoms in order to control the threshold voltage. With scaling of the channel width, channel length, and source/drain junction depth of the MOS transistor, the number of dopant atoms in the channel region reduces, and it has been estimated that in the sub-90 nm regime, the number of dopant atoms is only a few tens. Therefore, the number of dopant atoms in the channel region of a scaled MOS transistor is a statistical quantity with the probability of occupying any random location.

Definition 9 (RDD Effect): The random variations in the number and placement of dopant atoms in the channel region of a MOS transistor varies between two identically designed devices within the same die. Such variation causes the threshold voltage to vary between a pair of identical devices and is referred to as the random discrete dopant effect.

12.2.2.2 Line Edge/Line Width Roughness

The wavelength of light used to pattern the critical dimensions of a MOS transistor is scaled to sustain the scaling of the MOS transistor. However, at the 180 nm technology node and even at the 45 nm technology node, wavelength scaling is ceased at 193 nm due to the increased cost of the lithography process.

Definition 10 (LER/LWR Effect): The difference between the wavelength of the light used and the critical dimensions of a MOS transistor leads to a stochastic variation of the edge of the channel length of a MOS transistor across the entire width of the transistor. These undulations of channel length and channel width of a MOS transistor are referred to as line edge/line width roughness (LER/LWR).

The direct impact of the LER/LWR phenomenon lies in the stochastic variability of the threshold voltage of a MOS transistor.

12.2.2.3 Oxide Thickness Variations

In 180 nm CMOS technology, the typical value of gate oxide thickness is around 3-4 nm, while in 65 nm CMOS technology, the same is around 2nm. On the other hand, the atomized size of silicon is around 0.2 nm. Thus the value of oxide thickness is equivalent to the length of a few Si atoms with a comparable interface roughness of about one to two atomic layers.

Definition 11: The interface roughness between gate material and gate dielectric, as well as between silicon and gate dielectric, leads to variations in the nominal thickness of the oxide layer, referred to as the oxide thickness variation. This interface roughness in turn leads to fluctuation of the potential drop across the oxide layer causing threshold voltage variation.

12.2.3 Numerical Measures of Descriptive Statistics

For characterization methodology used in the analysis of data, the branch of statistics which is mostly used is known as descriptive statistics. In a data set, the main problem is to describe and extract information from a large mass of data. Some numerical measures which describe the main features of data are (i) central tendency, (ii) dispersion, (iii) skewness, and (iv) kurtosis. These are defined as follows [8].

Definition 12 (Arithmetic Mean): If $x_1, x_2, ..., x_k$ be k measures of a variable and $f_1, f_2, ..., f_k$ be their respective frequencies or weights, then the arithmetic mean is defined by

$$\bar{x} = \frac{1}{N}\sum_{i=1}^{k} x_i f_i \tag{12.4}$$

where $N = \sum_{i=1}^{k} f_i$

Definition 13 (Variance): Variance is defined as the average of the squares of the deviation taken from the mean and is derived from being computed through

$$\sigma^2 = \frac{1}{N}\sum_{i=1}^{k} x_i^2 - \bar{x}^2 \tag{12.5}$$

Definition 14 (Skewness): Skewness signifies lack of symmetry in a statistical distribution. A frequency distribution is said to be skewed if it is not symmetrical. The moment measure of skewness is expressed as

$$g_1 = \frac{m_3}{\sigma^3} \tag{12.6}$$

where m_3 is the third order central moment and is expressed as

$$m_3 = \frac{1}{N}\sum_{i} (x_i - \bar{x})^3 f_i \tag{12.7}$$

Skewed distributions are of two types: Positive (skewed to the right) and negative (or skewed to the left) distributions.

Definition 15 (Kurtosis): Kurtosis refers to the peakedness of a frequency curve. It measures the convexity of the curve. Pearson measurement of kurtosis is written *as*

$$b_2 = \frac{m_4}{\sigma^4} \tag{12.8}$$

where the fourth-order central moment is defined as

$$m_4 = \frac{1}{N}\sum_{i} (x_i - \bar{x})^4 f_i \tag{12.9}$$

and σ is the standard deviation. A frequency distribution having moderate peakedness is known as mesokurtic; it is called leptokurtic if it is more peaked than the moderate one, and platykurtic if it is less peaked than the normal curve. A normal curve is supposed to have moderate kurtosis.

12.3 STATISTICAL CHARACTERIZATION METHODOLOGY FOR STOCHASTIC PROCESS VARIABILITY

The various steps involved in the methodology for statistical characterization are illustrated in Figure 12.3, and the steps are discussed in the following subsections.

12.3.1 Identification of the Berkeley Simulation IGFET Model (BSIM) Compact Model Parameters to Stochastic Process Variabilities

As discussed earlier, in Figure 12.1, the relationship between the circuit parameters and the circuit performance parameters is evaluated through the SPICE simulation process. An important input component of this process is the compact model, of which the BSIM [9] is the most widely used compact model. Therefore, it is essential to identify the appropriate BSIM compact model parameters, the variations of which may appreciably represent the stochastic process variability effects. The various device parameters, corresponding BSIM model parameters, and associated stochastic process variability effects are summarized in Table 12.1 [10].

12.3.2 Sample Design

Statistical characterization deals with studying some of the characteristics of a larger whole of which all the aspects cannot be observed for practical reasons. Sampling

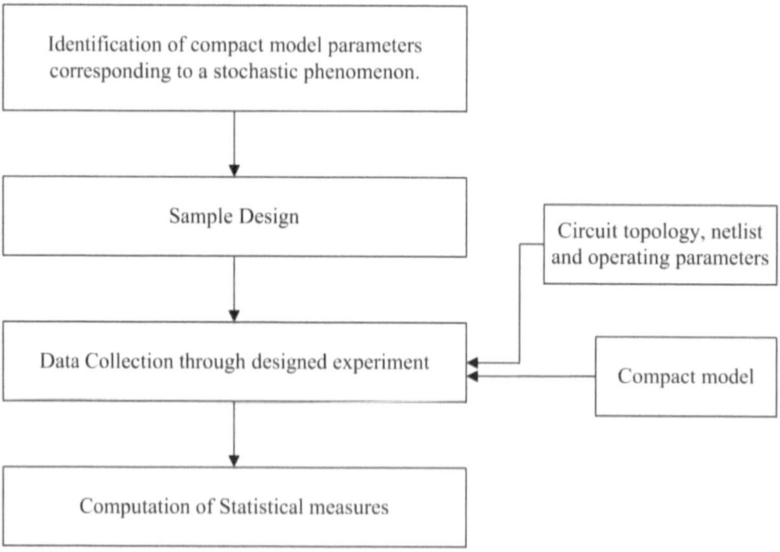

FIGURE 12.3 Illustration of the characterization methodology.

TABLE 12.1
Mapping of BSIM Compact Model Parameters to Stochastic Process Variability Effects and Correspondence With Device Parameters

Device Parameter		Model Parameter		
Symbol	**Definition**	**Symbol**	**Definition**	**Variability**
Vth_0	Threshold Voltage	V_{TH0}	V_T @ $V_{BS}=0$	*RDD*
L	Channel length	XL	Length offset	*LER*
W	Channel width	XW	Width offset	*LWR*
Tox	Gate oxide thickness	T_{OXE}	Equivalent Oxide thickness	*OTV*

theory consists of selecting a small representative part from the larger whole under study [11]. The larger whole as mentioned above is known as the population. The small part which is selected from the larger whole in order to have some idea about the larger whole is known as the sample. If a sample of size n is drawn from a population of size N in such a way that each and every member of the population has an equal chance of being selected in the sample, then the sampling is called simple random sampling. Simple random sampling is widely used in Monte Carlo simulation analysis of integrated circuits.

Probability sampling is the more generic sampling technique useful for studying the stochastic process variability effects. In probability sampling, each unit in the population has a known probability of sampling and a chance mechanism is used to choose the specific units to be included in the sample. The advantage of using a probability sample is that a relatively small sample set can be used to infer the parameters of a large population. It may be noted that simple random sampling is a special case in probability sampling [11], where each and every population unit has the same chance of being included in the sample. An equal probability sampling scheme is easy to design and explain. However, sometimes it is not practicable to use an equal probability sampling design, so an unequal probability sampling design is used.

The sampling strategy will be based on normal probability distribution when sampling is from a normal population. However, when the distribution nature of the population is unknown, the sample size plays a critical role. When the sample size n is small, the shape of the distribution will depend largely on the shape of the parent population, but as n gets larger the shape of a sampling distribution becomes more like a normal distribution. This is the essence of the central limit theorem [11]. Therefore, the sampling scheme will be based upon normal distribution for the statistical characterization of the stochastic process variability effects.

12.3.3 Data Collection Through Designed Experiment

The data required for statistical characterization is collected by conducting a designed experiment [12]. The experiment is conducted through the following two steps:

- Generate the sample set corresponding to the mapped compact model parameters from the probability distribution of the parameters.
- Through the SPICE circuit simulation process, the values of the performance parameters of the circuit under examination are computed.

This procedure of data collection has a few advantages. The sample size is independent of the number of circuit parameters, and the nature of the probability distribution of the circuit performance parameters is not assumed. In addition, it does not make any assumption on the nature of the relationship between the circuit parameters and the circuit performance parameters.

12.3.4 Sample Statistics and Measures

An important purpose of selecting a sample is to obtain estimates of certain population characteristics or population quantities. The population characteristics are referred to as population parameters. On the other hand, the sample characteristic parameters are referred to as statistics or sample statistics. The description and extraction of the values of the sample statistic are extremely important for statistical characterization of the stochastic process variability effects. The tools and techniques of descriptive statistics, as described earlier will be utilized for this. In this chapter, we will limit ourselves to this. On the other hand, in a more advanced procedure, such information is used to predict or infer the population parameters using the tools and techniques of inferential statistics [11].

12.4 IMPLEMENTATION OF THE METHODOLOGY

12.4.1 Systematic Process Variability

The statistical characterization methodology as described above is implemented with 180 nm standard threshold voltage CMOS technology from SCL (Semiconductor Laboratory, India). Although in this chapter, our main focus is the characterization of stochastic process variability, for the sake of completeness, we mention characterizing the systematic process variability too. The inter-die process variability models are represented by the various corner models provided within the process technology development kit. These include the TT model (typical n-channel and typical p-channel), SS (slow n-channel and slow p-channel), FF (fast n-channel and fast p-channel), FS (fast n-channel and slow p-channel) and SF (slow n-channel and fast p-channel) corner model files. The four worst-case models are constructed from the typical model by off-setting a selected compact model parameter by $\pm\, dP = n\sigma$, where $n \approx 3-6$ is the number of σ for the model parameter [9]. The typical model parameter is selected for the deterministic simulation of the design. On the other hand, the worst-case corner models are used for simulation while characterizing the systematic process variability effects.

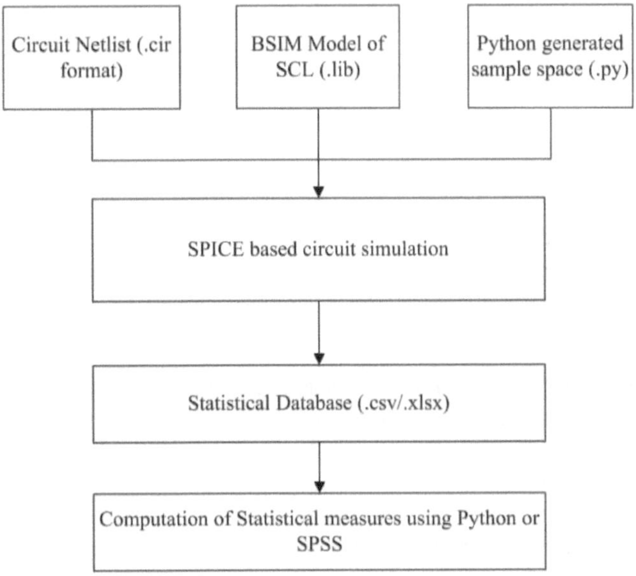

FIGURE 12.4 Implementation of the stochastic characterization methodology using Python and SPICE.

12.4.2 Stochastic Process Variability

On the other hand, implementation of the stochastic process variability effects is not straightforward, and the methodology described above needs to integrate a high-level language tool with SPICE simulation. At the University of Calcutta, Python language [13] is used to develop in-house software that works in tandem with SPICE simulation tool. The working principle is illustrated in Figure 12.4. This software performs the data generation process through designed experiments for a set of samples. The extraction of current/voltage data at the various node points obtained from SPICE is also handled through it. It may be noted that for SPICE simulation to be used for data generation, a slightly modified form of the compact model file is to be provided. A parameterized compact model file will be employed and the model parameters will be supplied externally, as obtained from the appropriate probability sampling technique. Finally the various statistical measures are computed by the software. For in-depth statistical characterization, third-party specialized software tools, e.g., SPSS [14] or R programming language, may be used.

12.5 APPLICATION EXAMPLE: CURRENT REFERENCE CIRCUIT FOR ULTRA-LOW POWER ANALOG APPLICATIONS

This section illustrates the statistical characterization methodology using a case study. The test circuit is a 1 nA current reference circuit, an important design component for ultra-low-power analog applications [15]. The simplified schematic circuit diagram is shown in Figure 12.5(a). The transistors operate in the weak inversion mode and are critically prone to process variability effects. The deterministic design of the circuit is based on a 0.18μm standard threshold voltage CMOS technology of SCL with a 1.8 V supply voltage. The compact model used corresponds to the Level 49 SPICE model or BSIM3v3. The variation of the output current I_{out} with respect to the variation of the supply voltage V_{DD} is shown in Figure 12.5(b). The circuit shows that beyond a certain minimum supply voltage, the reference current of 1 nA is almost independent of the variation of the supply voltage.

12.5.1 Characterization of Systematic Process Variability

In order to characterize the effects of systematic process variability on the performance of the circuit under test, we perform a worst-case corner analysis.

The current-voltage characteristics of the circuit under test at the various process corners are illustrated in Figure 12.6. The reference current at the supply voltage of 1.8 V for the various process corners is summarized in Table 12.2. We observe that the nominal value of the current has been shifted to different values at the various process corners. However, the dependency on the variation of the supply voltage does not change. The results, therefore, support our earlier discussions on systematic process variability.

12.5.2 Characterization of Stochastic Process Variability

The two chosen compact model parameters for illustration of the stochastic characterization methodology are (i) T_{ox} which is the oxide thickness parameter and (ii) v_{tho} which is the long channel threshold voltage parameter.

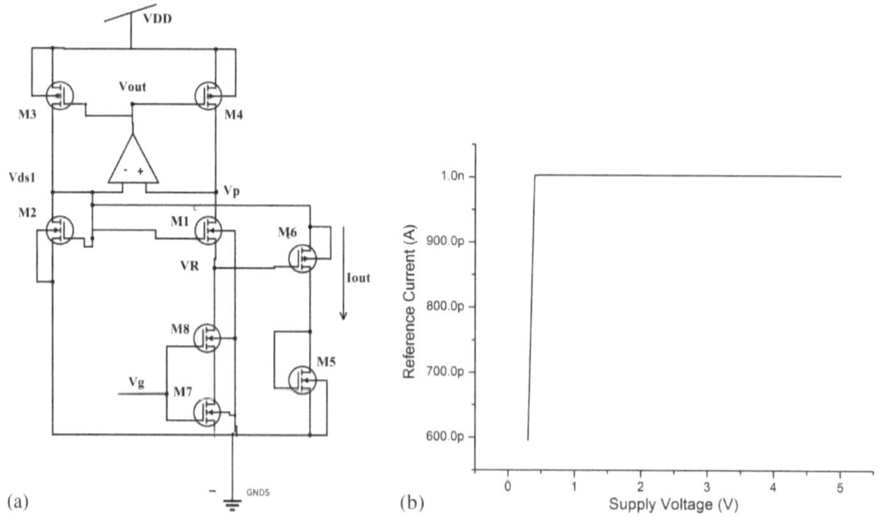

FIGURE 12.5 Schematic circuit diagram and current-voltage characteristics of the 1 nA current reference circuit considered as the circuit under test. (a) Schematic circuit diagram. (b) Current-voltage characteristics.

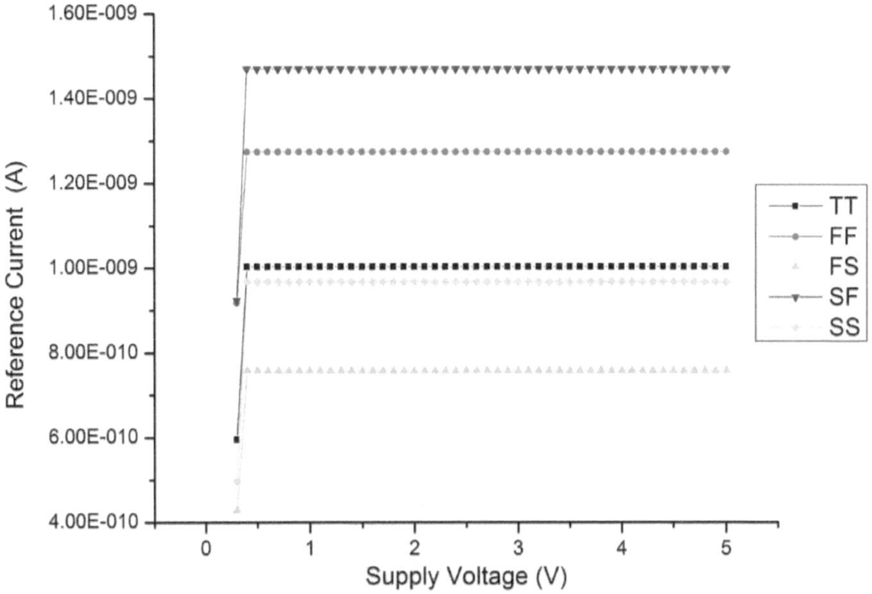

FIGURE 12.6 Process corner analysis results of the current-voltage characteristics of the circuit under test.

TABLE 12.2
Reference Current Values at the Various Corners

Voltages	TT	FF	FS	SF	SS
1.8 V	1.003 nA	1.274 nA	0.758 nA	1.469 nA	0.967 nA

12.5.2.1 Oxide Thickness Variations

A probability sampling technique employing Gaussian distribution was used. The process variability effects were studied for variation of $3\sigma/\mu=9\%$, where σ was the standard deviation and μ is the statistical mean of the samples. We selected 200 samples for the study. The descriptive statistical measures for the sampled values of oxide thickness parameters are summarized in Table 12.3. The histogram plot is shown in Figure 12.7.

As a performance parameter, we measured the value of the reference current at a 1.8V supply voltage. The descriptive statistical measures for the sampled values of the reference current are summarized in Table 12.4. The histogram plot is shown in Figure 12.8.

12.5.2.2 Threshold Voltage Variations

Similar to the oxide thickness variations, we also studied the variations of the threshold voltage parameters on the performance of the circuit under examination. We

TABLE 12.3
Descriptive Statistical Measures of the Sampled Values of the Oxide Thickness Parameter

Parameters	Values
Sample Number	200
Mean	4.0008 nm
Median	3.9881 nm
Mode	3.6194 nm
Std. deviation	0.12528 nm
Skewness	–0.013961
Kurtosis	0.180195
Range	0.74075 nm
Minimum	3.6194 nm
Maximum	4.3601 nm

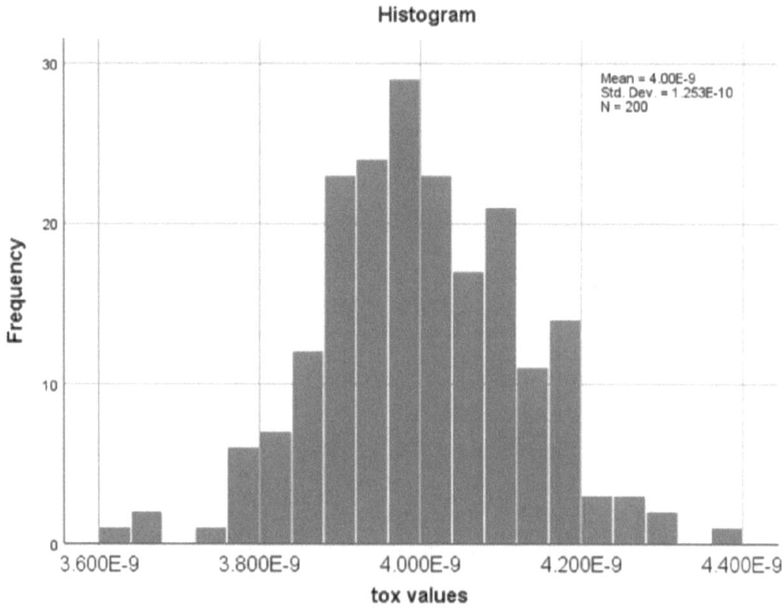

FIGURE 12.7 Histogram plot of the distributions of oxide thickness parameter sample values.

selected 200 samples with $3\sigma/\mu=9\%$. The descriptive statistical measures for the sampled values of the reference current are summarized in Table 12.5. The histogram plot is shown in Figure 12.9(a), and Figure 12.9(b) shows the I-V characteristics of the various samples.

TABLE 12.4
Descriptive Statistical Measures of the Sampled Values of the Reference Current For Variations of the Oxide Thickness Parameter

Parameters	Values
Sample Number	200
Mean	1.0147 nA
Median	1.0194 nA
Mode	0.603 nA
Std. deviation	0.173 nA
Skewness	0.518
Kurtosis	0.836
Range	1.054 nA
Minimum	0.603 nA
Maximum	1.658 nA

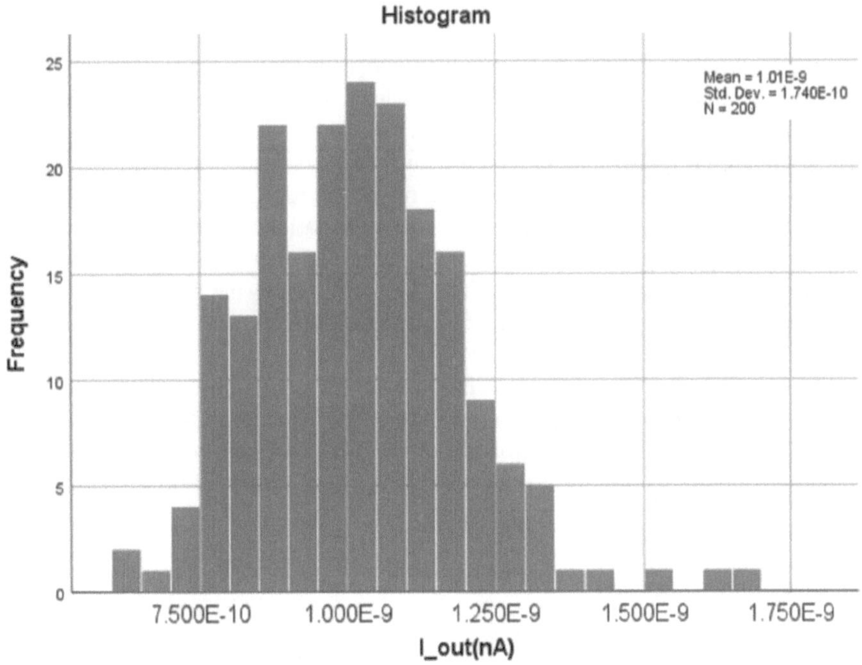

FIGURE 12.8 Histogram plot of the distributions of the sample values of the reference current for variations of the oxide thickness parameter.

TABLE 12.5
Descriptive Statistical Measures of the Sampled Values of the Reference Current For Variations of the Threshold Voltage Parameter

Parameters	Values
Sample Number	200
Mean	1.030 nA
Median	0.961 nA
Mode	0.396 nA
Std. deviation	0.370 nA
Skewness	0.814
Kurtosis	0.520
Range	1.798 nA
Minimum	0.396 nA
Maximum	2.195 nA

12.6 CONCLUSION

Statistical variations of the different process parameters play a significant role in determining the performance of a circuit when fabricated over batch processing mode. Deterministic design methodology, which is used for the design of integrated circuits, must be supplemented with statistical design methodology. Of the latter, statistical characterization plays an important role in understanding process variabilities' effects on performance parameter variations. In this chapter, we provided a description of the statistical characterization methodology for systematic and stochastic process variabilities. The former is somewhat standard, while the latter needs to be developed through concepts of research methodology and statistical quality assurance. In addition, several specialized computation tools need to work in tandem. We illustrate the methodology through a case study of an ultra-low current source generator, where the transistors operate in the weak inversion mode. The effects of variations of the oxide thickness parameter and threshold voltage parameter on a set of 200 samples have been illustrated through statistical results.

ACKNOWLEDGMENT

The author thanks the Special Manpower Development Program for VLSI, Chips to System Design, University of Calcutta, sponsored by MeitY, Govt. of India for providing the necessary computational facilities to carry out the research works reported herein. In addition, he acknowledges the contributions of his undergraduate students T. Sau, S. Upadhyay, S. Mitra, and A. Bhowmick for developing the Python-based software.

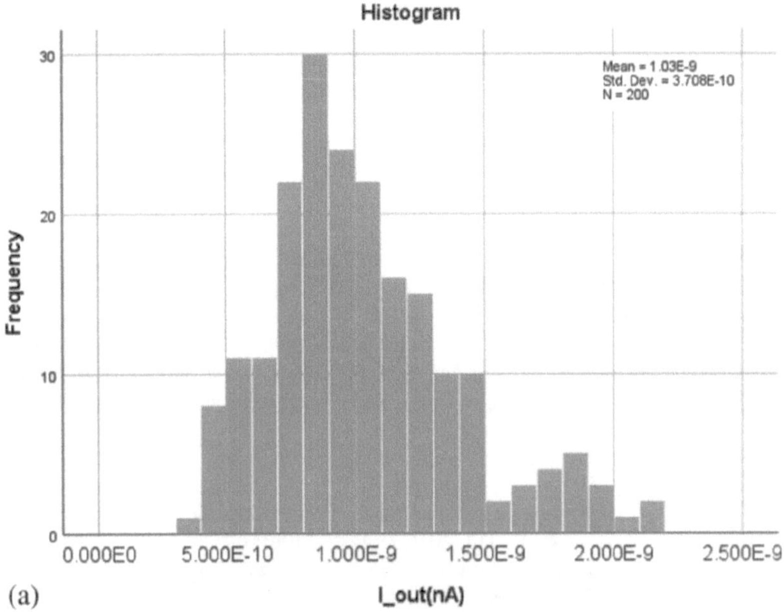

(a)

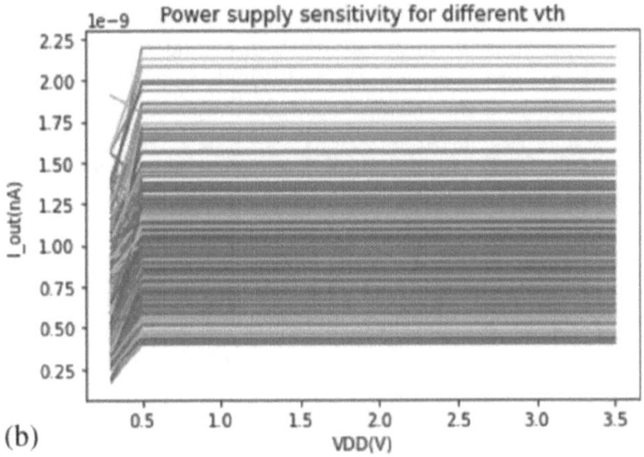

(b)

FIGURE 12.9 Statistical characterization of the reference current of the circuit under test for variations of the threshold voltage parameter. (a) Histogram plot of the distributions of the sample values of the reference current for variations of the threshold voltage parameter. (b) Statistical variations of the current-voltage characteristics for variations of the threshold voltage parameter.

REFERENCES

1. S. Pandit, and (Editor) C.K. Sarkar. *MOSFET Characterization for VLSI Circuit Simulation in Technology Computer Aided Design Simulation for VLSI MOSFET.* CRC Press, 1st edition, 2013.

2. R.J. Baker. *CMOS: Circuit Design, Layout, and Simulation*. Wiley- IEEE, 4th edition, 2019.
3. S. Pandit. Nanoscale MOSFET: MOS Transistor as Basic Building Block. In A. Sengupta and C.K. Sarkar, editors, *Introduction to Nano. Engineering Materials*. Springer, 1st edition, 2015.
4. S. Pandit. Nanoscale silicon MOS transistors. In S. Roy, C.K. Ghosh, and C.K. Sarkar, editors, *Nanotechnology: Synthesis to Applications*, pages 259–284. CRC Press, 2017.
5. S. Pandit, C.R. Mandal, and A. Patra. *Nano-Scale CMOS Analog Circuits Models and CAD Techniques for High-Level Design*. CRC Press, 2014.
6. S. Sengupta, and S. Pandit. Variability in Nanoscale Technology and Eδ DC MOS Transistor. In B.K. Kaushik, editor, *Nanoscale Devices Physics, Modeling, and Their Application*. CRC Press, November 2018.
7. S.K. Saha. Modeling Process Variability in Scaled CMOS Technology. *IEEE Design and Test of Computers*, 27:8–16, March/April 2010. DOI:10.1109/MDT.2010.50.
8. D. Bhattacharya, and S. Roychowdhury. *Statistics: Theory and Practice*. U.N.Dhur and Sons Private LTD., 3rd edition, 2019.
9. S.K. Saha. *Compact Models for Integrated Circuit Design Conventional Transistors and Beyond*. CRC Press, 1st edition, July 2016.
10. S.K. Saha. Compact MOSFET Modeling for Process Variability-Aware VLSI Circuit Design. *IEEE Access*, 2:104–115, 2014. DOI: 10.1109/AC- CESS.2014.2304568.
11. D. Bhattacharya and S. Roychowdhury. *Probability and Statistical Inference: Theory and Practice*. U.N.Dhur and Sons Private LTD., 3rd edition, 2017.
12. C.R. Kothari and G. Garg. *Research Methodology: Methods and Techniques*. New Age International Publishers, 4th edition, 2019.
13. A. Kar Gupta. *Scientific Computing in Python*. Techno World, 2nd edition, 2021.
14. L. Jasrai. *Data Analysis Using SPSS*. Sage Publications India Pvt. Ltd, 1st edition, 2020.
15. S. Pandit. Design Methodology for Ultra-Low-Power CMOS Analog Circuits for ELF-SLF Applications. In R. Dhiman and R. Chandel, editors, *Nanoscale VLSI: Devices, Circuits and Applications*, pages 23–43. Springer, 2020.

13 Versatile Single Input Single Output Filter Topology Suitable for Integrated Circuits

Bhartendu Chaturvedi,
Jitendra Mohan, and Jitender

CONTENTS

13.1 INTRODUCTION

The current advancements in the technology sector have been driven by state-of-the-art VLSI devices and circuits. A small chip can now perform such a wide range of complex tasks. This has been possible due to the miniaturization of devices and components up to the nano-scale and beyond. So, any new configuration of a signal processing module, such as an analog filter that employs nano-scale transistors, will always be welcomed. All-pass (AP) and notch functionalities of analog filters are extensively utilized in various electronics and communication engineering applications [1–5]. In an AP filter, the pass band comprises all frequency components without any suppression, whereas in the notch filter a particular range of frequencies is suppressed. Based on the pass band characteristics of AP and notch, these filters have a distinguished set of applications in the communication engineering domain. An AP filter is primarily used as a phase shifter or phase equalizer to adjust the phase delays of the received signals, whereas a notch filter is used to nullify noise

DOI: 10.1201/9781003155751-13

occurrences over a particular frequency range by suppressing those noise frequencies. Another difference between the AP and notch filters lies in their filter order. An AP filter can be designed from first-order onwards, whereas a notch filter is of second-order minimum.

The literature associated with the AP and notch filter is relatively vast. Some useful and relevant works related to first-order AP filters, second-order AP filters, and second-order notch filters are surveyed in this chapter to explore the scope of improvement from a performance point of view [6–27]. A common factor accounting for the design complexity of all the reviewed works [6–11, 14, 19, 21–26] is the use of more than one active building block (ABB). First-order AP circuits reported in [8, 12, 13, 15–17] employ more than three passive components and the ones presented in [7–14, 17] employ floating capacitor(s). This increased passive component count and the floating nature of the capacitor are not favorable from an IC fabrication point of view.

Second-order works reported in [23, 26, 27] are also affected by the increased passive component count as they employ more than four passive components. Among the reviewed works, the designs in [18–21] only offer second-order multi-functional behavior with the same circuit topology, and the design in [18] is capable of realizing first-order AP, second-order AP, and second-order notch responses with the same circuit topology, but employs two ABBs.

In order to mitigate the above-mentioned limitations and incorporate multi-functional behavior, a novel voltage-mode filter topology is presented in this chapter. The topology has a single input and a single output (SISO), and its versatility lies in the realization of different filtering functionalities on the appropriate selection of the generic impedances. These filtering functionalities are first-order AP filter, second-order AP filter, and second-order notch filter. Another advantageous feature offered by the circuit topology is the high impedance nature of the input port, which helps to eliminate the buffer stage while cascading the voltage-mode configurations. In the case of current-mode configurations [28, 29], impedance requirements for cascadability are the reverse of the voltage-mode circuit's requirements. The presented filter structure is based on an extra-X second-generation current conveyor (EXCCII) [30], which uses only 21 MOS transistors in its CMOS structure.

13.2 THEORETICAL DETAILS OF THE PROPOSED FILTER

This section describes the working principle of the employed ABB, i.e. EXCCII, and the structure of the proposed versatile filter topology.

13.2.1 Working Principle of EXCCII

Similar to the dual-X second-generation current conveyor (DXCCII) [31, 32], an EXCCII [30] is also an efficient variation of the conventional second-generation current conveyor (CCII). Compared to CCII, which has one X-terminal, both DXCCII and EXCCII have two X-terminals. A DXCCII has one inverting X-terminal and one non-inverting X-terminal, whereas an EXCCII has two non-inverting X-terminals.

The symbol depicting various EXCCII ports is shown in Figure 13.1. The voltage conveying action takes place from Y to X ports, whereas the current conveying action takes place from X to Z ports. In the case of the proposed filter topology, only one Z-terminal is required with inverted current conveying action, so the CMOS structure of the EXCCII is slightly modified to meet the requirements, as depicted in Figure 13.2. The port relations between the EXCCII's terminals are shown in matrix form in Eq. (13.1).

$$\begin{bmatrix} I_Y \\ V_{X1} \\ V_{X2} \\ I_{Z1n} \end{bmatrix} = \begin{bmatrix} 0 & 0 \\ 1 & 0 \\ 1 & 0 \\ 0 & -1 \end{bmatrix} \begin{bmatrix} V_Y \\ I_{X1} \end{bmatrix} \tag{13.1}$$

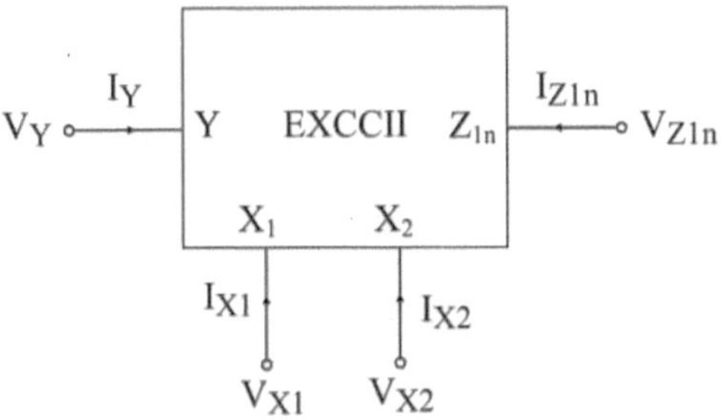

FIGURE 13.1 EXCCII's block-level representation.

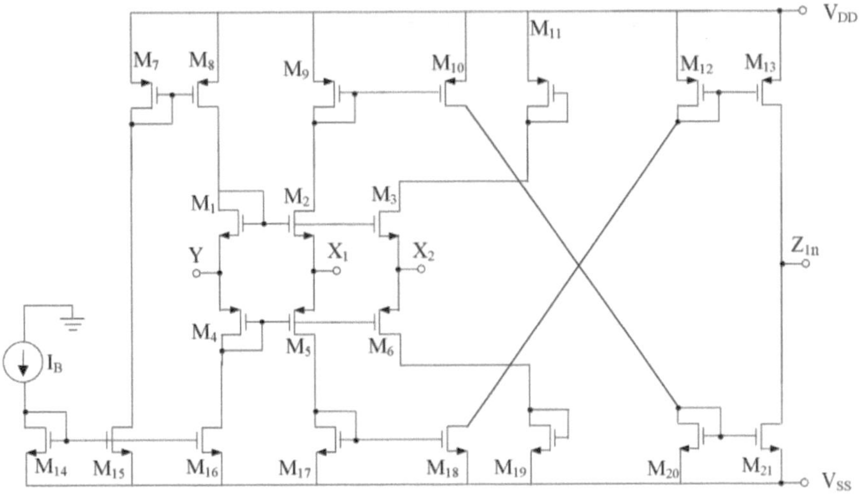

FIGURE 13.2 CMOS structure of EXCCII [30].

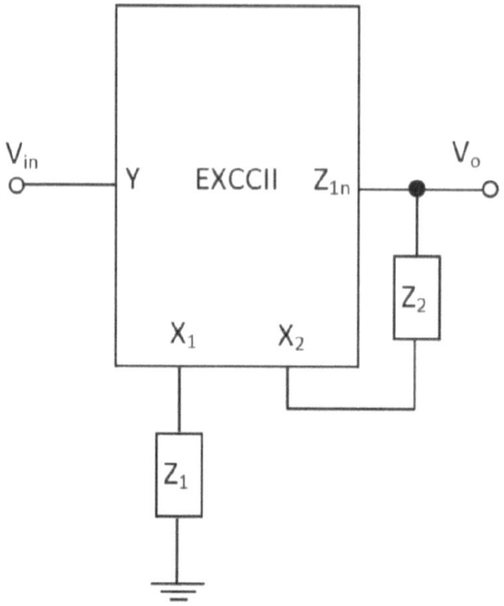

FIGURE 13.3 Proposed SISO filter topology.

13.2.2 Proposed SISO filter topology

The proposed SISO filter topology is based on EXCCII [14] and supports voltage-mode operation. EXCCII has featured in various signal processing applications [17, 33] in recent years. Two general impedances, Z_1 and Z_2, are connected alongside the EXCCII to design the filter topology. A voltage input V_{in} is applied at the high impedance port of the EXCCII, and output is taken across the V_o terminal, as shown in Figure 13.3. The transfer function obtained by using nodal analysis and EXCCII's port relations is expressed as follows:

$$\frac{V_o}{V_{in}} = \frac{Z_1 - Z_2}{Z_1} \tag{13.2}$$

13.3 FILTER FUNCTIONALITIES DERIVED FROM PROPOSED SISO FILTER TOPOLOGY

Suitable selections of the generalized impedances Z_1 and Z_2 in the form of resistors and capacitors result in distinct filtering functionalities. These functionalities can be stated as first-order AP filter, second-order AP filter, and second-order notch filter. All these functionalities are described below, one by one.

13.3.1 First-Order AP Filter

A first-order AP response is achieved by choosing Z_1 as a series combination of a resistor and a capacitor, i.e. $Z_1 = R_1 + (1/sC_1)$ and Z_2 as a passive resistor, i.e. R_2. The

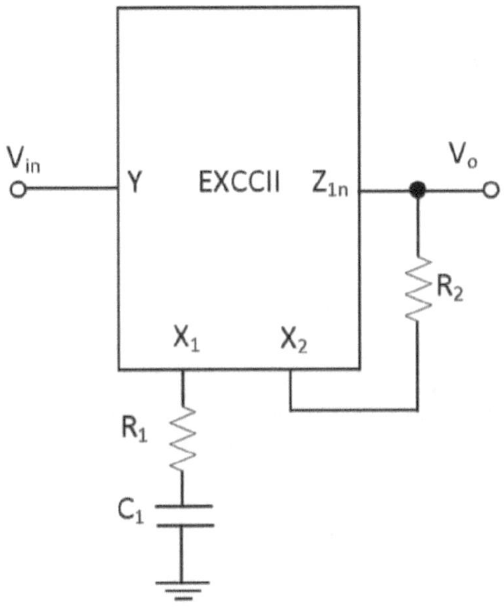

FIGURE 13.4 First-order AP filter derived from the circuit in Figure 13.3.

structure of the derived first-order AP filter is shown in Figure 13.4. Proper replacements of Z_1 and Z_2 in Eq. (13.2) yields the following updated transfer function:

$$\frac{V_o}{V_{in}} = \frac{1 + sR_1C_1 - sR_2C_1}{1 + sR_1C_1} \tag{13.3}$$

Choosing a value of resistor R_2 as double of the resistor R_1 value and denoting C_1 by C, convert the above transfer function into a standard first-order AP transfer function as expressed in Eq. (13.4).

$$\frac{V_o}{V_{in}} = -\frac{s - \frac{1}{RC}}{S + \frac{1}{RC}} \tag{13.4}$$

Observing Eq. (13.4), the expressions for AP gain, pole frequency, and phase angle are expressed as:

$$H_{AP} = -1 \tag{13.5}$$

$$\omega_0 = \frac{1}{RC} \tag{13.6}$$

$$\phi(\omega) = -2\tan^{-1}(\omega RC) \tag{13.7}$$

The sensitivity of the pole frequency of the obtained first order AP response with respect to passive components can be expressed as follows:

$$S_{R,C}^{\omega_0} = -1 \tag{13.8}$$

A good sensitivity performance can be ensured by observing the above sensitivity expression.

13.3.2 Second-Order AP Filter and Notch Filter

To obtain the second-order AP and notch response from the proposed versatile SISO filter topology, impedance Z_1 is chosen as a series combination of a resistor and a capacitor, whereas Z_2 is chosen as the parallel combination of a resistor and a capacitor, i.e. $Z_1 = R_1 + (1/sC_1)$ and $Z_2 = R_2 \parallel (1/sC_2)$ as depicted in Figure 13.5. Putting these impedance equivalents into Eq. (13.2) gives the following transfer function.

$$\frac{V_o}{V_{in}} = \frac{1 + s^2R_1R_2C_1C_2 + sR_2C_2 + sR_1C_1 - sR_2C_1}{s^2R_1R_2C_1C_2 + sR_1C_1 + sR_2C_2 + 1} \tag{13.9}$$

Using the matched resistor and capacitor values of $R_2 = 4R_1 = 4R$ and $C_1 = 4C_2 = 4C$ in Eq. (13.9) gives the updated transfer function equivalent to a second-order AP response, as expressed in Eq. (13.10).

$$\frac{V_o}{V_{in}} = \frac{16s^2R^2C^2 - 8sRC + 1}{16s^2R^2C^2 + 8sRC + 1} \tag{13.10}$$

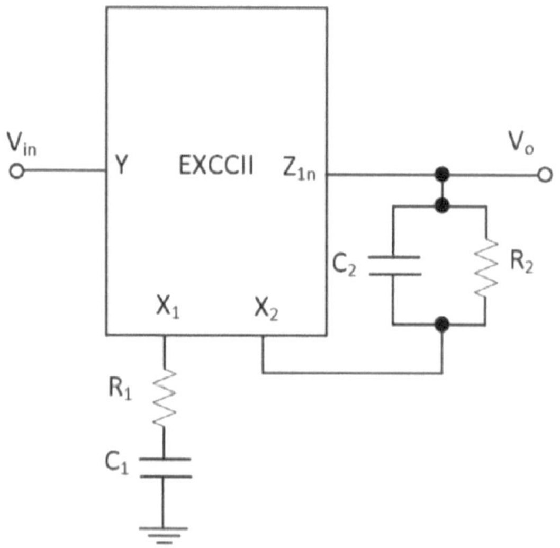

FIGURE 13.5 Second-order filter derived from the circuit in Figure 13.3.

Using the matched resistor and capacitor values of $R_2 = 2R_1 = 2R$ and $C_1 = 2C_2 = 2C$ in Eq. (13.9) gives the updated transfer function equivalent to a second-order notch response, as expressed in Eq. (13.11).

$$\frac{V_o}{V_{in}} = \frac{4s^2R^2C^2 + 1}{4s^2R^2C^2 + 4sRC + 1} \tag{13.11}$$

Expressions for pole frequency and quality factor for the second-order AP response are expressed in Eqs. (13.12) and (13.13), whereas, for the second-order notch response, pole frequency and quality factor are expressed by Eq. (13.14) and (13.15).

$$\omega_{0(AP)} = \frac{1}{4RC} \tag{13.12}$$

$$Q_{(AP)} = \frac{1}{2} \tag{13.13}$$

$$\omega_{0(N)} = \frac{1}{2RC} \tag{13.14}$$

$$Q_{(N)} = \frac{1}{2} \tag{13.15}$$

Sensitivity of the pole frequency with respect to resistor and capacitor for both second-order AP and second-order notch are identical and are unity in magnitude, as expressed by Eq. (13.16). This also signifies a good sensitivity performance of the designed second-order filters.

$$S^{\omega_0}_{R,C(AP,N)} = -1 \tag{13.16}$$

13.4 NON-IDEAL ANALYSIS OF THE PROPOSED SISO FILTER TOPOLOGY

To perform the non-ideal analysis of the proposed SISO filter topology, the ideal port relations between various EXCCII ports are replaced by the non-ideal ones as expressed in Eq. (13.17).

$$\begin{bmatrix} I_Y \\ V_{X1} \\ V_{X2} \\ I_{Z1n} \end{bmatrix} = \begin{bmatrix} 0 & 0 \\ \beta_1 & 0 \\ \beta_2 & 0 \\ 0 & -\alpha \end{bmatrix} \begin{bmatrix} V_Y \\ I_{X1} \end{bmatrix} \tag{13.17}$$

Where β_1 and β_2 are the non-ideal voltage-transfer gains between the Y and X ports and α is the non-ideal current transfer gain between X_1 and Z_{1n} port. Using this non-ideal port relation matrix while performing the nodal analysis the updated generalized transfer function is expressed as:

$$\frac{V_o}{V_{in}} = \frac{Z_1\beta_2 - Z_2\alpha_1\beta_1}{Z_1} \tag{13.18}$$

Choosing appropriate passive component combinations for Z_1 and Z_2 results in different filter orders. When $Z_1 = R_1 + (1/sC_1)$ and $Z_2 = R_2$ are chosen with matching conditions of $R_2 = 2R_1 = 2R$ and $C_1 = C$, a first-order AP response is obtained. Transfer functions observed before and after applying the matching conditions are expressed by Eqs. (13.19) and (13.20), respectively.

$$\frac{V_o}{V_{in}} = \frac{\beta_2 + s(R_1C_1\beta_2 - \alpha_1\beta_1R_2C_1)}{1 + sR_1C_1} \tag{13.19}$$

$$\frac{V_o}{V_{in}} = \frac{\beta_2 + sRC(\beta_2 - 2\alpha_1\beta_1)}{1 + sRC} \tag{13.20}$$

Similarly choosing $Z_1 = R_1 + (1/sC_1)$ and $Z_2 = R_2 \parallel (1/sC_2)$ results in second-order transfer function as follows:

$$\frac{V_o}{V_{in}} = \frac{s^2R_1R_2C_1C_2\beta_2 + sR_2C_2\beta_2 + sR_1C_1\beta_2 - sR_2C_1\alpha_1\beta_1 + \beta_2}{s^2R_1R_2C_1C_2 + sR_1C_1 + sR_2C_2 + 1} \tag{13.21}$$

Applying the matching conditions of $R_2 = 4R_1 = 4R$, $C_1 = 4C_2 = 4C$, result in a second-order AP transfer function as follows:

$$\frac{V_o}{V_{in}} = \frac{16s^2R^2C^2\beta_2 + 8sRC\beta_2 - 16sRC\alpha_1\beta_1 + \beta_2}{16s^2R^2C^2 + 8sRC + 1} \tag{13.22}$$

Applying matching conditions of $R_2 = 2R_1 = 2R$, $C_1 = 2C_2 = 2C$, result in a second-order notch transfer function as follows:

$$\frac{V_o}{V_{in}} = \frac{4s^2R^2C^2\beta_2 + 4sRC\beta_2 - 4sRC\alpha_1\beta_1 + \beta_2}{4s^2R^2C^2 + 4sRC + 1} \tag{13.23}$$

Careful observation of Eqs. (13.20), (13.22), and (13.23) shows that these transfer functions are closely correlated with the ideal transfer functions expressed in Eqs. (13.4), (13.10), and (13.11). The expressions for pole frequency and quality factor also remain identical even if the non-ideal behavior of EXCCII is considered.

13.5 SIMULATION RESULTS

Simulations are carried out on PSPICE using 250 nm CMOS technology. The supply voltage is chosen to be ± 1.25 V, and a bias current of 22 μA is applied. Aspect ratios of the MOS transistors used in EXCII's CMOS implementation are given in

Table 13.1. To simulate the first-order AP filter circuit of Figure 13.4, resistance R_1 and R_2 are chosen to be 10 kΩ and 20 kΩ, respectively and the capacitor value is chosen to be 10 pF. This results in a theoretical pole frequency value of 1.59 MHz. Figure 13.6 shows the phase and gain response of the derived first-order AP filter from which the simulated value of pole frequency is found to be 1.5 MHz. Time-domain waveforms for the first-order AP filter are shown in Figure 13.7, which highlights the 90° phase shift between the output and input voltages. Further, Monte

TABLE 13.1
Aspect Ratios of the MOS Transistors Used in the CMOS Structure of EXCCII [30]

Transistors	M_1–M_3, M_7–M_{13}	M_4–M_6	M_{14}–M_{21}
W(µm) / L(µm)	10/0.5	16/0.5	6/0.5

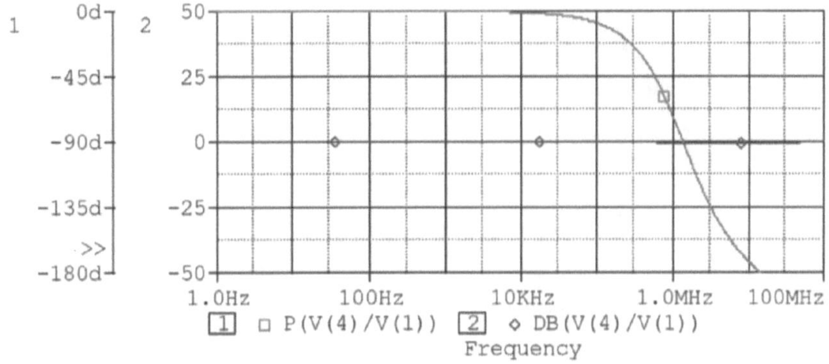

FIGURE 13.6 Phase and gain response of the first-order AP filter.

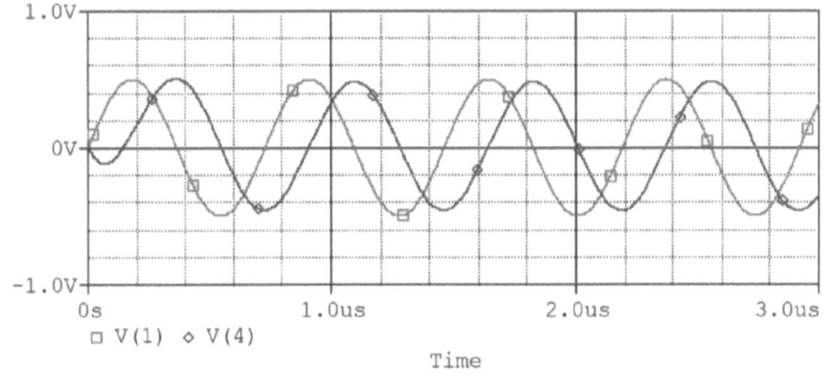

FIGURE 13.7 Time-domain waveforms of the first-order AP filter.

Carlo (MC) histograms for capacitor and threshold voltage variations of 10% and 2% Gaussian deviation, respectively, are plotted for the first-order AP response period. The MC histograms are shown in Figures 13.8 and 13.9 and indicate that the filter's performance is almost unaffected by these variations.

To observe the second-order AP response through the proposed SISO filter topology, the value of the passive components R_1, R_2, C_1 and C_2 are chosen to be 5 kΩ, 20 kΩ, 20 pF, and 5 pF, respectively, for the circuit in Figure 13.5. The AC response for the second-order AP response is shown in Figure 13.10. The value of the pole frequency observed in Figure 13.10 is found to be 1.54 MHz, which is quite close to the theoretical value of 1.59 MHz. The transient response for the second-order AP filter is shown in Figure 13.11. Both the AC and transient response highlight the 180° phase

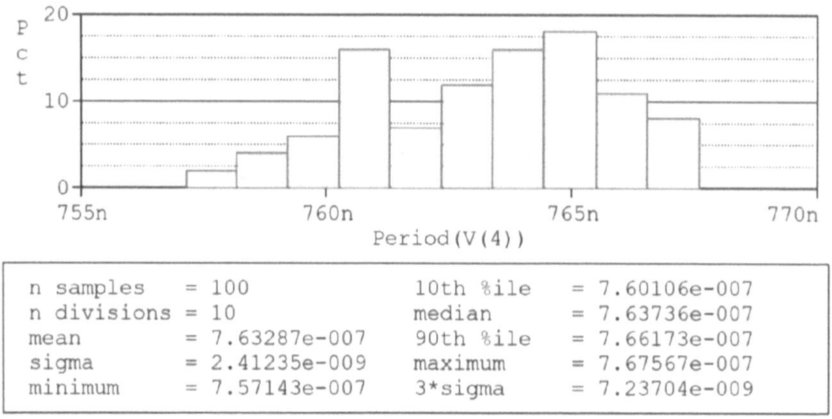

FIGURE 13.8 MC histograms for the time period of the first-order AP output for 10% capacitor variations.

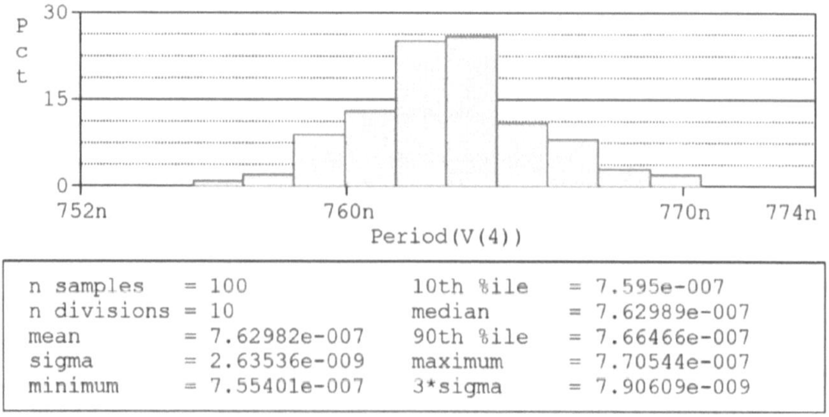

FIGURE 13.9 MC histograms for the time period of the first-order AP output for 2% threshold voltage variations.

change in the output at the pole frequency, which is an essential characteristic of the second-order AP filter. Similarly, choosing the passive component values as R_1 = 10 kΩ, R_2 = 20 kΩ, C_1 = 10 pF, and C_2 = 5 pF provides the second-order notch response as shown by the gain and phase plots in Figure 13.12. The simulated value of the pole frequency is 1.56 MHz. The performance of all three filters derived from the

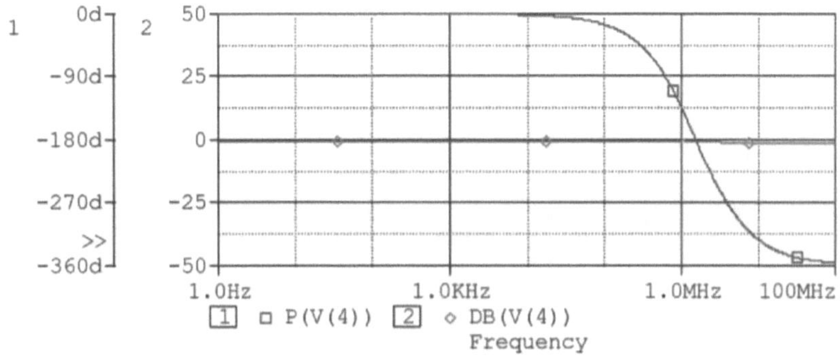

FIGURE 13.10 AC response for the second-order AP filter.

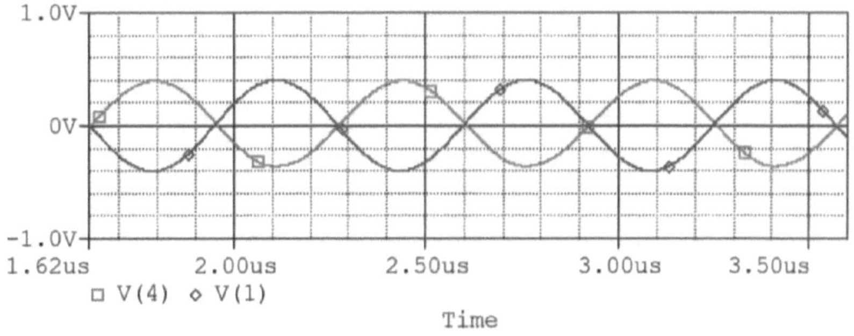

FIGURE 13.11 Transient response for the second-order AP filter.

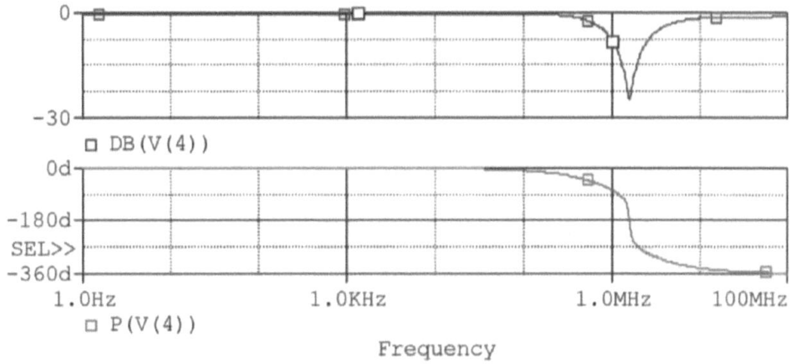

FIGURE 13.12 Gain and phase plots for a second-order notch filter.

proposed topology is also checked for temperature variations in the range of –25 °C to 125 °C. These variations negligibly affect the performance of first-order AP, second-order AP, and second-order notch filters, as shown in Figure 13.13. The value of power dissipation observed from simulations of the first-order AP and second-order notch filters is 0.4 mW, whereas the value for the second-order AP filter is 0.5 mW.

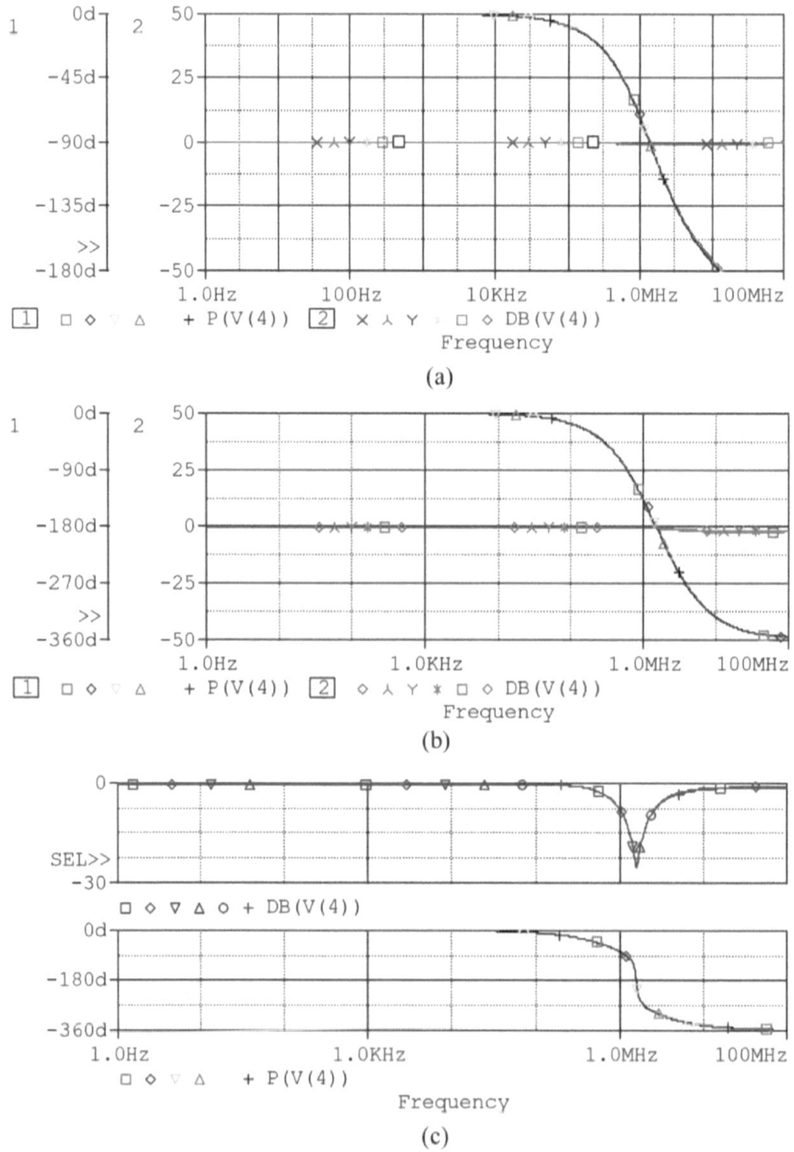

FIGURE 13.13 Effect of temperature variation on the performance of a) first-order AP filter, b) second-order AP filter, and c) second-order notch filter.

The value of total harmonic distortion for first-order AP, second-order AP, and second-order notch responses are found to be 1.5%, 1%, and 2.5%, respectively.

13.6 CONCLUSION

A highly versatile filter topology capable of delivering three different filtering responses, namely first-order AP, second-order AP, and second-order notch, was presented. The presented topology was based on EXCCII and had a single input and single output structure. Diverse functionalities were obtained by appropriately selecting the circuit impedances and passive component values associated with these impedances. Single ABB-based realization was the most attractive feature of the presented filter topology. The circuit had high input impedance, which goes well with the cascadability requirements of a voltage-mode circuit. Another noteworthy point about the topology was that it used only 21 MOS transistors in the CMOS implementation of the EXCCII. A set of simulation results justified the versatility of the presented filter topology and its theoretically claimed performance.

REFERENCES

1. B. Chaturvedi and J. Mohan, Single DV-DXCCII based voltage controlled first-order all-pass filter with inverting and non-inverting responses, *Iranian Journal of Electrical and Electronic Engineering.* **11** (2015) 301–309.
2. P. Kumar and K. Pal, High input impedance band pass, all pass and notch filters using two CCIIs, *HAIT Journal of Science and Engineering.* **3** (2006) 2–13.
3. B. Chaturvedi, A. Kumar and J. Mohan, Low voltage operated current-mode first-order universal filter and sinusoidal oscillator suitable for signal processing applications, *AEU-International Journal of Electronics and Communications.* **99** (2019) 110–118.
4. A. Kumar and B. Chaturvedi, Realization of novel cascadable current-mode all-pass sections, *Iranian Journal of Electrical and Electronic Engineering.* **2** (2018) 162–169.
5. B. Chaturvedi, A. Kumar, Electronically tunable first-order filters and dual-mode multiphase oscillator, *Circuits, Systems and Signal Processing.* **38** (2019) 2–25.
6. S. Minaei and E. Yuce, Novel voltage-mode all-pass filter based on using DVCCs, *Circuits, Systems Signal Processing.* **29** (2010) 391–402.
7. S. Minaei and O. Cicekoglu, A resistorless realization of the first-order all-pass filter, *International Journal of Electronics.* **93** (2006) 177–183.
8. I. A. Khan, M. I. Masud and S. A. Moiz, Reconfigurable fully differential first order all pass filter using digitally controlled CMOS DVCC, Proceedings of 8th IEEE GCC Conference and Exhibition, Muscat, Oman, 2015, pp. 1–5.
9. K. Pal and S. Rana, Some new first-order all-pass realizations using CCII, *Active and Passive Electronic Components.* **27** (2004) 91–94.
10. B. Chaturvedi and S. Maheshwari, An ideal voltage-mode all-pass filter and its application, *Journal of Communication and Computer.* **9** (2012) 613–623.
11. P. Kumar, A. U. Keskin and K. Pal, Wide-band resistorless all-pass sections with single element tuning, *International Journal of Electronics.* **94** (2007) 597–604.
12. F. Kaçar and Y. Özcelep, CDBA based voltage-mode first-order all-pass filter topologies, *Journal of Electrical and Electronics Engineering.* **11** (2011) 1327–1332.

13. C. Cakir, U. Cam and O. Cicekoglu, Novel all-pass filter configuration employing single OTRA, *IEEE Transactions on Circuits and Systems-II.* **52** (2005) 122–125.
14. I. A. Khan and A. M. Nahhas, Reconfigurable voltage mode phase shifter using low voltage digitally controlled CMOS CCII, *Electrical and Electronic Engineering.* **2** (2012) 226–229.
15. E. Yuce, K. Pal and S. Minaei, A high input impedance voltage-mode all-pass/notch filter using a single variable gain current conveyor. *Journal of Circuits, Systems and Computers.* **17** (2008) 827–834.
16. B. Metin. and K. Pal, New all-pass filter circuit compensating for C-CDBA non-idealities, *Journal of Circuits, Systems and Computers.* **19** (2010) 381–391.
17. Jitender, J. Mohan and B. Chaturvedi, A novel voltage-mode configuration for first order all-pass filter with one active element and all grounded passive components, 2020 6th International Conference on Signal Processing and Communication (ICSC), Noida, India, 2020, pp. 235–239.
18. S. Maheshwari, J. Mohan and D. S. Chauhan, Cascadable all-pass and notch filter configurations employing two plus-type DDCCs, *Journal of Circuits, Systems, and Computers.* **20** (2011) 329–347.
19. S. Maheshwari, J. Mohan and D. S. Chauhan, Novel cascadable all-pass/notch filters using a single FDCCII and grounded capacitors, *Circuits Systems and Signal Processing.* **30** (2011) 643–654.
20. B. Metin, S. Minae and O. Cicekoglu, Enhanced dynamic range analog filter topologies with a notch/all-pass circuit example, *Analog Integrated Circuits and Signal Processing.* **51** (2007) 181–189.
21. P. Kumar and K. Pal, Variable Q all-pass, notch and band-pass filters using single CCUU, *Frequenz.* **59** (2005) 9–10.
22. J. Mohan, B. Chaturvedi and S. Maheshwari, Single active element based voltage-mode multifunction filter, *Advances in Electrical Engineering.* **2014** (2014) 514019.
23. E. Yuce and S. Tez, A novel voltage mode universal filter composed of two terminal active devices, *AEU - International Journal of Electronics and Communications.* **86** (2018) 202–209.
24. F. Yucel and E. Yuce, A new electronically fine tunable grounded voltage controlled positive resistor, *IEEE Transactions on Circuits and Systems II: Express Briefs.* **65** (2017), 451–455.
25. W. Tangsrirat, O. Channumsin and T. Pukkalanun, Resistorless realization of electronically tunable voltage mode SIFO type universal filter, *Microelectronics Journal.* **44** (2013), 210–215.
26. S. Maheshwari, J. Mohan and D. S. Chauhan, High input impedance voltage mode universal filter and quadrature oscillator. *Journal of Circuits, Systems, and Computers.* **19** (2010) 1597 –1607.
27. B. Chaturvedi, J. Mohan and A. Kumar, A new versatile universal biquad configuration for emerging signal processing applications, *Journal of Circuits, Systems, and Computers.* **27** (2018) 1850196.
28. J. Mohan, B. Chaturvedi, S. Maheshwari, Novel current-mode all-pass filter with minimum component count, *International Journal of Image, Graphics and Signal Processing.* **5** (2013) 32–37.
29. B. Chaturvedi, J. Mohan, A. Kumar and K. Pal, Current-mode first order universal filter and it's voltage-mode transformation, *Journal of Circuits, Systems, and Computers.* **29** (2020) 2050149.
30. S. Maheshwari and D. Agrawal, High performance voltage-mode tunable all-pass section, *Journal of Circuits, Systems, and Computers.* **24** (2015) 1550080.

31. S. Maheshwari and B. Chaturvedi, High input low output impedance all-pass filters using one active element, *IET Circuit Devices and Systems.* **6** (2012) 103–110.
32. Jitender, J. Mohan and B. Chaturvedi, All-pass frequency selective structures: application for analog domain, *Journal of Circuit Systems and Computers.* **30** (2021) 2150150.
33. A. Kumar and B. Chaturvedi, current-mode MOS only precision full-wave rectifier, *IETE Journal of Research.* (2020). DOI: 10.1080/03772063.2020.1830861.

14 Secured Integrated Circuit (IC/IP) Design Flow

Rahul Chaurasia, Anirban Sengupta, and Prasad Pradeeprao Kanhegaonkar

CONTENTS

DOI: 10.1201/9781003155751-14

14.1 INTRODUCTION

The consumer electronics (CE) industry has grown substantially in recent years. There are various types of devices in CE that are used for entertainment, communication, and recreation. In the home and office, we use many devices like refrigerators, televisions, set-top boxes, radios, mobile phones, laptops, printers, modems, various IoT devices, and so on (Thavalengal and Corcoran, 2016; Mohanty, 2015; Kim et al., 2015; Schneiderman, 2010). Many of these devices work on digital signal processing (DSP) technology, where such devices operate based on some form of signal processing. During the operation of such devices, the input captured by the sensors is converted into a digital signal, which is then processed by the hardware/processor which is present in it. It is therefore crucial to obscure the functionality and structure of the hardware architecture of such devices in order to ensure that such hardware is not compromised through cyber-attacks, such as piracy and Trojan insertion. At the same time, they should be reliable, practical, and cost-effective, and provide the correct output at all times.

In order to achieve these goals, the security of CE devices, both in terms of software and hardware, is essential (Sengupta, 2016). Security must be ensured for such devices, and they should be resilient to different types of hardware threats or attacks. If these things are not taken care of, then it may lead to huge business losses for the concerned companies. Also, it may destroy the brand equity of such companies, preventing them from being competitive and profitable in the CE industry.

Keeping these things in mind, the developers must ensure the hardware security of CE devices; there should be an inbuilt preventive mechanism to protect such devices from various hardware security threats or attacks, such as piracy, Trojans, malware, etc. (Castillo et al., 2007). The preventive mechanism against hardware security threats in CE devices should be such that it should be costly, infeasible, and impractical for the attackers to access the devices, and that they should not get any benefit from such attacks (Sengupta et al., 2017; Sengupta and Roy, 2018). There are various approaches present in today's digital world to ensure hardware security for CE devices; a few of them are discussed in the subsections below.

14.2 DISCUSSION OF CONTEMPORARY APPROACHES

Before starting the discussion on hardware security approaches, we would like to inform the readers about the design process of IC/IP cores used as computation engines and/or application-specific processors/coprocessors for their respective operations in modern CE devices. The design process of IP cores comprises various abstraction levels, such as system level, algorithmic level, register transfer level (RTL), gate level, physical level, etc. (arranged in top to bottom order).

High-level synthesis (HLS) is part of the design process. It converts the algorithmic or system-level specification of the concerned application into the respective register transfer level (Sengupta and Sedaghat, 2011). HLS has various basic substeps such as scheduling, allocation, and binding. Scheduling is used to partition the algorithm into various control steps or timestamps. It defines the various states of the finite state machine. Each control step contains a small portion of the algorithm performed in one clock cycle in the hardware. Allocation and binding are used to map the instructions and variables to the hardware components, multiplexers, registers, and wires of the data path.

Some of the hardware security approaches which are integrated within the HLS design flow of IP cores for DSP/multimedia applications, such as digital filters, image processing kernels, and image compression algorithms, include structural obfuscation (Sengupta et al., 2017; Lao and Parhi, 2015; Sengupta and Roy, 2017a,b), hardware watermarking (Sengupta and Roy, 2017; Bhadauria and Sengupta, 2016), and functional obfuscation (Sengupta and Roy, 2018). Each of these approaches addresses different hardware security threat models, and these approaches protect the original IP design from such threats.

14.3 STRUCTURAL OBFUSCATION OF IP CORES

14.3.1 The Basic Definition and Threat Model of Structural Obfuscation

Structural Obfuscation (SO) aims to secure the design of IP cores from standard hardware security threats, such as backdoor Trojan insertion. It achieves this by making the analysis or reverse engineering (RE) process difficult or complex by making the structural design of an IP core highly unobvious or significantly different from that of the original design while preserving its original functionality (Chakraborty and Bhunia, 2009). The adversary or attacker would not be able to comprehend the actual functionality and design architecture of the IP core due to the transformed structure after applying structural obfuscation. The goal of SO is to significantly alter the structural design of an IP core design so that it is highly obscured from an attacker while preserving its original functionality and thereby hindering the RE (Sengupta et al., 2017). SO is performed using techniques like transformations, arbitrary assignment of logic components, and inserting dummy wires or irregular routing. SO does not insert any additional components as key gates into the design. Therefore the design overhead does not add up during SO.

The threat model addressed by structural obfuscation is Trojan insertion. Due to Trojan insertion, the chip may not perform its intended operation, work unexpectedly, become damaged, or provide wrong outputs. So these threats lead to high business losses for the concerned companies.

14.3.2 Why Performing SO during HLS is Beneficial?

The design flow of an IP core includes several abstraction levels, as stated in Section 1.3.1. Structural obfuscation for complex DSP applications should be preferably performed at higher abstraction levels, such as the architecture level during HLS. Many recent advances in the electronic design automation (EDA) process exist, and various computer-aided design (CAD) tools are currently available. HLS performed during the higher abstraction levels in the design process enables the design of an

IP core to be completed in less time, with higher reliability, less errors, larger productivity, shorter time to market, and an easier design space exploration process for optimization of area, power, delay, temperature, etc., as well as making the circuit technology available to non-experts (Sengupta et al., 2017).

An IP core design, such as for DSP applications, can be made imperceptible in many ways during the design process. SO performed during HLS in the design process is beneficial. It is due to the fact that the changes made at higher levels are automatically reflected or percolated in the lower layers of the respective design. Due to this, the obfuscation performed at higher abstraction levels of the design automatically alters the corresponding data path and controller of the design. The design of such obfuscated IP cores is also altered at the gate-level structure or netlist. Due to the benefits of HLS performed at higher abstraction levels, as stated before, we can optimize the IP core design in terms of area, complexity, power, delay, performance, and fault tolerance. So, SO during HLS is capable of generating a strongly obscured design and thereby hindering the RE process (Sengupta et al., 2017).

It is noteworthy that SO may be used as a preventive mechanism against backdoor Trojan insertion by thwarting the RE process. If an attacker tries to RE an obfuscated IP core netlist or GDS file to analyse its functionality and design structure, it is harder for him to do so. A design is secured by applying a SO function, and the original structure is non-interpretable for an attacker (Sengupta et al., 2017).

14.3.3 Description of Multi-Stage Structural Obfuscation Process (Sengupta et al., 2017)

There are multiple compiler-driven high-level transformation techniques used for the structural obfuscation of DSP applications, such as redundant operation elimination (ROE), logic transformation (LT), tree height transformation (THT), loop unrolling (LU), folding, loop invariant code motion (LICM), and data flow graph partitioning. They can obfuscate both loop and non-loop-based IP cores for DSP applications. Loop-based transformations (e.g., LU and LICM) are not applicable to non-loop-based DSP applications. As a result of employing structural obfuscation, there will be many functionally equivalent structurally ambiguous RTL architectures in the same IP core. A low-cost structurally obfuscated design architecture may be further explored by integrating PSO-based design space exploration (DSE) during the design process to obtain an optimal solution in terms of cost, area, power, and delay (Sengupta et al., 2017).

The folding transformation generates various meaningful modes, which increases the complexity for an attacker to differentiate the correct functionality within the different modes. Different folding factors cause different equivalent circuits, which leads to ambiguity for an attacker. Due to folding transformation, circuits with different functions may have the same structure and circuits with the same functionality may have dissimilar structures. The folding factor has a crucial impact on the design cost. Therefore selecting an optimal folding factor is necessary to design an optimized IP core for DSP application. The inability to obtain an optimal folding factor would lead to a higher design overhead (Sengupta et al., 2017).

ROE removes the redundant or duplicate operations from a set of operations in the control data flow graph (CDFG) of the DSP application, and it then performs necessary adjustments in the graph structure. In the CDFG of the respective application, each operation is represented by a node. A node is considered redundant if another node with the same parents and operation type is present in the same CDFG structure. If redundant nodes are detected, the nodes with higher node numbers are eliminated from the CDFG. Some adjustments are made in the CDFG after ROE to preserve the original functionality. Therefore structurally obfuscated IP core can be generated using HLS. It can also be used as an intermediate output and fed to the next obfuscation stage (Sengupta et al., 2017).

LT modifies the structure of a DSP application keeping the functionality logically equivalent. A simple example of this is that multiplying a number by two is the same as adding a number to itself. In this, the multiplication operation is converted into addition. LT alters the nodes of the input CDFG so that its structure is different from the original and preserves the correct functionality of the application. This transformation is powerful in reflecting change at the gate level after HLS (Sengupta et al., 2017).

THT modifies the height of the CDFG by changing the critical path dependency of the application. Critical path computations are divided into multiple sub-computations in THT. The sub-computations can then be executed in parallel. So, it creates structurally diverse yet functionally similar graphs. A THT-based structurally obfuscated IP core may have a different controller structure at the gate level than the original controller structure. THT obfuscated IP core also affects the interconnectivity of the data path structure at the RTL and gate level. Therefore, stronger obfuscation strength can be achieved due to RTL/gate-level obfuscations of the data path and controller (5Sengupta et al., 2017).

LU unfolds or unrolls a loop-based DSP application. Unrolling the loop means executing the operation inside the loop body for a specific number of iterations in parallel. A LU-driven structurally obfuscated IP core is generated based on a predefined value of the unrolling factor. A PSO-driven DSE process can be used to find a low-cost unrolling factor for structural obfuscation. LU also minimizes the delay by introducing concurrency so that parallel computing can be implemented in the resulting structurally obfuscated IP core design. LU-based structural obfuscation may affect the RTL structure in terms of change in the number of functional units, size of interconnect resources, and number of storage elements (Sengupta et al., 2017).

LICM identifies the loop-independent operations inside the loop and moves them out of the loop body. The nodes in CDFG, which are independent of the loop, are shifted out from the loop body. This transformation due to LICM affects the data path of the processor (IP core) in terms of changing the size of interconnected resources and the number of storage elements. Further, due to this, the equivalent gate-level representation also structurally changes without affecting functionality (Sengupta et al., 2017).

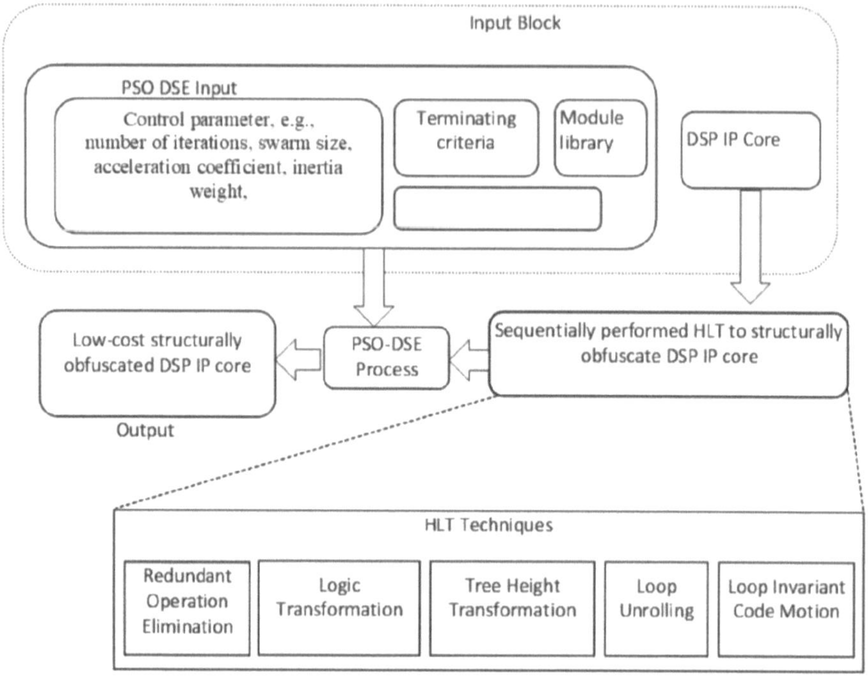

FIGURE 14.1 High-level view of low-cost multi-stage structural obfuscation methodology.

In the literature, multi-stage structural obfuscation using the above high-level transformations was proposed in (Sengupta et al., 2017), as shown in Figure 14.1. In this approach, five consecutive high-level transformations were employed in series on the DSP dataflow graph. The PSO-DSE module was integrated with the transformation block in order to generate a low-cost optimal hardware solution for the HLS design process (Mishra and Sengupta, 2014). The primary inputs of this approach included a module library for fitness evaluation during DSE, a pre-processing block for unrolling factors, and the original DFG of the DSP application; while the output of the approach was a low-cost structurally obfuscated DSP core (RTL representation). Low-cost obfuscated design is achieved with the aid of PSO-DSE. Optimization is required to develop an optimal IP core with low area, power, and delay. In the context of a low-cost IP core design solution, each particle represents a source configuration and unrolling factor value as a design solution. The first three particles are initialized with a minimum (min), maximum (max), and mid-resources configuration, along with its unrolling factor value. If the initial population size is more than three particles, then the rest of the particles are initialized randomly within the range of min and max resource configuration values. The fitness of each particle is calculated using the cost function based on normalized area/power and normalized delay of the design (Sengupta et al., 2017). An example of the transformed DFG and its corresponding low-cost obfuscated RTL data path of an IP core is shown in Figures 14.2 and 14.3 respectively.

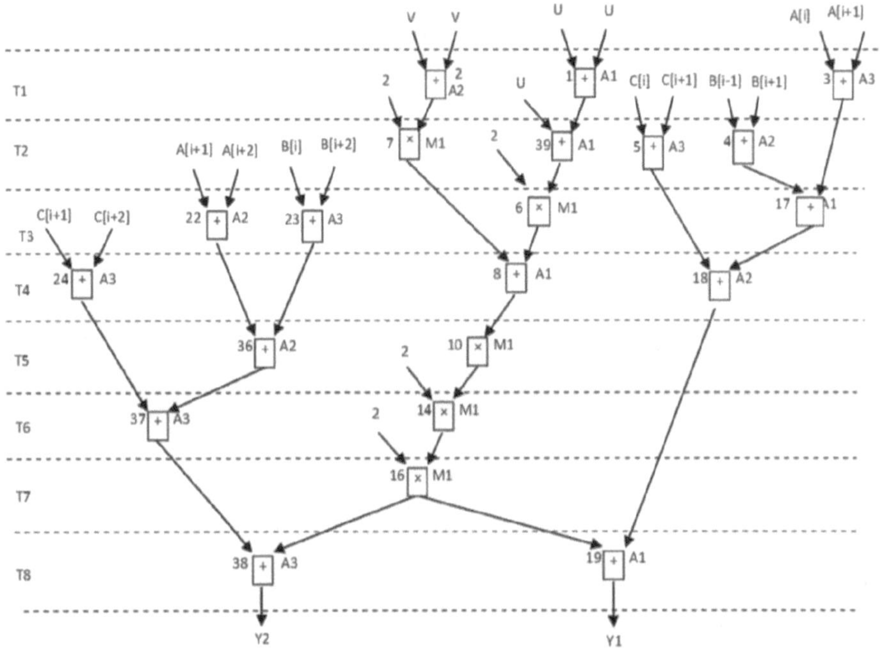

FIGURE 14.2 Scheduling of a structurally obfuscated IP core design post applying multi-stage obfuscation techniques.

14.4 FUNCTIONAL OBFUSCATION OF DSP IP CORES

Functional obfuscation is also known as logic obfuscation, logic locking or logic encryption (Rajendran et al., 2015; Roy et al., 2008; Zhang, 2016; Subramanyan et al., 2015). In the semiconductor or VLSI industry, functional obfuscation may be used as another line of defense, besides structural obfuscation for the enhanced security of DSP IP cores, against threats such as piracy and Trojan insertion. In modern business models in the semiconductor industry, different entities may be involved, such as developers, fabricators, and third-party IP (3PIP) vendors (Yasin et al., 2016). The latter two may become untrustworthy in certain situations. The rights of the original owners of DSP IP cores may be abused by such untrustworthy parties, leading to high business losses and damaging the reputation of the owners in the market. Piracy may cause revenue/profit loss. Trojan insertion may cause malfunctioning, leaking of confidential information, performance degradation, etc. of the DSP IP cores (Yasin et al., 2016; Sengupta and Roy, 2018).

14.4.1 Reverse Engineering Attacks during the Various Stages of IP Core Design

Reverse engineering is a legal process in many countries, unless it is used with malicious intent to cause harm to the original entity/owner of the DSP IP core

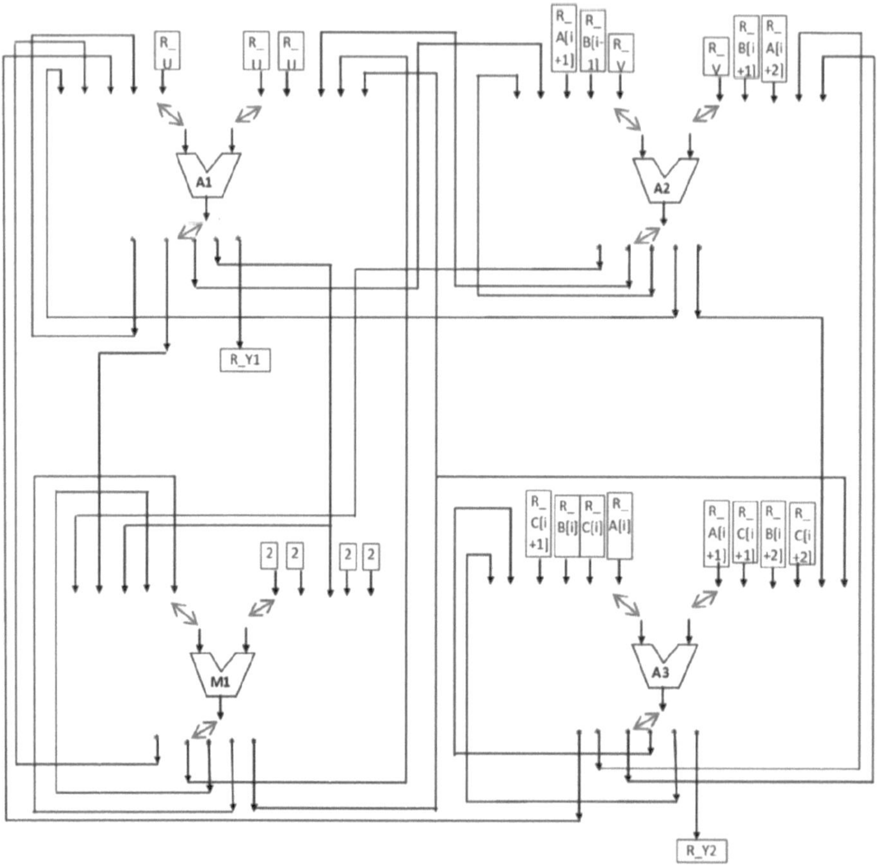

FIGURE 14.3 Structurally obfuscated RTL data path of the IP core design.

(Torrance and James, 2009). However, RE becomes an attack if the attackers use RE to uncover the structure, design, or functionality of DSP IP cores for piracy or Trojan insertion.

The DSP application in the initial stages is generally represented in the form of an algorithmic or system-level description of the respective application. HLS is performed to obtain its corresponding RTL description in the form of a netlist. Functional obfuscation may be performed on this netlist for security. After structural obfuscation, it may also be performed on the netlist for enhanced security. After functional obfuscation, the original netlist becomes locked/encrypted, which doesn't perform its intended operation unless the predefined correct key is applied. After the layout generation stage, the locked/encrypted gate-level netlist in consumer electronics devices is converted into a GDSII file. Further, the locked GDSII file is converted into a mask followed by the fabrication of a non-functional IC. After packaging and activation, the market-ready functional IC is generated. This is the common design cycle to produce an encrypted IP core (Yasin et al., 2016).

The attacker may perform RE on (1) the locked/functionally obfuscated RTL netlist, (2) the GDSII file after its layout stage, (3) the IC mask, (4) the non-functional IC, and (5) the market-ready functional IC after the packaging/activation stage, etc. If the attacker succeeds in the RE process, s/he obtains the encrypted netlist. Further, the attacker then tries to launch attacks (discussed in the next subsection) to get the deciphered netlist. Once s/he gets the deciphered netlist, s/he may perform IP piracy and Trojan insertion (Yasin et al., 2016).

14.4.2 Attack Scenarios

There are two main types of attacks in functional obfuscation, viz., key-sensitization (Yasin et al., 2016) and removal (Yasin et al., 2017), in the case of DSP IP cores. Functional obfuscation is used to lock or encrypt the circuit, as discussed in Section 14.4.1. To encrypt the circuit functionality of DSP IP cores, IP core locking blocks (ILBs) can be used. The circuit performs correct functionality only when provided with the correct key values. A key-sensitization attack makes use of controlled inputs to send key values to the output. In a removal attack, an attacker attempts to remove the ILBs from the circuit to nullify the functional obfuscation.

The incorrect placement of key gates in ILBs makes them prone to key-sensitization attacks. The incorrect placements include isolated key gates, run of key gates, and mutable key gates. In isolated key gates, there is no path from one key gate to another key gate. In a run of key gates, certain key gates are placed back-to-back. In mutable key gates, two or more key gates converge at a common gate but have no common path between them. The key value can easily be sensitized in isolated key gates. The correct key can easily be identified in a run of key gates due to an increase in the possible correct key combinations. Moreover, a single key gate can replace a run of key gate structures. A specific key bit can easily be traced from one key gate by muting another key gate in the case of mutable key gates. Such improper placement of key gates may lead to key-sensitization attacks. Once keys are sensitized by an attacker, s/he may obtain the correct functionality, structure, and design of the original netlist (Sengupta and Roy, 2018).

In the case of a removal attack, the attacker tries to remove all key components used to lock the circuit. This is called a removal attack. Key-sensitization attacks and removal attacks nullify or remove the functional obfuscation effect. Eventually, both attacks lead to piracy and Trojan insertion (Sengupta and Roy, 2018).

14.4.3 Functional Obfuscation Approach for DSP Cores

A low-cost PSO-DSE driven functional obfuscation approach, as shown in Figure 14.4, has been proposed by (Sengupta and Roy, 2018). As shown in the figure, the input to this approach is the algorithmic description (such as C/C++ code or transfer function) or intermediate representation (such as CDFG) of the DSP application. Further, the PSO-DSE process takes the module library and certain control parameters as input. ILBs are also fed as input. Functionally obfuscated DSP IP core design, along with optimized area, power, and delay requirements, leading to

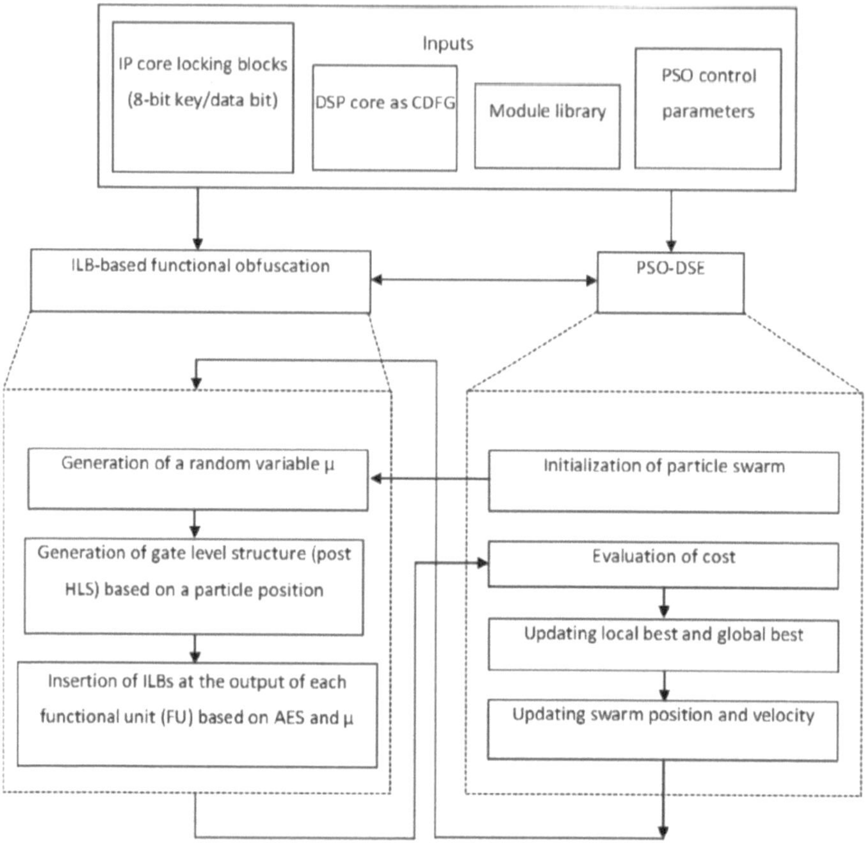

FIGURE 14.4 Framework for functional obfuscation of DSP cores.

minimal design overhead, is the final output of this approach. The PSO-DSE process was discussed in Section 14.3.3. The ILBs, their security features, and insertion techniques are discussed below.

ILBs are used as key components to functionally obfuscate the DSP IP core design. Each ILB activates by applying the correct value of the 8-bit key. The structure of ILBs is made reconfigurable and depends on the Advanced encryption standard (AES) block's fixed secret key values (128 bit), and the respective output generated (128 bit) by the AES hardware. Each ILB comprises different combinations of basic and universal logic gates. There can be a large number of possible ILB structures depending upon the combination of key bits. ILBs generate correct output only by providing correct key values. Figure 14.5 shows a sample ILB structure designed using either direct key values or the key values coming through the AES block.

The ILBs proposed in (Sengupta and Roy, 2018) have features, such as multi-pairwise security, prohibiting key-gate isolation, protection against the run of key gates, non-mutable convergent key gates, etc. Multi-pairwise security means that

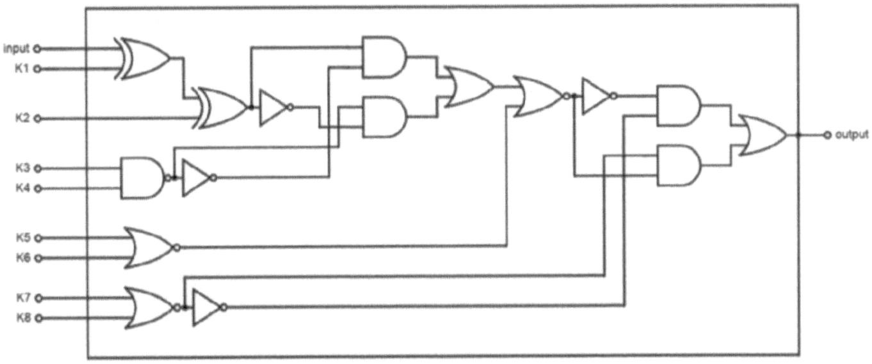

FIGURE 14.5 Eight-bit key value-based sample ILB structure (keys come either directly or from AES block).

an attacker is not able to sensitize any key bit without controlling or knowing the remaining key bit values. The attacker doesn't have access to other key bits in the locked netlist. In prohibiting key-gate isolation, key gates are arranged in the ILB structure, so there is no isolated key-gate. In protection against the run of key gates, the key gates of ILBs are mixed with each other. So, multiple correct key combinations are highly unlikely and replacing all key gates with one key gate is not feasible. In non-mutable convergent key gates, none of the key gates in the ILB structure can be muted by applying a direct value to a primary input. Since ILBs contain eight non-mutable key gates, to sensitize one key bit to the output the remaining seven key bits need to be muted, which again is highly challenging. Thus, all these features make the ILBs secure against key-sensitization attacks. Security against removal attacks is achieved as the ILB structures are reconfigured based on AES output and does not use a fixed ILB architectural template in the design netlist for encryption.

As per (Sengupta and Roy, 2018), a special technique is used for inserting ILBs into the design. In this technique, using the PSO-DSE process, a random number 'μ ' is generated, such that: $1 \leq \mu \leq T_{ILB}$, where T_{ILB} indicates the total number of ILB structures/templates available to insert in the design. Authors in (Sengupta and Roy, 2018) used $T_{ILB} = 4$. Once the gate-level data path structures are generated after HLS, based on the 'μ ' value to generate a locked netlist, ILBs are inserted in each output data bit of functional units. The same ILB is inserted 'μ ' times into the design. Then, the next ILB from T_{ILB} is selected. The process goes on until all the output data bits of functional units are connected to the ILBs. Here, no output data bits remain unobfuscated, ensuring stronger security. The use of PSO-DSE to find the optimal value of 'μ ' handles the area, power, and delay trade-offs, and generates an optimal design.

14.4.4 Handling Attacks

Figure 14.6 shows the generic interface of functionally obfuscated DSP core with AES hardware (Sengupta and Roy, 2018). A custom AES hardware in synthesized

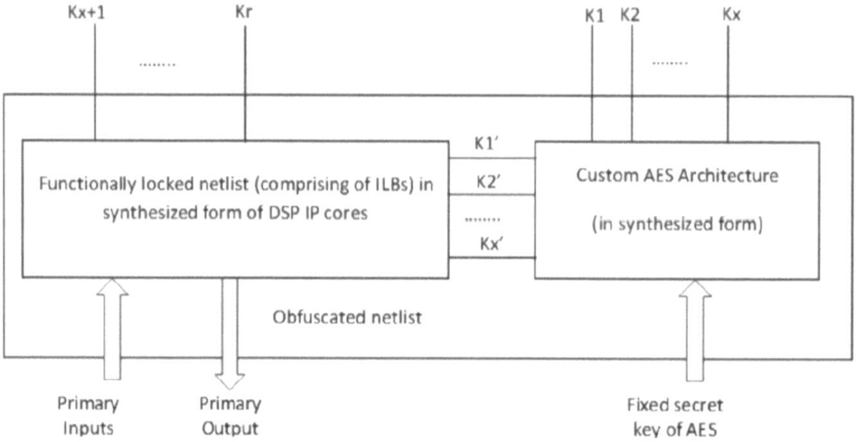

FIGURE 14.6 Interface of a functionally obfuscated DSP IP core design with a custom design AES block to hinder removal attack.

form is integrated (Mohanty and Sengupta, 2019). A sub-set of key bits is fed as input to the custom-designed AES block. This is called pseudo key input. The generated output of the AES block works as the actual key input to the reconfigured ILB structure, which is produced based on the 128-bit fixed secret key provided by the IP vendor. A functionally obfuscated netlist in synthesized form is composed of such reconfigured ILB structures. The resulting DSP IP core design can handle key-sensitization attacks, removal attacks, IP piracy attacks, and reverse engineering attacks. Below is a short description of each.

- *Key-sensitization attack*: can be launched by an attacker if key gates are isolated, not multi-pairwise secured, mutable, and connected in a back-to-back fashion. In (Sengupta and Roy, 2018), the ILBs are designed by interweaving several gate structures based on 8-bit key inputs. Here, the key gates are multi-pairwise secure, no isolated key gates are available in any ILB, and no key gates can be muted. Therefore, the attacker's complexity is increased to determine and replace the run of key gates using a single equivalent key gate. Thus, this approach provides resiliency against key-sensitization attacks.
- *Removal attack*: Few ILB structures are reconfigured in (Sengupta and Roy, 2018) based on the custom-designed AES encrypted output for a specific secret key (chosen by the IP vendor). The reconfiguration of ILB depends on the AES output, resulting in several possible structures of ILBs. No fixed template of ILBs is integrated with the obfuscated design netlist. If the number of key inputs fed through AES to the ILBs is large, then more ILB structures can be reconfigured. Henceforth, an attacker cannot deduce in advance the ILB gate structure for performing removal from the

obfuscated design netlist. A low-cost version of a custom-designed AES architecture is integrated with the obfuscated design, which is not publicly available. Therefore, it is not feasible for an attacker to launch a removal attack on the AES gate-level hardware design, as s/he is unaware of the respective structure. Further, post-synthesis, the individual components of the overall complex DSP core design netlist become indistinguishable from each other, thereby hindering removal attacks.

- *IP piracy and RE*: In (Sengupta and Roy, 2018), the 8-bit key input-based ILB structures are heavily interweaved among themselves. Because of the integration of innumerable possible ILB structures with the DSP design, the strength of the encrypted (locked) netlist is robust (a minimum of a 512-bit key input is required for decoding the output value). This makes it challenging for an attacker to decrypt the locked netlist. Further, due to the integration of AES with ILBs in the design netlist, deducing even a small 8-bit key value is challenging for an attacker. Thus for a practical design with several ILBs (with a minimum of 512-bit keys), deducing the exact key value is impossible. This additionally renders the DSP design stronger in terms of obfuscation and thwarts against decryption of the locked netlist. The locked netlist of the DSP IP core performs its intended operation only upon providing a correct key–value pair. In the absence of correct key–value pair, an attacker cannot use the functionality of the DSP IP core. Further this makes both piracy and Trojan insertion attacks infeasible for an attacker.

14.5 HARDWARE WATERMARKING

Hardware watermarking is a process of secretly inserting the owner's (authentic) signature into the design during the design synthesis process. There are different design synthesis phases where an owner's watermark can be inserted into his/her IP core design, such as behavioral synthesis (architectural synthesis), RTL/logic synthesis, and layout synthesis. In general a hardware watermark should satisfy the following properties:

a) Post embedding a watermark into a hardware design, the original functionality of the design must not be affected.
b) A watermark should not result in excessive design overhead.
c) It should be robust against typical attacks, such as finding ghost signatures and tampering.
d) Watermark detection should be seamless for the IP owner and law enforcement authority (such as specialized IP courts) upon providing relevant information.
e) It should be imperceptible to an attacker.
f) It should be fault-tolerant and capable of nullifying a fraudulent IP user's false claim of IP ownership.
g) Time to embed a watermark into the design should not be large.

14.5.1 Features of the Multi-Variable Watermarking Approach (Bhadauria and Sengupta, 2016) for Signature Insertion During High-Level Synthesis

The multi-variable watermarking approach proposed by (Bhadauria and Sengupta, 2016) has the following features

- The signature in the multi variable watermarking approach comprises the following four variables: 'i', 'I', 'T', and '!'.
- The multi-variable watermarking approach is a single-phase watermarking where the signature constraints are embedded during the register allocation phase of the HLS process.
- It exploits particle swarm optimization (PSO) driven design space exploration process to obtain optimal watermark, ensuring minimal overhead post watermarking.

14.5.2 Signature Encoding Mechanism in Multi-Variable Watermarking Approach (Bhadauria and Sengupta, 2016)

As discussed in the previous subsection, signature constraints are embedded during the register allocation phase of the HLS process. In order to embed signature constraints, register allocation in HLS is performed based on the concept of a colored interval graph where the nodes of the graph represent the storage variables, and the edges represent the existence of an overlapping lifetime between variables. Two nodes can have the same color only if their respective storage variables are not executing in the same control step. For embedding a watermark, additional constraints (hardware security constraints) in the form of edges are imposed into the CIG. These hardware security constraints are obtained by encoding the signature variables using the designer's encoding mechanism. In the multi-variable watermarking approach, the IP owner's signature is a combination of four variables, viz., 'i', 'I', 'T', and '!'. The encoding mechanism of the four variables mentioned above (I, I, T, and!) is as follows :

For 'i', 'I', 'T', and '!', the encoding of a signature variable is an edge (constraint) in the CIG with node pairs of (prime, prime), (even, even), (odd, even), and (0, any integer), respectively.

Adding constraints in the register allocation step may sometimes lead to conflict, and this can be resolved either by taking a brand new color (register) or from the existing ones; however, it should not result in future conflicts. The more the number of constraint edges, the greater the watermark's security.

14.5.3 Process of Watermark Creation for an IP Core Using a Multi-Variable Signature (Bhadauria and Sengupta, 2016)

The process comprises the following steps:

i. Schedule the control data flow graph based on the optimal resource configuration post-design space exploration.

ii. Draw the colored interval graph to calculate the minimum number of registers required for allocation and generate a register allocation table.
iii. Sort storage variables as per their number in increasing order.
iv. Generate the desired signature from the tuple comprising of 'i', 'I', 'T', and '!' based on the designer's choice.
v. Generate a list W[k] of additional node pairs corresponding to its encoded values by traversing the sorted storage variables in step (iii).
vi. Implant the additional edges (constraints) as a watermark into the CIG.
vii. Modify the register allocation table of the IP design on the basis of the implanted watermark.

The process of embedding the watermark can be illustrated in Figure 14.7

14.5.4 Motivation for Performing Design Space Exploration of an Optimal Watermark

Design space exploration is a process of performing a trade-off between the hardware area and latency of a watermarked design to explore its optimal implementation. For IP protection using watermarking, performing a trade-off is extremely critical for low-cost watermarked design implementation (i.e., a solution with minimum hardware area and latency). There is a trade-off between hardware area and latency because, with an increase in the number of resources and a possible register overhead, the area of the watermark solution increases, whereas latency decreases as it offers more parallelism in this case.

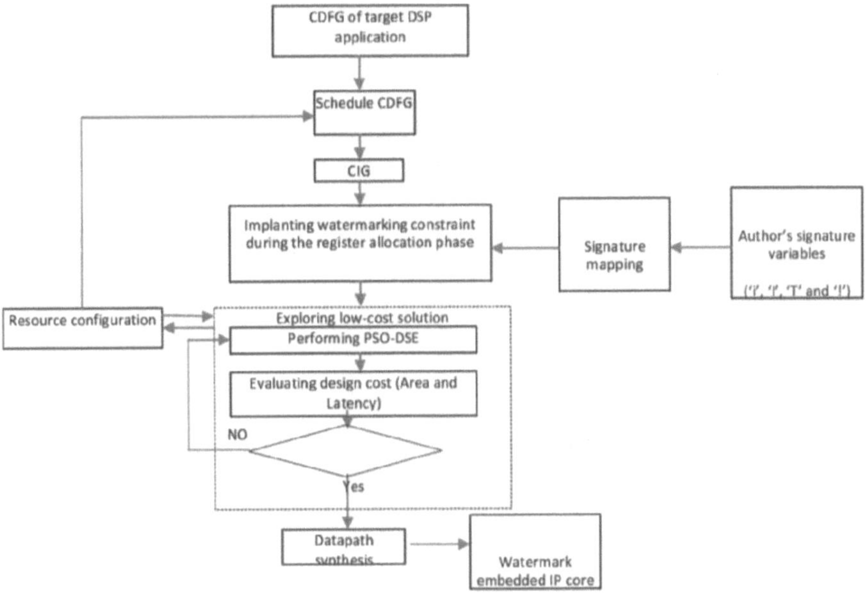

FIGURE 14.7 Single-phase watermarking-based IP protection technique.

14.5.5 Resolving Attacks and Ownership Problems through a Multi-Variable Watermark Approach (Bhadauria and Sengupta, 2016)

In this regard, we use the example of two entities, viz., 'Ram' and 'Ravan,' involved in the semiconductor business model, where Ram has a design 'D', which is watermarked with his signature. Ravan is the buyer of the design 'D'. Following IP ownership conflicts are possible:

Finding unintended signature: If Ravan successfully extracts a signature from the IP core design, Ram will be the victor because only he can prove his signature more strongly and in a meaningful sense.

Embedding unauthorized signature: If Ravan becomes successful in inserting his signature into the design, Ram will still be the winner. This is because the design claimed by Ravan will now carry two signatures (including Ram), whereas the original design only contains a single signature (Ram's signature).

Tampering signature: This is never an easy option for Ravan because the chances of affecting the quality of the design are very high.

14.5.6 Properties/Metrics of Watermark Generated through Multi-Variable Watermarking Approach (Bhadauria and Sengupta, 2016)

The watermark generated through the multi-variable watermarking approach satisfies the following properties:

Minimization of cost: The cost of the watermarked design is minimal because of the exploration of an optimal watermark using the PSO-DSE process.

Resiliency: Watermarking is resilient against attacks because the constraints are distributed throughout the IP design after HLS.

Fault tolerance: The watermark is fault-tolerant because IP ownership still remains preserved even if an attacker removes a part of the watermark.

Adaptability to any CAD tool: The watermark is added as a pre-processing step; hence, the watermarking approach can be integrated with any existing CAD tool.

Watermark creation time and signature detection time: Both must be less for the IP designer.

14.6 CONCLUSION

This chapter presented a detailed description of several hardware security and IP core protection methodologies, such as structural obfuscation, functional obfuscation, and hardware watermarking, in the context of data-intensive DSP cores. In the structural obfuscation section, details of the threat model, a formal definition of structural obfuscation, the importance of performing structural obfuscation during high-level synthesis, and an explanation of the multi-stage structural obfuscation process were explicitly discussed. Further, in the functional obfuscation section, details of the threat model, scenario of reverse engineering attack during various stages of IC design, possible attacks and threats, as well as an explanation of the functional obfuscation approach for DSP cores, were discussed. Finally, in the hardware watermarking section, an overview of the threat model, desirable properties of a watermark, details of multi-variable signature encoding-based hardware watermarking, as well as possible attack scenarios, were discussed. This chapter, therefore, lucidly touches upon the most important aspects of hardware security and IP core protection techniques for DSP/multimedia applications.

REFERENCES

A. Sengupta (2016), 'Intellectual property cores: Protection designs for CE products,' *IEEE Consumer Electronics Magazine*, vol. 5 (1), pp. 83–88.

A. Sengupta, and D. Roy (2017a), 'Antipiracy-aware IP chipset design for CE devices: A robust watermarking approach,' *IEEE Consumer Electronics Magazine*, vol. 6 (2), pp. 118–124.

A. Sengupta, and D. Roy (2017b), 'Protecting an intellectual property core during architectural synthesis using high-level transformation based obfuscation,' *IET Electronics Letters*, vol. 53 (13), pp. 849–851.

A. Sengupta, and R. Sedaghat (2011), 'Integrated scheduling, allocation and binding in high level synthesis using multi structure genetic algorithm based design space exploration,' International Symposium on Quality Electronic Design, Santa Clara, CA, pp. 1–9.

D. Kachave Sengupta, and D. Roy (2018), 'Low cost functional obfuscation of reusable IP cores used in CE hardware through robust locking,' *IEEE Transactions on Computer-Aided Design of Integrated Circuits and Systems*, vol. PP (99), pp. 1–1.

D. Roy, Sengupta, S. P. Mohanty, and P. Corcoran (2017), 'DSP design protection in CE through algorithmic transformation based structural obfuscation,' *IEEE Transactions on Consumer Electronics*, vol. 63 (4), pp. 467–476.

E. Castillo, U. Meyer-Baese, A. Garcia, L. Parrilla, and A. Lloris (2007), 'IPP@HDL: Efficient intellectual property protection scheme for IP cores,' *IEEE Transactions on Very Large Scale Integration Systems*, vol. 15 (5), pp. 578–591.

J. Kim, E. S. Jung, Y. T. Lee, and W. Ryu (2015), 'Home appliance control framework based on smart TV set-top box,' *IEEE Transactions on Consumer Electronics*, vol. 61 (3), pp. 279–285.

J. Rajendran et al. (2015), 'Fault analysis-based logic encryption,' *IEEE Transactions on Computers*, vol. 64 (2), pp. 410–424.

J. Zhang (2016), 'A practical logic obfuscation technique for hardware security,' *IEEE Transactions on Very Large Scale Integration System*, vol. 24 (3), pp. 1193–1197.

J. A. Roy, F. Koushanfar, and I. L. Markov (2008), 'EPIC: Ending piracy of integrated circuits,' *DATE*, vol. 43, pp. 1069–1074.

M. Yasin, J. J. Rajendran, O. Sinanoglu, and R. Karri (2016), 'On improving the security of logic locking,' *IEEE Transactions on Computer-Aided Design of Integrated Circuits and Systems*, vol. 35 (9), pp. 1411–1424.

M. Yasin, B. Mazumdar, O. Sinanoglu, and J. Rajendran (2017), 'Removal attacks on logic locking and camouflaging techniques,' *IEEE Transactions on Emerging Topics in Computing*, vol. PP (99), pp. 1–1.

P. Subramanyan, S. Ray, and S. Malik (2015), 'Evaluating the security of logic encryption algorithms,' IEEE International Symposium on Hardware Oriented Security and Trust (HOST), Washington, DC, USA, pp. 137–143.

R. Schneiderman (2010), 'DSPs evolving in consumer electronics applications,' *IEEE Signal Processing Magazine*, vol. 27 (3), pp. 6–10.

R. Torrance and D. James (2009), 'The state-of-the-art in IC reverse engineering,' Proceedings of Workshop on Cryptographic Hardware and Embedded Systems (CHES) (Springer, Berlin, Heidelberg), Lausanne, Switzerland, pp. 363–381.

R. S. Chakraborty, and S. Bhunia (2009), 'HARPOON: An obfuscation-based SoC design methodology for hardware protection,' *IEEE Transactions on CAD of Integrated Circuits and Systems*, vol. 28 (10), pp. 1493–1502.

S. Bhadauria, and A. Sengupta (2016), 'Exploring low cost optimal watermark for reusable IP cores during high level synthesis,' *IEEE Access*, vol. 4, pp. 2198–2215.

S. Thavalengal, and P. Corcoran (2016), 'User authentication on smartphones: Focusing on iris biometrics,' *IEEE Consumer Electronics Magazine*, vol. 5 (2), pp. 87–93.

S. P. Mohanty (2015), *Nanoelectronic Mixed-Signal System Design*, McGraw-Hill Education, New York, no. 0071825711.

S. P. Mohanty, and A. Sengupta (2019), 'Advanced encryption standard (AES) and its hardware watermarking for ownership protection', *Book: 'IP Core Protection and Hardware-Assisted Security for Consumer Electronics*, e-ISBN: 9781785618000, pp. 317–335.

V. K. Mishra, and A. Sengupta (2014), 'MO-PSE: Adaptive multi-objective particle swarm optimization based design space exploration in architectural synthesis for application specific processor design,' *Advances in Engineering Software*, vol. 67, pp. 111–124.

Y. Lao, and K. K. Parhi (2015), 'Obfuscating DSP circuits via high-level transformations,' *IEEE Transactions on Very Large Scale Integration System*, vol. 23 (5), pp. 819–830.

Index